More Reader Comments about

What Causes Human Behavior—
Stars, Selves, or Contingencies?
by Stephen F. Ledoux

Stephen Ledoux's book, What Causes Human Behavior—
Stars, Selves, or Contingencies? *is a strong, non–compromising,
theoretical and philosophical argument that the answers come from
behaviorology, the natural science of behavior, that the answers
do not come from astrology, theology, etc., or from psychology, the
mentalistic unnatural science of the mind. And he supports his
argument with examples of effective, science–based applications of
applied behaviorology (applied behavior analysis) and with analyses
of human behavior in everyday life, going from simple behaviors,
to complex verbal behavior, with suggestions that behaviorology is
crucial to the solutions of the world problems of overpopulation,
sustainability, and global warming. But also, he's not afraid to
make these complex topics more readable by using an occasional
contraction, an informal expression, and even a little humor, i.e.
he's way cool.*
❧Richard Malott, ᴘʜ.ᴅ. (Professor, Western Michigan University)

*Professor Ledoux has written a primer on a newly emerging
discipline:* behaviorology. *It is the natural science of environment–
behavior relations and an intellectually challenging subject, one
that variously intersects with astrology, psychology, philosophy,
education, and physiology plus other biological and behavioral
sciences. Ledoux's discussion of explanatory fictions and a variety
of other explanatory fallacies alone, however, is worth the price of
admission. And there is so much more!*
❧John Stone, ᴘʜ.ᴅ. (Professor, East Tennessee State University,
Johnson City, and President, Education Consumers Foundation at
www.education-consumers.org)

Dr. Ledoux has written a book that is accessible to all readers willing to take the time to peruse its pages. He clearly explains the principles of behavior in this informative primer, with the last chapter, in particular, tied to applying them to saving humankind from disasters resulting from human behavior or the lack thereof.
❧Michael Shuler (Engineer, Peru, IN, retired)

In the prevailing academic order of 21st–century society, the complex subject of the behavior of organisms, particularly regarding the causes and effects of human behavior, is controlled by the explanatory myths that are inherent to psychology. This book presents a threat to that prevailing academic order. So every person prepared to actually discover why people do what they do must read it.
❧Michael Rauseo, Psy.D. (Faculty, Los Angeles Unified School District)

This book concisely covers a vast amount of ground in its mere 400 pages, preparing readers well for further, and more in–depth, examination of the discipline. Although the book is primarily written to elucidate human *behavior, the laws and principles of behavior are used not only by behaviorologists but also by animal behavior technologists who study the behavior of non–humans. Dr. Ledoux has presented a complex discipline in an accessible format that is reinforcing to read and introduces a natural–science alternative to the various mystical and pseudoscience disciplines so prominent in society today. A natural–science approach brings with it highly effective and efficient engineering strategies and tactics, which are just what our world needs. This timely book deserves a Nobel Prize in my opinion!*
❧James O'Heare, DLBC (Companion Animal Sciences Institute, Ottawa, Ontario, Canada)

As this book shows, we, and the next and younger cohorts of behaviorologists, must continue to make abundantly clear, to our fellow natural scientists as well as to the general public, the philosophically incommensurable differences between what B. F. Skinner called… "radical behaviorism," and the traditional position with its stars, selves, minds, souls, and psychés.
❧Werner Matthijs, M.A. (Team Coördinator van de Toegepaste Gedragsologie, Universitair Psychiatrisch Centrum Sint Kamillus, Bierbeek, Belgium, retired)

Many of the principles of behavior, and their cultural implications, threaten received conceptions of behavior, including dearly–held pre–scientific assumptions about the nature and causes of our "selves." This book gently walks readers through these principles and implications, allowing readers to overcome any attendant threats and to appreciate fully the subject matter at hand. In this primer Ledoux renders it all accessible, inviting, and eminently relevant.
❧Bruce Hamm, M.A., BCBA (Director, Blackbird Academy of Childhood Education, Vancouver, British Columbia, Canada)

Currently, students reaching the end of their formal educations suffer a terrible gap in coping skills, never having had an opportunity to study behavior in the formal mode of an uncompromised natural science. This easy–to–read primer presents some elementary behaviorological principles and examines important implications of bringing, and failing to bring, behavioral phenomena under the scrutiny of a pure natural science.
❧Lawrence Fraley, ED.D. (Professor, West Virginia University, Morgantown, retired)

If all children, youths, and adults, knew why they and all those around them behave as they do, the history of our cultures could have been much more positive. Taking a wonderful leap in the right direction, this book helps all readers to acquire such pivotal understanding as well as practical tools for changing our ways toward a better world.
❧Comunidad Los Horcones, Mexico

Stephen Ledoux's book is an accessible and thorough primer to a modern science of behavior. It is very difficult to find both the breadth of topics and in–depth detail that Ledoux offers in this very readable and important contribution to the field.
❧Michael Clayton, PH.D. (Professor, Missouri State University, Springfield)

This book talks about the journey into a science of behavior and why there is still so much resistance to it. The author explains why pre–scientific mystical, agential explanations are so slow to be replaced by natural–science explanations. Our daily language is filled with vestiges of this pre–scientific world including, as the author points

out, terms like "mind," "self," and even the pronouns we use. The author includes our misguided tendency to use descriptions as explanations. Anticipation of future events, and other "fictional accounts," result in cessation of searching for real causes.
❧Norman Peterson, PH.D. (author, and former professor, Western Michigan University, Kalamazoo)

Occasional blank pages provide space for:

Reader's Notes

What Causes Human Behavior—Stars, Selves, or Contingencies?

Toward a better future

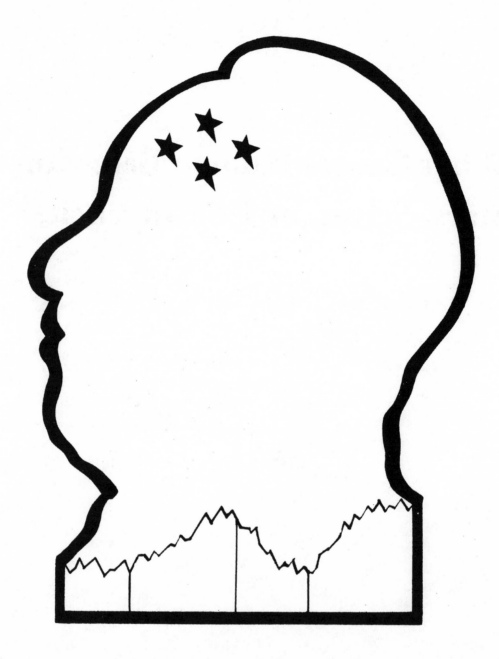

Overt and Covert Behaviors and Contingencies

What Causes Human Behavior—Stars, Selves, or Contingencies?

Stephen F. Ledoux

Published by **BehaveTech Publishing**
Ottawa, Ontario, Canada

What Causes Human Behavior—
Stars, Selves, or Contingencies?

Stephen F. Ledoux, Ph.D.

Published by **BehaveTech Publishing** of Ottawa, Ontario, Canada

ISBN 978–1–927744–14–7 (soft–cover edition)   Electronic versions available

Cover design by James O'Heare & Stephen Ledoux on an image by Janka Dharmasena
    of the *Vitruvian Man* by Leonardo da Vinci

Book layout by Stephen Ledoux   Printed in the United States of America

---

Prospective **cataloging–in–publication data (under "natural science" or "biology" but *not*—since behaviorology is not any kind of psychology—under psychology or "BF"; such cataloging would be like placing a biology book under creationism):**

Ledoux, Stephen F., 1950–.
        What Causes Human Behavior—Stars, Selves, or Contingencies?
        by Stephen F. Ledoux
        xx, 432 p. ill. 26 cm.
        Includes appendix, glossary, bibliographic references, index.
        1. Behaviorology. 2. Biology. 3. Behaviorism. 4. Consciousness. 5. Natural science.
            6. Human behavior. 7. Operant behavior. 8. Global warming solutions.
        I. Title.
        ISBN 978–1–927744–14–7 (acid–free paper)

This edition is printed on acid–free paper to comply with the permanent
paper z39.48 standard of the American National Standards Institute.

Improved printing number:	>10	9	8	7	6	5	4	3	2	1	0
Nearest year of printing:	>2050		2040		2030		2025		2020		2017

Dedication

To all my students, past and passed:

Your efforts, questions, and reactions became part of the contingencies that shaped my repertoire and thus this book.

Other Books by Stephen F. Ledoux

Study Questions for Paul De Kruif's Microbe Hunters (1972)

Grandpa Fred's Baby Tender, or Why and How We Built Our Aircribs
with Carl Cheney (1987)

The Panda and Monkey King Christmas—A Family's Year in China
with first author Nelly Case (1997)

Eight other books of *Study Questions* for various texts, one with other
authors (1999–2015; see BOOKS at www.behaviorology.org)

Behaviorology Majors Make a Difference
with 11 student authors (2013; a reassembly of a 1977 book)

*Running Out of Time—Introducing Behaviorology to Help Solve
Global Problems* (2014)

An Introduction to Verbal Behavior—Second Edition
with first author Norman Peterson (2014)

Origins and Components of Behaviorology—Third Edition (2015)

*Beautiful Sights and Sensations—Small Collections of Native American and
Other Arts* (2016)

Contents

Acknowledgements

$\mathcal{M}$any people deserve thanks and praise for catching various "typos" and other errors that crept into my massaging of a wide range of related materials into the manuscript that became an earlier book and then this book. While any residual problems are of course my own, I want to express my deepest appreciation to a subset of all those deserving folks. These friends and colleagues went above and beyond the usual levels of reviewer assistance. They provided extensive, careful, detailed and exacting suggestions and commentary. Their efforts particularly improved the clarity, readability, and topical coverage of the larger book from which this book directly derived. That larger book was my 2014 book, *Running Out of Time—Introducing Behaviorology to Help Solve Global Problems.* In alphabetical order (and from the USA unless otherwise specified) these people are John Ferreira, Lawrence Fraley, Bruce Hamm (from Canada), Philip Johnson, Werner Matthijs (from Belgium), James O'Heare (from Canada), Jón Sigurjónsson (from Iceland), and William Trumble. For similar assistance directly with the current book, in addition to all those who provided supportive comments (e.g., see the back cover) the people who deserve thanks and praise include Traci Cihon and her graduate students (at the University of North Texas), Nelly Case, Chris Cryer, Jaymes Farrell, Miles Ledoux, Doug Ploehn, Mike Shuler, and Susannah Sudborough. They each and all have my sincerest appreciation (and I hope we are still friends).

The current book not only developed from, but also serves as a primer for, my earlier, more comprehensive and technical book, *Running Out of Time…* From these books I hope all humanity derives benefits. $\mathcal{SFL}$ ∾

Foreword

*M*y work includes serving as the faculty advisor for the Teaching Science Lab (TSL) in the Department of Behavior Analysis at the University of North Texas (UNT). The TSL is responsible for the design and implementation of six sections of a two–course Introduction to Behavior Principles sequence for undergraduate students. The TSL is also a teaching and research lab in which graduate and undergraduate students in Behavior Analysis and related disciplines learn about Behavioral Systems Analysis, Instructional Design, Precision Teaching, the Constructional Approach, and behavior analytic applications to education through their work in the TSL.

Each semester somewhere between 100 and 200 undergraduate students enroll in the Introduction to Behavior Principles classes at UNT. Many of these students have never heard of Behavior Analysis let alone Behaviorology. Most of the students enroll in the course because it is a core course in the Social/ Behavioral Sciences. Others take the course because they know someone who is affected by an Autism Spectrum Disorder (ASD) or because they are encouraged by faculty in other discipline areas such as Psychology, Education, or Speech Language Pathology to learn about Applied Behavior Analysis (ABA). Many others think they are going to learn how to do criminal profiling. Perhaps the most challenging part of our job as course instructors is that the majority of the students who enroll in our courses have a long history of looking anywhere except the environment for the causes of behavior.

The students who enroll in these courses have the misfortune of being part of a generation that is tasked with the job of contributing solutions to some of the world's most challenging problems. However, these students are also some of the most fortunate undergraduate students because they are enrolled at a university that has a Department of Behavior Analysis that is separate from the Departments of Psychology and Education. They can choose to major or minor in Behavior Analysis, a choice many undergraduate students at universities all over the world do not get to make, because there are not many undergraduate programs in Behavior Analysis/Behaviorology. But how does one encourage them to make that choice? How do you encourage a student who has never contacted behavioral science, or the philosophy of behaviorism, to see that taking a natural–science approach to understanding behavior may be the best option they have for contributing to solving some of the most pressing problems facing humanity? How do you tell them that very few behavior analysts are gainfully employed working in areas other than autism intervention but that there is much we can contribute? I do not have the answers to these questions but I do have the opportunity to find out.

In the fall of 2014, the members of the TSL and I became disenchanted with our current instructional design and student learning outcomes. We decided to take on a massive course redesign effort; we wanted a course in which the undergraduate students could learn about Behavior Analysis by experiencing Behavior Analysis as applied to their educational experience. We wanted our students to be able to shift in and out of a behavior analytic world view, to be able to make an informed choice to declare a major or minor in Behavior Analysis, and for those who did not choose to major or minor in Behavior Analysis, to know under what conditions they should call a Behavior Analyst/ Behaviorologist to their interdisciplinary and multidisciplinary teams. We also wanted to help our students to acquire the repertoires that would help them to contribute to more effective strategies to address societal issues.

So we started to look for a new book for the course. We considered Skinner's 1953 book, *Science and Human Behavior,* and reviewed nearly every popular introduction to behavior analysis text available. After settling on one and beginning to construct our course materials, I happened upon a book Dr. Ledoux had recently published: *Running Out of Time—Introducing Behaviorology to Help Solve Global Problems.* I ordered a copy and brought it to the TSL. I knew after a cursory review that this was the textbook I wanted to adopt for the class but the members of the TSL were already six months into preparing course materials based on the text we had previously selected. I passed the book around to the TSL members and within a couple of weeks everyone was on board. We scrapped everything we had done and started again. We had found a text that introduced behavior analysis/behaviorology as a natural science, a text that spelled out how behavior analysis/behaviorology was different from other disciplines, a text that introduced basic principles and techniques of behavior analysis/behaviorology and did it in the context of why behavior analysis/behaviorology was important to contributing more effective solutions to global issues. It was different from other textbooks. It included chapters on the history of behavior analysis/behaviorology. It immediately tied basic principle into applications that are less commonly explored by behavior analysts. But it was written for an audience of more traditional natural scientists (e.g., biologists, physiologists) and natural science practitioners (e.g., engineers). We knew it would be a difficult text for our undergraduate students but we bravely adopted it and we rolled out our first redesigned courses in the fall of 2015.

I cannot remember how or when in this process I came to meet Dr. Ledoux—maybe it was the ABAI conference in the spring of 2015—but I do remember that he was eager to visit UNT and to see our redesigned courses. Dr. Ledoux came to UNT in the fall of 2015. He visited all six sections of our Introduction to Behavior Principles courses, our TSL meetings, signed books for our students, and spent several hours talking to our students and TSL members about the science of behavior (something he now does at least once each year). Dr. Ledoux and I also started an ongoing dialogue. We would talk about the history of our discipline, about the future of our discipline, about how much more we needed to do, and about how to best reach the next generations of behaviorologists. These discussions and our first semesters of outcome data sparked a discussion about the possibility of a new book, one that better suited the aforementioned audience characteristics of our student population. Dr. Ledoux readily agreed to take on this challenge and has worked diligently to produce this book, *What Causes Human Behavior—Stars, Selves, or Contingencies?*

In *What Causes Human Behavior—Stars, Selves, or Contingencies?* Dr. Ledoux brings the natural science of behavior to a broader audience, which is ideal for an undergraduate–level Introduction to Behavior Analysis course. Beginning with why a natural science of behavior is not only important but also necessary for our survival, and ending with a plea for behaviorological solutions that decrease the overconsumption of the limited resources available in our global community, Dr. Ledoux has written a primer to *Running Out of Time...*, preparing his readers for further study, reflection, and application of the natural science of behavior.

At times, readers may become frustrated with the subject matter of this book; it is more science than they may have ever thought was involved with behavior. They may resist the challenges Dr. Ledoux poses to his readers as they are pressured to question their beliefs about why behavior occurs. Others will marvel about why they have not been introduced to behaviorology before coming into contact with this book. Perhaps most importantly, the readers of this book will be an integral part of the next generation of society who will decide what role, and how much of a role, the natural science of behavior will have in contributing to solutions to the world's most pressing social issues.

Traci M. Cihon, Ph.D., BCBA–D
Associate Professor
Department of Behavior Analysis
University of North Texas
May 2017

On Typography & Related Resources

*T*his book is set in the Adobe Garamond, Adobe Garamond Expert, and Tekton collections of typefaces. In addition, a valuable basis for the typographic standards of this work deserves acknowledgment. As much as possible, this book follows the practices described in two highly recommended volumes by Ms. Robin Williams (both of which Peachpit Press, in Berkeley, CA, USA, publishes). One is the 1990 edition of *The Mac is Not a Typewriter.* The other is the 1996 edition of *Beyond the Mac is Not a Typewriter.* For example, on page 16 of the 1990 book, Williams specifies practices regarding the placement of punctuation used with quotation marks, an area in which some ambiguity has existed with respect to what is "proper." In addition the present book follows the advice in these books about avoiding "widows" (which is the name for leaving less than two words on the last line of a paragraph) and "orphans" (which is the name either for leaving the first line of a paragraph alone at the bottom of a page, or for leaving the last line of a paragraph alone at the top of the next page). Also, since some confusing alternatives remain regarding the use of hyphens and dashes, this book would simply limit hyphens to separating the parts of words that break at a line end, although this book never breaks words at line ends. Then, "en dashes" most commonly separate the whole words of compound adjectives, and "em dashes" set off multiple–word—a compound adjective with an en dash—phrases or clauses. (Note that ebook formatting typically destroys most of these easy–reading characteristics that developed across humanity's centuries of successful printing–press experience.)

You can address correspondence regarding this book to the author (at ledoux@canton.edu). For more information, visit www.behaviorology.org where you can find all the back issues of the journal of TIBI (The International Behaviorology Institute). Previously named *Behaviorology Today* (ISSN 1536–6669), TIBI renamed it *Journal of Behaviorology* (ISSN 2331–0774) in 2013.

Some related books may also interest the reader. One is my 2014 text, *Running Out of Time—Introducing Behaviorology to Help Solve Global Problems.* Another, first published in 1997, is a 2015 book of my thematically related readings, *Origins and Components of Behaviorology—Third Edition,* which also contains contributions by other TIBI founders, including Glenn Latham and Lawrence Fraley. Also, consider Fraley's 2012 *Dignified Dying—A Behaviorological Thanatology* and his 2013 *Behaviorological Rehabilitation and the Criminal Justice System.* To order the books by Fraley, write the publisher, ABCs, at ledoux@canton.edu. Order *Running Out of Time...* or *Origins and Components...* directly from the distributor, **Direct Book Service, Inc.,** at 800–776–2665. They will likely answer the phone with "Dogwise," because one of their long standing and most popular specialities involves books about our canine friends; several of these books already specifically apply the laws of behavior that *Running Out of Time...* systematically introduces in detail.☙

"Introduction"
(Traditionally, the *Preface*)

Starting in the last decades of the twentieth century, traditional natural scientists (e.g., physicists, chemists, and biologists) and engineers turned their attention to solving new, major (and minor) problems around the globe. They quickly discovered that both the problems and the solutions involve human behavior and changes in human behavior. Yet most of them were generally unaware that a 100–year–old natural science of behavior, now called behaviorology, was available to address its part in the solutions through its engineering counterpart, "Applied Behavior Analysis" (ABA). The 100th year of behaviorism—its centenary year in 2012—provided an occasion to review this discipline for them, if ever so briefly, in the form of an abridged article. Entitled "Behaviorism at 100," the article appeared in the first issue (January 2012) of the centenary volume of the journal *American Scientist* (available at www.americanscientist.org). The unabridged, peer–reviewed version appeared in volume 15, number 1, of the journal *Behaviorology Today* two months later (available at www.behaviorology.org). To support the teamwork of all the natural sciences in solving local and global problems, this book provides some elaboration of the basic science content of the unabridged article, using more examples and less technical terminology.

The focus of this book, however, resides elsewhere rather than solely in the value of behaviorology for helping solve global problems. This is not because such a focus lacks merit, but because another book already supports that focus. My 2014 textbook, *Running Out of time—Introducing Behaviorology to Help Solve Global Problems* (often simply called the *ROOT* book or, for the cover color, "The Green Book") makes such connections throughout its pages.

The present book also connects with "help humanity save itself and the planet." However, for most readers, this book likely represents their first behavior–related *scientific* book. So it limits this connection mostly to a later chapter. Instead this book focuses simply on enabling some basic understanding of many fascinating points about human behavior in ways that not only interest people but also help them provide answers for their questions about the topic.

In the process of fulfilling that focus, this book must also address the many traditional, and pre–scientific, views about behavior that have proven so misleading for people, such as many views in astrology and psychology. To neither of these does the label, "natural science," apply. Using pre–scientific phrasing, they ask the traditional question of "Why do people do what they do?" And the answers that these disciplines provide for this question have proven inadequate. If we rephrase the question scientifically, then we can ask it of behaviorology: "Why does human behavior happen?" Basic answers to either form of the question, from these three unequal disciplines, could be paraphrased this way: Human behavior occurs, "because of our stars (astrology)" or "because of our selves (psychology)" or "because of our contingencies (behaviorology)."

Behaviorology answers with the term, "contingencies," because this term refers to the "if–then" relations between behavior and its causes. That is, *contingencies* encompass, in general, all the dependencies, all the interrelationships, between behavior, as an effect, and all of its various and interacting natural–science causes. These include "nature" and "nurture" (i.e., genes and physiology and environment) as we shall later explore.

Some comment on the astrology and psychology alternatives is pertinent first. Historically astrology basically claimed that star or planet positions cause our behavior. And currently psychology claims that non–physical agents inside us, such as selves, cause our behavior. Spontaneously, the selves supposedly tell the body what to do. This repeats in secular (i.e., non–religious) terms a long–standing claim of theology (i.e., the study of religion) that souls cause our behavior by telling the body what to do. Today, astrology maintains a certain entertainment value (e.g., daily newspaper horoscopes). Psychology, however, is taken much more seriously (as is theology). For that reason, this book includes comparisons and contrasts particularly between behaviorology and psychology.

Alluding to historical points that we cover in early chapters, both the first 100 years of the natural science of behavior, and its more recent (in 1987) complete separation and independence from psychology, *as behaviorology,* are now behind us. Thus we must clarify that this natural science of behavior, behaviorology, is neither a part of, nor any kind of, psychology, which means that this book is *not* a book in psychology or for it. The term, behaviorology, names the basic natural science behind a range of applications, including Applied Behavior Analysis, which is so vital in the treatment of autism.

With that kind of background, several options competed for the title of this book. Some possibilities still pushed the "help solve global problems" focus, like *"So Many People, So Little Time,"* or *"Behaviorology—Advancing Sustainability."* Other nice titles, like *"Saving the Planet Through Cultural Shift,"* pushed a bigger scope than the book covers. This book, however, mainly provides a primer for behaviorology and its scientific answers for the "why behavior happens" questions. So, in the context of dealing with traditional, taken–for–granted, pre–scientific and contradictory answers for these questions, the title *"What Causes Human Behavior—Stars, Selves, or Contingencies?"* works well.

This book can, however, provide a couple of curious challenges for some people. For one, reading it begins the process of countering the current effects of the previous 50,000 years or more of pre–scientific cultural conditioning about human nature and human behavior, something that we all grow up experiencing. This includes all the myths and legends and explanations, from multiple cultures, that may still hold life lessons but whose foundations can no longer bear close, especially scientific, scrutiny. Nevertheless, we all grow up surrounded and cultured—the usual term is *conditioned*—by these views about human nature and human behavior, which induces our taking them for granted. Should something come along that calls them into question, the result can be intellectually and emotionally challenging. Happily, reasonable exposure to the realities of the natural science of behavior can meet such challenges.

A different kind of challenge concerns some gradual changes in the way we talk and write. This book implements talking and writing changes that help move us away from our reliance on sentence phrasings that inherently contradict our scientific knowledge about behavior. Instead we shift to phrase forms and word usages that typically get criticized as non–standard forms and usages. We then use them in ways that better support our scientific knowledge about behavior, thereby helping make them into standard forms and usages. For example occasionally we use nouns as adjectives, and we use adjectives as nouns for the subjects of sentences. This reduces the reliance on pronouns as subjects of sentences. The result may start out seeming awkward. But it moves us toward a more efficient grammar that better supports scientific realities. This shift in phrase forms and word usages contains fewer inherent implications of ghosts (i.e., mystical—as in untestable or unmeasurable—behavior–directing agents) residing inside bodies, a point that also receives more complete treatment in various chapters. I beg the reader's indulgence in support of these grammatical shifts. As readers will discover as we move through the chapters of the book, the benefits of these shifts gradually increase, and far outweigh the risks of maintaining traditional aspects of grammar that have been harming us for millennia by coincidently supporting anti–science superstitions.

The chapters of the book divide into two parts. Part i pursues history, assumptions, concepts, principles, and practices. Part ii pursues more complex developments, along with the beginnings of some natural–science answers to some of humanity's long–standing questions. You see these topics briefly described under the chapter entries in the Table of Contents.

This book describes many of the discoveries and applications of the first 100 years of behaviorological science, particularly regarding *why* human behavior occurs. The undercurrent throughout the book concerns the relevance of this natural science not only to helping build a sustainable society in a timely manner but also to addressing the wide range of personal, local and global issues confronting individuals and humanity. All these issues have definitive behavior components. So addressing these issues benefits from a comprehensive natural science of behavior. These issues range from how to train a pet to the problem of overpopulation, which is perhaps humanity's most fundamental practical problem. No book, including this one, solves such problems by itself. But the more humanity understands about the causes of behavior, which is the topic of this book, the more success each of us, and humanity in general, will have at solving such problems. We can even solve them in the timely manner that will prevent us from having to experience the worst effects of major problems like global warming.

Some folks feel that finding out "why people do what they do" (or, more scientifically, "why human behavior happens") is so important and fascinating that everything about the topic must *immediately* appear, now, in this chapter! While that cannot occur, their enthusiasm is praiseworthy. Meaningful coverage of a topic like human behavior takes time, and space, *and pages.* We need to address our topic not only fairly and reasonably but also in a manner that can leave readers able to provide appropriate answers to their own additional questions, including how to further apply this knowledge. Such topic coverage requires at least this whole book.

Along those lines, this book follows the common, natural–science educational pattern. This pattern consists of first briefly mentioning a topic, or series of topics, some of which might be related. Then, we revisit these topics, developing them in greater depth and with more details and examples. At the same time, we also continue to mention new, additional and possibly related topics, at least briefly. And these cycles repeat, chapter after chapter, with a bit of review but always with greater depth and more details and examples and interconnections. Thus, you may feel that our coverage of some principle or concept seems inadequate at some early point. Fear not. We will return to it, especially after covering newer related points that help us put things together. We have even already taken the first steps for some cycles by mentioning some topics in this introductory chapter. You will see the topics of these cycles return soon. Meanwhile, you may sometimes run across one or another non–technical word that seems new to you. Your English dictionary can then serve you well. On the other hand, you can find some technically defined behaviorological terms in the Glossary. Or, you may read the Glossary as a sort of sleep therapy.

Lastly, and by design, this book invokes a vast range of examples from ordinary, everyday, and mostly human behavior. As a result virtually every reader can surely find something to dislike in the book, something which actually only reflects the vast and wonderful natural complexity of life. While for most readers this book serves as a first book about behaviorology, some readers may find that it does not go far enough for them into the systematic substance and components of the behaviorology discipline. For example, this book says little about experimental methodology or other particularly complicated technical topics in the discipline, such as the topic of "equivalence relations" with its vast implication, including for regular public education. In such cases the easiest next step takes readers to my 2014 *ROOT* book, which does cover these areas. The *…Related Resources* material, which appeared just before this chapter, contains additional, pertinent information about the 2014 book. Beyond that, a wide selection of books and articles, listed in the bibliography, can satisfy much remaining curiosity.

Addendum. Before closing this chapter, let's *briefly* consider three advanced topics (or you can skip them). Touching on them helps demonstrate, by example, why this book leaves advanced topics for the next–level book.

The three topics involve a certain omission in this book, and two characteristics of this book. The omission concerns "pragmatism," while one characteristic concerns the similarity between this book and its predecessor,

ROOT, and the other characteristic concerns the connection between the discipline of behaviorology and the discipline of physiology.

Let's begin with the similarity between books, because it stems from the historical events that produced this book. While my earlier *ROOT* textbook proved appropriate for advanced undergraduate majors and beyond, it seemed a bit too thorough or difficult for first–semester students still unsure of their major. So a professor who mostly had first–semester students in class asked me to downsize the *ROOT* book, and quickly (i.e., for the following fall semester, a time frame that precluded writing a completely new book). She also suggested using less technical terminology for this different, more general audience. The book you hold is the result. Of course, this led to many sets of pages containing much the same content between this book and the *ROOT* book, although in the *ROOT* book, topics receive more thorough coverage with more emphasis on technical details. As a result some worries might arise about so–called "self plagiarism." To me, however, this term is somewhat "self contradictory" in that it contradicts the main meaning of plagiarism (i.e., to steal and use *another's* ideas or writings as one's own). One cannot really "steal" one's own ideas or writings. So allow me to make two suggestions to anyone worried about the similarities between some parts of these two books. (a) Enjoy them and benefit from them, as these similarities are part of why these two books make a great reading sequence. More importantly (b) let's all worry instead, and work in a timely manner, on helping humanity save Planet Earth. At present this needs to become the highest priority, and our science has much to contribute.

In addition this new book (and even to some extent the *ROOT* book) is not really for those of us who are already familiar with the natural science of behavior, although the occasional review of the discipline does provide us with benefits. Instead this book is principally a primer for helping to bring our science to the attention of everyone else. Indeed, consider that *many* good books exist that *cover applications* of this science while just providing a *summary* of the science *to help cover the applications better* (e.g., in my opinion the best one is Whaley & Malott's 1971 book, *Elementary Principles of Behavior*). Since the appearance of Skinner's 1953 book, *Science and Human Behavior*, however, how many books cover this *science per se,* while just providing some applications *to help cover the science better?* This and the *ROOT* book try to follow the latter, using some applications *mainly to support the science coverage.* The *ROOT* book probably does this better, but the culture has inadequately prepared everyone for reading it, which is why that professor asked me to down–size it.

Next let's briefly consider this book's omission of pragmatism. As a typical definition, pragmatism refers to the value of a course of action residing in its measurable consequences. This book omits any detailed discussion, because the topic is more complicated than topics appropriate for a primer. Nevertheless, in the opinion of some disciplinary professionals, Skinner's pragmatism is his most important contribution to the science. For example Dr. Guy Bruce describes Skinner's pragmatism this way:

By pragmatism, I mean his preference for effective and efficient methods for discovering effective and efficient explanations of behavior change and procedures for changing behavior. By way of illustration, consider his pragmatic definition of respondent and operant behavior (as adaptive interactions between organism and environment), his interest in discovering orderly relationships between organism actions and environmental events, both of which are natural, not supernatural, events directly observable and measurable, specifically the type of environmental events [that scientists can study] allowing the discovery of useful explanations and procedures for changing behavior. Taking pragmatism as the starting point, we can see how all aspects of the science of behavior that Skinner created are consistent with that pragmatism, how we define behavior, how we measure it, within and across subject experimental manipulations, explanations, and procedures (personal communication).

Much merit resides in that view. Still, in my opinion, Skinner's analysis of verbal behavior likely represents his most important contribution. (Books mentioned in this *"Introduction" chapter/Preface* appear in the Bibliography.)

Finally, briefly consider this book's characteristic of including the overlapping interests of physiology and behaviorology. This inclusion causes reactions from "pause" to consternation among some of those who are already natural scientists of behavior. But such reactions seldom occur among traditional natural scientists (e.g., physicists, chemists, biologists) who see this inclusion as appropriately clarifying our intersection with the traditional natural sciences. It helps explain, for them, our natural–science status.

Without thoroughly including these connections in "first–contact" books like this one, people contacting our science for the first time have difficulty seeing that, and accepting our pointing out that, we are not part of psychology. All their conditioning has been to see psychology as *the* behavior discipline (beyond theology). But, as we later discuss in detail, psychology cannot maintain the natural functional history of events across multiple disciplinary levels, as expected and demanded by traditional natural sciences, because psychology breaks that history when it includes its spontaneously acting, mystically invented inner agents in the causal sequence. But our science does maintain the natural functional history of events across multiple disciplinary levels, rather than break it, because the natural variables that we address *are already parts of that history.*

For such reasons I include our discipline's connections with physiology in this book, not that our discipline cannot stand alone, because it can. But we need not, and cannot afford to, stand alone. All the natural sciences must stand together to solve the planet's problems in the required timely manner. Such is the direction that this book takes.

Stephen F. Ledoux   Los Alamos NM USA   June 2017

What Causes Human Behavior— Stars, Selves, or Contingencies?

As we shift from pre–scientific understandings to helping build a sustainable society in a timely manner, answering this *matters:*

Are the causes of our behavior in our stars, in our selves, or in our contingencies?

Stephen F. Ledoux

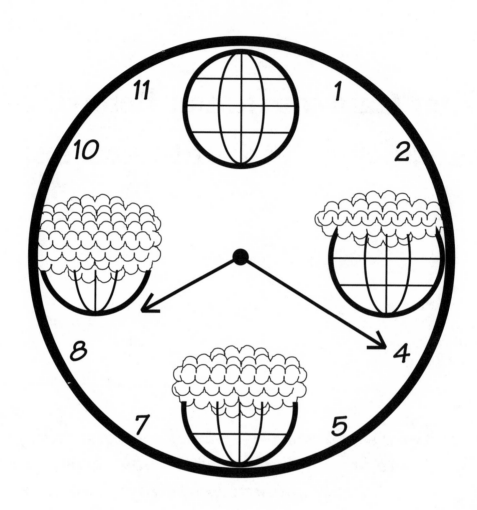

Running Out of Time

PART I

Basic Considerations and Contingencies, including some bits of History, and Philosophy of Science, plus some Concepts, Principles, & Practices

Reader's Notes

Chapter 1
Reasons for Interest in the Natural–Science Causes of Behavior:

Why should humans bother about behavior? With the *Law of Cumulative Complexity*, contingencies cause behavior rather than stars or selves. …

*W*elcome aboard! Sit back. Relax. We are beginning a journey—sometimes uncomfortable, often fun, always revealing—to explore some fascinating and recent (in the last 100 years) scientific discoveries, and the benefits we derive from them. These discoveries concern nature, human nature, the nature of human behavior, and even consciousness and reality. Additional topics also have direct and significant effects on human existence and experience, although at this point many will surely sound technical. These topics include shaping, fading, evocation, emotions, feelings, reinforcement, extinction, personhood, verbal behavior, values, rights, ethics, morals, robotics, the alternatives to coercion and punishment, and even life and death. The behaviorology discipline covers far greater detail for many more topics. Many of these constantly change little by little as new research data and conclusions accumulate. In comparison, what appears in these pages provides some stable highlights as a primer for the story of behavior and its scientific causes. Of course, the first chapter always seems hardest, because it overviews so much. But it is worth the effort, although it is also the only chapter you could skip, if you return to it later. As I implied in the *Introduction* (which we would traditionally have called the "Preface"; you did not skip it, I hope) this book aims to appeal to a broad, general audience. To that end, "plain English" appears in place of technical jargon wherever possible.

Having introduced our use of the terms "stars," "selves," and "contingencies" in the Introduction, the time has come to expand the term "contingency" a little. As an if–then relationship, if one variable happens, then the other variable happens. Each relationship involves a dependency between variables. One variable causes behavior while the other variable *is* the behavior as the effect of the cause. A dependency exists between the effect and the cause. The occurrence of the effect *depends* on the occurrence of the cause. Given the title role that "contingency" holds in this book, while gradually getting used to it, try substituting "dependency" for it. As we will see in the next chapter, science generally avoids the various bits of philosophical baggage that come with words like "cause" and "effect." We can, however, still employ them.

We approach the story of behavior and its causes as natural scientists. Traditional natural sciences cover the basic subject matters of energy, matter, and life forms as well as their related extensions. These thus include physics, chemistry, and biology, along with astronomy, geology, physiology, and so on. We can see the beginnings of all these with Galileo's telescopic observations in the early 1600s. The natural science we examine in these pages, emerging over the last 100 years, concerns the subject matter that here we can call "life functions." It not only covers but clarifies and expands the content of the ancient dictum, "know thyself." This natural science encompasses all behavior, especially human nature and human behavior, the behavior of you and me, of friends and trouble makers, of the ordinary and the extreme, of the shy and the obvious. Thus focused on behavior, we call this science "Behaviorology," and we explore its historical, conceptual, philosophical, analytical, and engineering aspects as part of discovering many of its principles, laws, and applications.

We begin this chapter by looking at the big picture. This orients us to some crucial historical, philosophical, and scientific premises behind the natural science that constitutes our subject matter. This chapter lays out some lines of progress, occurring over the last 100 years in the natural science of human nature and human behavior, for which later chapters provide some details.

As part of this discovery quest, you may like to access additional resources that provide other details and verifications. In this book such sources initially appear in the citation form of "author, year–of–publication" (usually in parentheses). The full reference then appears at the end of the chapter. To avoid the interference with reading that mentioning many citations can cause, most chapters contain only a few citations, relying instead on readers following up through the extensive resources listed in the bibliography.

Preview

Many human intellectual endeavors start out on the foundation of some basic assumptions. We use the term *philosophy* to encompass the combined variety of assumptions behind any endeavor, and we find that those assumptions actually exert a vital controlling element that helps keep the behavior of the concerned parties within appropriate boundaries. This control occurs in both scientific and non–scientific activities, and even underlies many emotional activities. We describe this control by saying that the philosophy "informs" the endeavor. In our case some familiarity with the philosophy behind science in general, which means natural sciences, and behind the particular science that interests us, behaviorology, thus becomes necessary.

While the next chapter provides some details, we begin here. With scientific activities we use the term *philosophy of science* to denote the set of informing assumptions. We cover philosophy of science just enough to clarify its role and value, especially for behaviorology, the natural science of behavior, which is filling a chair at the roundtable of the basic natural sciences along with physics, chemistry, and biology. We use the term *naturalism* for the general philosophy of science of all these natural sciences.

Some sciences also develop additional assumptions that extend naturalism to further inform their work. The natural science of behavior is one such science, because studying human nature and human behavior scientifically presents concerns that are specific to particular aspects of behavioral phenomena. We use the term *behaviorism* to denote the current extensions of naturalism that inform the natural science of behavior. Actually we use the term *"radical" behaviorism*. It is not radical in the sense of extremist. It is radical in the sense of comprehensive. We will say more about radical behaviorism in due time.

To lay a solid foundation upon which to cover our behavior subject matter, we devote some of this first chapter to introducing some of the assumptions, and the rationale behind them, that inform the natural–science approach to human nature and human behavior. Furthermore, to paint the big picture for the larger context encompassing our behavior–science study in this book, this chapter also introduces the historical perspective by summarizing the first 100 years of behaviorism. This includes mentioning some of the discoveries and developments in the natural science of behavior that behaviorism (i.e., radical behaviorism) informs. (For articles that further cover the first 100 years of behaviorism, see Ledoux, 2012a and 2012b.)

Here is a quick abstract of this chapter that should help: Summarizing 100 years of the philosophy of science that we know as behaviorism, and its impact, starts with a description of B. F. Skinner's 1963 article, "Behaviorism at fifty," which covered the first 50 years (1913–1962). Then, a review of the second 50 years mentions interrelated developments in the philosophical, scientific, historical and organizational domains: (a) We consider Skinner's radical behaviorism, the philosophy that extends naturalism to inform the *natural* science of behavior that today we call behaviorology, after its separation from the non–natural, fundamentally mystical discipline that defines itself as studying "behavior and the mind." (b) We ponder some of the advances of behaviorological science. And (c) We mention some historical and organizational developments in this discipline. All these contribute to reducing global superstition, extending global science, and solving global problems.

Some Background for this Book

Part of what prompted this book was the natural science of behavior reaching 100 years old. The 2012 centenary year of behaviorism coincided with several interrelated circumstances: Many traditional natural scientists were working to solve problems like overpopulation and global warming within the limited time frame available before we must experience their worst effects. In this process they noticed the substantial human–behavior components of both the problems and the solutions. Yet they lacked both knowledge of, and full access to, the natural science of behavior, which makes natural scientists in general another audience for this book. Then, in 2012, *American Scientist,* the journal of a major natural–science organization—Sigma Xi, the Scientific Research Society—celebrated its centenary as well. This prompted the editor of *American Scientist,* David Schoonmaker, to accept my "Behaviorism at 100" article (Ledoux, 2012a), and to introduce it to readers with the comment, "Over its second 50 years, the study of behavior evolved to become a discipline, behaviorology, independent of psychology." Two months later *Behaviorology Today* (now called *Journal of Behaviorology)* published a longer, fully peer–reviewed version entitled "Behaviorism at 100 Unabridged" (Ledoux, 2012b). The interest that these two articles generated then prompted my efforts to produce book–length introductions to behaviorology (Ledoux, 2014, and the book you are reading). While other available books emphasize *applications* of this science to particular concerns, these books emphasize the science itself.

Some Behavior Science Parameters

Behaviorism, which actually refers to a philosophy of science, officially began with an article by John B. Watson in 1913. Over the next several decades, behaviorism developed some increasingly specific flavors. Across his career (from the early 1930s through the late 1980s) B. F. Skinner, along with his colleagues and students, built not only the natural science of behavior but also the current most developed behaviorism flavor, radical behaviorism. This informs a basic, separate *natural* science of behavior (i.e., a science of the *functions of life* involving energy exchanges between bodies and environments). This recently emerged, independent discipline of life functions complements traditional natural sciences of energy, matter, and life forms. It also shares in solving local and global problems by showing how to discover and effectively control the variables that cause behavior. Such a topic affects everyone.

In 1963, Skinner published his "Behaviorism at Fifty" article. In it he reviewed the varieties of behaviorism and the directions of natural behavior science. By the 1960s so–called common wisdom saw experimentally discovered laws of behavior as largely irrelevant to normal humans, and limited applications mostly to treatments for "psychotic" individuals and training for other animals. On scientific as well as philosophical grounds, Skinner's article emphasized the fallacy (i.e., error) of such views. Subsequently, data accumulating over the next 50 years validated his position that the natural laws governing behavior are relevant to *all* behavior of humans and other animals.

The 1960s were also a time when natural scientists of behavior continued their attempts to turn the discipline of psychology, where many of them had historically worked, into a natural science. Over the next 50 years, psychologists always adamantly refused these attempts; they liked the contingencies of their social and economic status, and no amount of data–based facts seemed able to change such contingencies.

So, natural scientists of behavior gradually took their discipline outside psychology. They recognized a separate independence for their established natural–science discipline, and some of them acknowledged this independence formally in 1987 using the name *behaviorology*. That name is synonymous with "the natural science of behavior" and is conveniently shorter. At present quite a number of possible natural scientists of behavior still use an older label for this discipline. They use the label, "behavior analysis." But psychology also claims this label, which costs behavior analysts some scientific credibility among traditional natural scientists. Meanwhile, only the behaviorology name definitively indicates the behavior–related discipline that remains completely separated from psychology and any discipline that accepts or espouses superstitious or mystical (i.e., non–scientific) accounts for behavior.

Different Behavior Disciplines

Actually, several different disciplines claim behavior as a subject matter, including theology, psychology, and behaviorology. They are not equal. While the distinction between theology and science got established centuries ago, some confusion lingers over the other two, so we should start by clearing this up. Behaviorology—and radical behaviorism—are neither any part of, nor any kind of, psychology!

The relation between behaviorology and psychology approximates the relation between biology and creationism. Both psychology and creationism claim to be sciences, because they use scientific methods. However, neither of them qualifies as *natural* science, because they both appeal to entities or events *outside* of nature (i.e., to superstitious or mystical or supernatural or anatural or otherwise non–natural events). Also, natural scientists fail to see the *secular* mystical accounts of psychology as any sort of improvement over *theological* mystical accounts.

The implications of those relations deserve consideration. If discussing the implications brings about appropriate action, the results can elevate the status of naturalism and the natural sciences, lead to solving more human problems, reduce susceptibility to superstition and mysticism (both theological and secular), and improve human intellectuality, rationality, and emotionality.

Naturalism, the general philosophy of science in the natural sciences, has among its characteristics one that particularly helps achieve those outcomes: Natural scientists respect what they call the "continuous, natural, functional history of events." This refers to the sequential, "causal" connections that events have with the events occurring before and after them. Respecting these connections enables scientific analyses to show parallels across disciplinary lines, and that enhances mutual respect among sciences. In contrast, ignoring the natural functional history of events can lead to unnecessary, even dangerous compromises between some natural sciences and non–scientific disciplines that make claims of mystical origination of events. An example is the 400–year–old, ultimately unhelpful, and yet still extant, give–away to theology of human nature and human behavior considerations in exchange for religion leaving science alone as it covers nearly everything else. Yet all the present global problems *require* some input from the *natural science* of human behavior for their solutions. Without this input we likely cannot solve our problems, which leads to species extinction. This need provides extreme contingencies that make humanity unable any longer to afford this compromise, if it ever could.

Related Natural Sciences

Conversely, here is a two–part example of cross–discipline parallels that stems from respecting the natural functional history of events: Behaviorology accounts for a stimulus evoking a behavior; for example the approach of a fast and close moving object evokes ducking the object. Meanwhile, in parallel, biology provides details at the physiological level about how an energy change at receptor cells affects behavior; for example the light reflected from the approach of a fast and close moving object strikes the retina and so sets off a cascade of changes through the nervous system that mediates—*not* originates—a behavior, like ducking. The occurrence of the energy caused the ducking by causing the nervous system activity. In addition, chemistry works on details at the cellular and sub–cellular level about the physiological events, while physics strives for atomic and sub–atomic details about the chemical events.

All four disciplines together provide a comprehensive account of the events. But if natural scientists instead compromise by allowing claims that behavior in general—or ducking in this particular case—results from the spontaneous, willful act of some putative inner agent, then they lose the whole subject matter of human behavior—a subject matter whose application is likely vital for human survival—to purveyors of non–science. Such compromises give undeserved status to mystical accounts, which causes natural science to lose ground, reducing its benefits. The resolution now revolves around understanding behaviorology better so as to avoid falling into such compromises.

Foundations in the First 50 Years

Skinner's 1963 treatment of behaviorism began by describing the primitive origins of mentalistic or psychic explanations of human behavior. His concern, however, was not the primitive origins of these explanations, but their continued use in psychology, a current discipline. This discipline began in the 1800s, when its original philosopher members wanted to be "scientific." If one restricts the term *scientific* to the use of scientific methods, as they did, then much of the psychology discipline could be construed as scientific. But *science,* as we typically understand the term, also includes adhering to the facets of naturalism. Perhaps the most basic facet maintains that science deals solely with natural—real, measurable—events as independent and dependent variables. No place remains, or is allowed, for non–natural (i.e., mystical, fictitious) events.

Such facets of naturalism tie the natural sciences together, and distinguish them from other disciplines.

As a discipline, however, psychology maintains a range of non–natural, even anti–natural, alternatives among its disciplinary schools. This alone excludes it from the natural sciences. In addition psychological explanations derive from common, traditional cultural biases by ultimately tracing to some type of mystical inner being or entity. Sometimes these remain little more than secular versions of the theological soul, like a behavior–originating mind or a behavior–initiating self agent. Supposedly these agents spontaneously consider various factors, decide what to do, and tell the body to do it. With these processes, or sub parts like id, ego, motive, choice, or trait, we cannot trace the behaviorological, physiological, chemical, or physical links of a natural functional history chain; we can only trace the causal chain back to the supposed spontaneous wilful act of the self agent. This breaks the chain of events in the natural functional history and so further excludes psychological analyses from natural science. However, the competition that these psychological analyses represent absorbs societal resources that could otherwise expand natural–science functional analyses and their related beneficial behaviorological–engineering applications. Such were some of the ongoing problems from the first 50 years.

Also in his 1963 paper, Skinner dealt explicitly with the question of "mind" (the quintessential self agent). His answer began with Charles Darwin's continuity of species. With humans and other animals qualitatively similar, some researchers looked for, and claimed to have found, human characteristics in other animals including consciousness and reasoning. Testable accounts that were simpler (i.e., more parsimonious) than claiming an animal "mind," however, could explain such findings. And if that was so with other animals, then more parsimonious accounts could explain such characteristics for humans also. Trying to discover parsimonious accounts prompted the natural, experimental science of behavior that began with Skinner in the 1930s. The point was not merely to discover the naturalistic explanations of human behaviors, including complex behaviors such as consciousness, language, reasoning, imaging, and thinking (both verbal or visual). The point was also to discover these naturalistic explanations initially as alternatives to, and ultimately as replacements for, superstitiously (i.e., mystically) grounded explanations, and then to apply these explanations for humanity's benefit.

Early Natural Behavior Science

Behavior, a natural phenomenon, happens—and changes—because independent variables affect the particular body structures that mediate it. No mysterious inner self agent *does* the behaving or instructs the body to behave. Instead the experimental literature describes two basic and continuously operating conditioning processes that we call respondent (i.e., Pavlovian) conditioning and operant (i.e., Skinnerian) conditioning. We cover both in later chapters, but a quick overview can set the stage for readers.

Both types of conditioning involve energy transfers between the environment (internal and external) and the body. These transfers occur in ways that, as our physiology colleagues can show, trigger cascades of neural firings. Such firings induce the greater energy expenditures involved in bodily movements. Such firings also induce small but crucial changes in neural structures that make the body a different body. That different body then mediates (not initiates) behavior differently on future occasions. Many people call the process "learning," but no inner agent—no "learner"—is present to "do" the learning (so we seldom use the term learning).

Early in the 1900s, Ivan Pavlov discovered respondent conditioning. It involves the "pairing" (i.e., the overlapping or successive occurrence) of neutral stimulus events with stimulus events that already (due to genetically determined neural structures) elicit responses. These pairings transfer energy to the body resulting in nervous–system changes that alter the way in which the neutral stimulus events function; they come to elicit responses also (because the body has changed, not the stimulus). Emotion and feelings and other reflexes and reactions involve respondent processes.

Some years after Pavlov's discovery, B. F. Skinner discovered operant conditioning. He called it "operant," because it involves stimulus–evoked responses affecting—as in "operating on"—the environment in ways that produce, as consequences of those responses, changes in the environment. These changes transfer energy back to the nervous system where they produce structural nervous–system changes that make similar future evocative antecedent stimuli more functional (e.g., more effective or efficient; see Seo & Lee, 2009).

The involvement of antecedent stimuli (i.e., stimuli occurring before a behavior) increases the complexity of operant behavior. During episodes of operant behavior, a wide range of such antecedent stimuli are usually present at any given time. We are always surrounded by sometimes vast numbers of individually identifiable stimuli. (See for yourself. Look around and try to count all the individually identifiable stimuli in just your immediate surroundings. Be careful, however, because such efforts might induce sleep.) In any case some of these antecedent stimuli work singly, but more often they work in combination. At other times they seem to compete with each other to affect us. Still, some among these antecedent stimuli evoke, or come to evoke, operant behavior. This happens based on current neural structures that occurs from species and personal conditioning history. After they evoke the behavior, additional stimuli that the behavior usually produces may then immediately consequate the behavior and so alter the neural structures that mediate the behavior. This alteration changes how readily future similar antecedent situations or stimuli will evoke the behavior. Walking, talking, singing, dancing, loving, thinking, studying, working, fighting, planning, partying, publishing, and problem solving—including engineering and scientific research and writing—are all among the typical and ongoing results of operant processes (i.e., contingencies).

With several later chapters elaborating many details and examples, we will get to respondent and operant conditioning processes separately. This is because the kinds of consequences occurring in operant processes appear to have little effect on the behaviors produced through respondent processes.

Places for Physiology and Consciousness

The physiology that mediates behavior, and the genetic basis of that physiology, are both real. However, we generally have little if any access to them when the time comes for practical behaviorological engineering interventions to alter environments in ways that produce differences in behavior. For example when the efforts of teachers make their students' behavior repertoires expand, the efforts focus not on physiology or genetics but on changing the instructional environment. With respect to behavior, the variables of physiology and genetics remain essentially inaccessible even to teachers.

Physiology and genetics remain real, as in measurable and testable. However, homunculi (literally, little beings), minds, or wills, remain unmeasurable and untestable. Thus they cannot relate to the basic point of all sciences, which is the prediction and control (i.e., experimental control) of the subject matter. Nothing in the nature of behavior requires a homunculus, mind, or will in accounting for any behavior. Indeed, the spontaneous production of responses by such self agents not only violates the conservation of energy principle but also mimics the spontaneous generation hypothesis in early biology that Pasteur put to rest. Other real variables, the variables in contingencies, cause behavior.

In his 1963 article, Skinner also described the expansion of naturalistic explanations toward a more complete scientific account of behavior. In this expansion Skinner discerned that the question of consciousness not only holds a central challenge but also attracts substantial attention and must be taken into account. Looking ahead a bit, in an excessively brief manner, this science begins accounting for consciousness in terms of *neural* behaviors such as awareness, recognition, observation, thinking, and comprehension. While muscle behavior (actually, neuro–muscular behavior) is more familiar to us, the behaviors of consciousness happen as *pure neural processes*. These lack neuron–stimulated muscle–contraction components. In agreement with Skinner, most people find consciousness high on their interesting–topic list, and we devote a whole later chapter to accounting for it.

For now, though, let's dig just a little deeper in our preview. To help grasp consciousness as pure neural responses, let's try describing consciousness behavior on several increasingly elaborate levels that take into account behaviors, as real events, serving as stimuli for further behaviors. Speaking first in traditional (i.e., agential, non–natural) terms, we note this: People can observe and report to themselves or others the occurrence of some of their behaviors. *Restating* this point with simple behaviorological phrasing (i.e., naturalistically, without implying self agents) gives us this: Certain of peoples' behaviors *evoke* subsequent behaviors of observing and reporting the earlier behaviors. *Embellishing* this point behaviorologically gives us this: The external and internal environments of peoples' daily existence feature the contingent occurrence of energy exchanges. These condition people so that some behaviors, including neural behaviors, like observing, function as real, independent variables evoking further neural and motor behaviors, like reporting, as real, dependent variables.

Actually many, perhaps most, moment to moment human behaviors play little or no part in evoking any consciousness behaviors. This may make the occurrence of consciousness behaviors even more notable (i.e., more capable of evoking further behavior). And if one's past conditioning history has made supposed agential accounts reinforcing, then the occurrence of otherwise ordinary, natural consciousness behavior can seem particularly impressive as supposedly being, or showing, the activity of one or another inner agent.

Through those kinds of processes, the "common wisdom" of traditional cultural contingencies builds inaccurate repertoires with large numbers of individuals. Such groups become the organized forces of theological or secular superstition and mysticism that oppose so much natural science today. This opposition occurs in spite of the opposers' thorough reliance on the vast array of products, from the opposed natural sciences, that enable living quality lives while rendering their opposition. For example the opposers take modern transportation to anti–science meetings held in beautifully designed and expertly engineered buildings with comfortable accommodations including flush toilets—and other personal–hygiene amenities—as well as healthy air quality and efficient food preparation areas. The opposers would not want to live without all of these products of the opposed sciences.

Elaborating the natural–science analysis of behavior, including consciousness, helps counter those trends. The elaboration of the consciousness behavior of observing/reporting some *other* behavior can also accommodate the physiological and genetic levels of analysis. While genes originally produce nervous–system structure, ongoing respondent and operant conditioning processes, occurring throughout life, continually change that structure as the internal and external environments exchange energies with the body on a moment by moment basis. Some of those changes to nervous–system structure are such that some structures now react in ways naturally mediating neural, consciousness behaviors that we describe as observing and reporting, and that differ from the observed and reported behaviors. Later chapters gradually clarify how all of this happens.

All entirely natural. When happening, all these intertwined environmental/neural/behavioral processes can move along at such a rapid pace that they may seem magical or undetermined. Events can quickly outpace our measurement technology, particularly when we try to encompass behavior in general across a time frame beyond a few moments. That, however, is a problem not with nature, which naturalism takes as lawful, but with our residual ignorance (what we still don't know and don't like not knowing). Continued research gradually increases how much we know, understand, and apply. Nevertheless, consistent with Skinner's behaviorism, those processes are *still all entirely natural*. Including self agents in accounts for behavior is not only redundant and misleading but also dangerous and irresponsible, because the resulting reduced effectiveness in problem solving can cause people harm.

Inadequate behaviorisms. In his 1963 paper, Skinner emphasized a range of concerns surrounding the question of consciousness, because this topic in particular seems to focus attention on the difference between science and non–science with respect to behavior. While Freud had assisted the behavioristic perspective by showing that consciousness was not required for other behavior to occur, the nature and occurrence of the consciousness itself presented a more difficult problem, namely the reduced access to real but private events (i.e., events that can function as stimuli only for the body in which they occur).

Some early behaviorists simply denied the existence of private events. For example, Skinner (1963) reports that Watson "tangled with introspective psychologists by denying the existence of images" (p. 952). Others accepted the public/private distinction but disallowed the inclusion of private events in scientific deliberations, because "science is public." Still others, while also accepting the public/private distinction, allowed such events in scientific discussion but only after defining out the private aspect. The private event of "hunger" can exemplify the approach to privacy of each of these types of behaviorism. (a) Behaviorists under "Watson's *original* radical behaviorism" might simply deny that hunger exists. (b) Behaviorists under "methodological behaviorism" accept that hunger exists but disallow its study due to its being private. (c) Behaviorists under "operational behaviorism" accept that hunger is private but study it only after defining away its privateness such as by defining it as some number of hours of food deprivation. All three give unsatisfactory accounts, because they sidestep the reality of private events, and thereby fail to deal with those events, particularly the events called consciousness.

Calling his 1963 article a "restatement of radical behaviorism" (p. 951), which contributed to calling his philosophy of science *radical behaviorism,* Skinner resolved the privacy problem by pointing out that, with the skin as a scientifically unimportant boundary, the physical dimensions of public and private events are the same. A natural science of behavior makes no assumptions that events inside the skin are of any special nature, or that we need to know them in any special way, that differs from how we know the rest of nature. Instead, an adequate natural behavior science deals with private events as part of behavior itself, as covert behaviors under the same natural laws that overt behaviors are under. In the bulk of his article, Skinner went on to discuss the results of experimental science that had already accumulated in support of this privacy–problem solution, and many of the implications and ramifications.

That experimental science, however, was largely a natural science of the behavior of other animals in its early years. In the subsequent 50 years, it became both a natural science, and an applied engineering technology, emphasizing *human* behavior.

Developments in the Second 50 Years

The solution to the privacy problem that capped the first 50 years prompted one of the major developments bearing on the question of consciousness in the years since 1963. This involved a greater appreciation of the overlap between the separate yet complementary natural sciences of physiology and behaviorology. For example, to deal scientifically with emotion requires the different analytical levels of these two disciplines. *Emotion* refers to a change in body chemistry due to the release of chemicals into the bloodstream (an area of physiology) that external or internal stimuli elicit (an area of behaviorology). That changed body chemistry produces the reactions called *feelings*. Perhaps even more importantly, that changed body chemistry produces effects on other responses. When a bear startles you, you run faster than you run under more ordinary circumstances. Or, excising the fictitious inner agent that the word "you" can mistakenly imply, the sudden appearance of a big brown bear from behind a boulder only a meter away evokes faster running—due to the elicited emotional body chemistry change—than the running that more ordinary circumstances evoke, such as a clock showing the time that a jogging session begins.

The Value of Physiology

Behaviorology accounts for specific functional relations between real causes (independent variables) on both sides of the skin, and real effects (dependent variables) of behavior changes on both sides of the skin. Meanwhile brain physiology accounts for the structural changes that are occurring as those behaviorological independent and dependent variables interact. Do not confuse brains, which are physical organs, with minds, which are mystical inventions. Brains mediate behavior that occurs as a function of other real variables. Brains neither originate nor initiate behavior. Thus, the more brain physiologists work to account for *the mediation of behavior,* particularly *the mediation of neural behavior,* rather than for *mind* or *mystical mental operations,* the greater success they experience, and the more valid and valuable their work becomes.

Essentially then, behaviorology is not a natural science *of how* a body mediates a behavior (e.g., of how striated muscle contractions are a function of neural processes); physiology studies *how* behavior occurs. Instead, behaviorology is a natural science *of why* a body mediates a behavior (i.e., of the cause–effect relations between independent variables, such as a boulder blocking a forest path, and the dependent variables of body–mediated *behavior,* such as the muscle contractions that the obstacle evokes thereby taking the body around the boulder). All the events at both levels of analysis are entirely natural; no mystical inner agent is considering options and then willing the body/physiology to do the action that it has decided to take. "Considering" and "decision–making" are naturally occurring neural behaviors that happen as a function of real independent variables.

Also naturalistically, behaviorology has addressed some ancient and fundamental questions, leading to some exciting outcomes. Again, while later chapters address complex topics in detail, here we just begin to scratch the surface, not as some sort of tease, but to support your anticipation and show why humans should bother about behavior.

Some ancient questions deal with knowledge and its sources. Since what scientists and philosophers and other knowers "do" *is* behavior—often verbal and with inner agents stripped away—behaviorology, as the natural science of behavior, provides scientific analyses of science and philosophy. More specifically, by the 1990s such natural–science analyses covered values, rights, ethics, and morals, with important implications for a range of engineering concerns including sustainability and robotics. These kinds of scientific extensions of behaviorology led Lawrence Fraley, in Chapter 29 of his *General Behaviorology* (2008), to conclusions about reality that parallel the conclusions that Stephen Hawking and Leonard Mlodinow, in their book, *The Grand Design* (2010), reached through the logic of naturalism in physics. Their shared conclusion is that our neurally behaving reality is the sole source of knowledge (i.e., conscious/neural responding) about reality, because we can get no closer to reality than the behaviors that the firings of sensory neurons evoke.

Past and future. A related question arises, both on its own merits and due to its relevance to accounting for consciousness: How can events that seem to be in the past or future affect our behavior? The basic answer is that past or future stimulus events *cannot* directly evoke or consequate responses. Both responses and stimuli occur only in the present. This raises the important implication that all behavior is *new* behavior (with details later on responses grouping into response classes for experimental analysis). Every behavior occurs under the functional control of *current* evocative—which includes eliciting—stimuli regardless of the complexity, multiplicity, or interactivity of the stimuli and responses. Even memories are not *stored responses.* They are new responses that current stimuli evoke and that *current* neural structures mediate, neural structures that have their current structure because conditioning processes changed them both at the time of, and since the time of, the original instance.

And more. The available, fascinating realities go on and on. But we cannot put the whole book in the first chapter. We must leave many such topics for later chapters (e.g., while you can review the Table of Contents for more, here are just a of couple of the complex topics related to consciousness that we cover later: [a] interactions between consciousness and the verbal community, and [b] the connections among genes, conditioning, and response chains).

The Law of Cumulative Complexity. As always, in all those complex topics that we have touched on so far, no capricious inner agent makes responses occur; any responses that occur are the only responses that can occur under the present independent variables. This applies even to chains of responses. As complexity increases, contingencies cause response chains to form. Given their status as real events, responses themselves serve as stimuli for further responses, gradually building chains of responses connected by the stimulation of each previous response. As natural scientists, we respect the natural functional history even of extremely complex and multiply–controlled response chains such as the text–composition responses that the physiology of this "complex carbon unit" (i.e., of the author) mediated under the then current conditions during the original moments (hours, weeks, years) of writing this book. It would be quite economically wasteful to bother with the expense of the detailed analysis to identify and describe the range of variables compelling the present wording. However, B. F. Skinner (1957a) as well as Lawrence Fraley (2008, pp. 949–1094), Stephen Ledoux (2014, Ch. 20), and Peterson and Ledoux (2014) have provided, in their textbooks, the foundations for making such an analysis, and it will occur under appropriate contingencies (e.g., see Fraley, 2014).

All that complexity in behavioral and physiological events can seem amazing, wondrous, awesome, even overwhelming, *but it is all entirely natural.* Perhaps summarizing the origin of such emerging complexity in the context of *all* the natural sciences as a theory or law will help. If it rates the status, let's call it the **Law of Cumulative Complexity:** *The natural physical/chemical interactions of matter and energy sometimes result in more complex structures and functions that endure and naturally interact further, resulting in an accumulating complexity.* Any origin of life is an outcome of the Law of Cumulative Complexity. On this planet other examples of this law include the interrelations of physiology and behaviorology in the contingencies responsible for the vast range of complexity in human behavior. *All these are cumulatively complex; all are entirely natural.* (The Glossary entry for *Law of Cumulative Complexity* has more examples.)

All the developments that we have discussed involve extensions of the philosophy of science that Skinner called *radical behaviorism.* It is, again, radical in the sense of comprehensive or fundamental. Furthermore, it informs both the natural science that experimentally studies human nature and human behavior, and the derivative engineering technologies of that science. These technologies, under "contingency engineering," effectively address accessible independent variables in ways that bring about improvements in behavior at home and work, in education and diplomacy, in interpersonal relationships, and indeed in all applied behavior fields from advertising to zoo keeping. This philosophy, and the science and technology that it supports, first arose among a thoroughly naturalistic group of researchers and academics, working in early twentieth–century psychology, that Skinner and his colleagues and their students best represent. However, this natural philosophy, science, and technology ultimately proved to be fully incompatible (i.e., "incommensurable") with the more commonly available, non–natural, fundamentally mystical, agential perspectives of certain fields that popular culture currently supports, including psychology. As a result, a separation of disciplines had to occur.

Organizational Developments

Such full incompatibility with natural science begins when any discipline studying any phenomenon rejects the assumptions of naturalism that provide the foundation for success in all natural sciences. This is because *all* real aspects of the universe—including energy, matter, life forms, *and life functions* (e.g., human behavior)—are potentially within the reach of objective scientific accounts and applications. To qualify as *natural* science though, natural philosophy—rather than any sort of mystical assumptions—must inform the framing of research questions and the interpretation of experimental results.

That is basic to natural science, but difficult to maintain in a culture steeped in superstition in which some insist that accounts based on mystical assumptions are adequate. Of course, with much scientific activity involving methods, anyone using scientific procedures, even mystical people, can objectively collect data on any real phenomenon. But this is not enough, because mystical assumptions inherently contradict natural science. Any variety of mystical assumptions that informs real, measurable data collection usually leads both to misdirected research questions and to misinterpreted experimental results. Either or both of these often get aimed mainly at somehow proving the mystical assumptions.

No one, however, can prove or disprove assumptions. The supporters of mystical assumptions either induce them off–the–cuff or, with allusions to tradition or common lore, induce them from various kinds of faith beliefs that also actually result from researchable natural processes. In contrast, supporters of the assumptions of the philosophy of naturalism mainly induce them from the results and developments of the last several hundred years. These results and developments include the accumulation of both experimental evidence and effective engineering practices regarding the wide range of phenomena that the various natural–science and engineering disciplines cover. Thus, these two types of assumptions, naturalistic/scientific and mystical/superstitious, are *not* equal in any meaningful way. They are incommensurable (i.e., incompatible).

That incommensurability, and the growing pressure of expanding experimental and applied research, provided the principal driving forces behind the 1987 reorganizing of the natural science of behavior as a separate and independent discipline. The general result of this development is a foundation natural science related to all other natural sciences, not at the discredited level of body–directing self agents, but at the level of a body's physics–based interactions with the external and internal environments. Working in this natural–science tradition, Skinner's treatment of behaviorism in his 1963 article was well rounded but necessarily minimal. A decade later his book *About Behaviorism* (Skinner, 1974) provided details and helped pave the way for the sometimes controversial steps in this reorganization. In a long paper, Lawrence Fraley and I (Fraley & Ledoux, 2015 [originally, 1992]) thoroughly discussed the issues and history of this transition, some highlights of which appear in a later chapter.

Scientific Developments

During the decades of gradual emergence as an independent discipline, natural scientists of behavior have reported a still growing range of experimental research and engineering applications. Consider, for example, this tiny sample of behaviorological science articles from that period: William Baum (1995) further analyzed radical behaviorism and the concept of agency. Joseph Cautela (1994) elaborated on the importance of the concept of the General Level of Reinforcement (GLR). Carl Cheney (1991) summarized the controlling relations and sources of behavior. John Eshleman and Ernest Vargas (1988) promoted application technologies for verbal–behavior analysis. Philip Johnson (2012) and John Ferreira (2012) each analyzed aspects of clinical successes from Progressive Neural Emotional Therapy (PNET). Lawrence Fraley discussed the cultural mission of behaviorology (1987), naturalistic consideration of uncertainties about determinism (1994a), behaviorological thanatology and medical ethics (2006), naturalistic analysis of the legal doctrine of *Mens Rea* (1983), and behaviorological penal corrections (1994b). Stephen Ledoux reported results on an experimental procedure to control simultaneously evocable and simultaneously occurring, complex human behaviors with multiple, even simultaneous, selectors, like reinforcers (2010), and summarized 25 years of behaviorology curriculum development successes (2015a). And Jack Michael (1982) separated discriminative and motivational functions of stimuli.

Also, consider this small sample of the many behaviorological science books from the last 70 years: Charles Ferster and B. F. Skinner (1957) published the data from their extensive laboratory research on schedules of reinforcement. Robert Epstein (1996) and B. F. Skinner, through their *Columban Simulation Project,* reported the phenomena that they called *recombination of repertoires.* Murray Sidman provided works (a) on research methods (1960/1988), (b) on stimulus equivalence relations (1994, a work with considerable implications for the nature of research *as behavior)* and (c) on the detrimental effects of coercion and punishment at all levels of social interaction including families, classrooms, workplaces, and international relations (2001). Glenn Latham contributed works on positive practices for raising (1994, 1999) and educating (1998, 2002) children. Cathy Watkins (1997) clarified the value of *Project Follow Through* for improving regular education. Catherine Maurice, Gina Green, and Stephen Luce (1996) detailed best practices for work with autistic children. Aubrey Daniels (1989) addressed performance management in business and industry. Lawrence Fraley (2008) published his graduate level behaviorology course materials, from which he had been teaching for 25 years, thereby providing the first extensive and systematic book detailing the separate and independent natural–science discipline of human behavior. Fraley also provided other book–length works, one on behaviorological thanatology and dignified dying (2012), and another on behaviorological rehabilitation and the criminal justice system (2013). Stephen Ledoux provided an entry level behaviorology textbook (2014) and a third edition (2015b) of his book of thematic readings on the origins and components of behaviorology. Norman Peterson and Stephen Ledoux also provided a second edition (2014) of Peterson's text for introducing B. F. Skinner's 1957 book, *Verbal Behavior.* The philosophy and science discussed here even prompted two popular and educational works of fiction, one each by B. F. Skinner (1948) and W. Joseph Wyatt (1997).

Those sample lists could have included a far larger number of the thousands of available articles and books. The particular selections that they include derive as much from the author's familiarity with them as from various opinions regarding their relative importance. The same point applies to the selection of topics, examples, and applications throughout this book.

Conclusion

In its second 50 years, the value and legacy of behaviorism broadened substantially. The natural science that Skinner's radical behaviorism supports and informs has emerged as an extensive, multi–faceted discipline, although its independence as behaviorology began only about a quarter–century ago. Its academic homes will continue to expand due to the repeatedly documented effectiveness and benefits of approaching human behavior naturalistically. Other disciplines also faced similarly difficult circumstances in their early histories and prevailed. The astronomical discoveries of Galileo 400 years ago helped move our human home, the Earth, beyond superstitious, mystical accounts. The biological discoveries of Darwin 150 years ago helped move the human body beyond superstitious, mystical accounts. And, based on the naturalism of Skinner's radical behaviorism, the current discoveries of behaviorological science help move human nature and human behavior beyond superstitious, mystical accounts. On this basis our continuing efforts not only improve effective scientific thinking about all subject matters including human nature and human behavior, but also reduce reliance on superstition across the worldwide culture, expand reliance on naturalism, science, and engineering, and increase success in solving personal, local, and world problems.❧

Endnotes

Many of the papers and books that we mentioned in this chapter, plus additional ones, along with links for many of the organizations and journals related to this chapter, appear on www.behaviorology.org (the web site of TIBI, The International Behaviorology Institute) or on www.americanscientist.org (the web site of *American Scientist,* the journal of Sigma Xi, the Scientific Research Society). Others appear on www.behavior.org (the web site of the Cambridge Center for Behavioral Studies) or on www.bfskinner.org (the web site of the B. F. Skinner Foundation).

References (with some annotations)

Baum, W. M. (1995). Radical behaviorism and the concept of agency. *Behaviorology, 3* (1), 93–106.

Cautela, J. R. (1994). General level of reinforcement II: Further elaborations. *Behaviorology, 2* (1), 1–16.

Cheney, C. D. (1991). The source and control of behavior. In W. Ishaq (Ed.). *Human Behavior in Today's World* (73–86). New York: Praeger.

Daniels, A. C. (1989). *Performance Management (Third Edition, revised).* Tucker, GA: Performance Management Publications. (This publisher, now in Atlanta, GA, released a fourth edition in 2006 with J. E. Daniels as coauthor.)

Epstein, R. (1996). *Cognition, Creativity, and Behavior.* Westport, CT: Praeger.

Eshleman, J. W. & Vargas, E. A. (1988). Promoting the behaviorological analysis of verbal behavior. *The Analysis of Verbal Behavior, 6,* 23–32.

Ferreira, J. B. (2012). Progressive neural emotional therapy (PNET): A behaviorological analysis. *Behaviorology Today, 15* (2), 3–9.

Ferster, C. B. & Skinner, B. F. (1957). *Schedules of Reinforcement.* Englewood Cliffs, NJ: Prentice–Hall.

Fraley, L. E. (1983). The behavioral analysis of *Mens Rea* (doctrine of culpable mental states). *Behaviorists for Social Action Journal, 4* (1), 2–7. (See Fraley, 2013.)

Fraley, L. E. (1987). The cultural mission of behaviorology. *The Behavior Analyst, 10,* 123–126. Expended versions appear in Fraley, 2012 and 2013.

Fraley, L. E. (1994a). Uncertainty about determinism: A critical review of challenges to the determinism of modern science. *Behavior and Philosophy, 22* (2), 71–83.

Fraley, L. E. (1994b). Behaviorological corrections: A new concept of prison from a natural–science perspective. *Behavior and Social Issues, 4,* 3–33. (See Fraley, 2013.)

Fraley, L. E. (2006). The ethics of medical practices during protracted dying: A natural–science perspective. *Behaviorology Today, 9* (1), 3–17. (See Fraley, 2012.)

Fraley, L. E. (2008). *General Behaviorology: The Natural Science of Human Behavior.* Canton, NY: ABCs.

Fraley, L. E. (2012). *Dignified Dying—A Behaviorological Thanatology.* Canton, NY: ABCs.

Fraley, L. E. (2013). *Behaviorological Rehabilitation and the Criminal Justice System.* Canton, NY: ABCs.

Fraley, L. E. (2014). Behaviorological science and the complexity of unfathomable variation. *Journal of Behaviorology, 17* (1), 13–18.

Fraley, L. E. & Ledoux, S. F. (2015 [originally appeared in 1992]). Origins, status, and mission of behaviorology. In S. F. Ledoux. *Origins and Components of Behaviorology—Third Edition* (pp. 33–169). Ottawa, CANADA: BehaveTech Publishing. This multi–chapter paper also appeared across 2006–2008 in these five parts in *Behaviorology Today* (the precursor to *Journal of*

Behaviorology): Chapters 1 & 2: *9* (2), 13–32. Chapter 3: *10* (1), 15–25. Chapter 4: *10* (2), 9–33. Chapter 5: *11* (1), 3–30. Chapters 6 & 7: *11* (2), 3–17.

Hawking, S. & Mlodinow, L. (2010). *The Grand Design.* New York: Bantam. (Especially see Chapter 3.)

Johnson, P. R. (2012). A behaviorological approach to management of neuroleptic–induced tardive dyskinesia: Progressive neural emotional therapy (PNET). *Behaviorology Today, 15* (2), 11–25.

Latham, G. I. (1994). *The Power of Positive Parenting.* Logan, UT: P & T ink.

Latham, G. I. (1998). *Keys to Classroom Management.* Logan, UT: P & T ink.

Latham, G. I. (1999). *Parenting with Love.* Salt Lake City, UT: Bookcraft.

Latham, G. I. (2002). *Behind the Schoolhouse Door: Managing Chaos with Science, Skills, and Strategies.* Logan, UT: P & T ink.

Ledoux, S. F. (2010). Multiple selectors in the control of simultaneously emittable [evocable] responses. *Behaviorology Today, 13* (2), 3–27. Also, under the title "Multiple selectors in the control of simultaneously evocable responses," this paper appears in S. F. Ledoux. *Origins and Components of Behaviorology—Third Edition* (pp. 205–242). Ottawa, CANADA: BehaveTech Publishing. This paper includes material about a procedure that further develops behaviorological research areas. See Ledoux, 2013, for an equipment update.

Ledoux, S. F. (2012a). Behaviorism at 100. *American Scientist, 100* (1), 60–65. *American Scientist* published this paper with introductory excerpts, on pages 54–59, from: Skinner, B. F. (1957b), which the editor listed as an "*American Scientist* Centennial Classic 1957." (Skinner's complete 1957 paper is also available online with this paper.)

Ledoux, S. F. (2012b). Behaviorism at 100 unabridged. *Behaviorology Today, 15* (1), 3–22. This peer–reviewed version of the paper included the material that the *American Scientist* editor had set aside at the last moment to make more room for the Skinner article excerpts that accompanied the original article (Ledoux, 2012a). (*American Scientist* also posted this unabridged version of the paper online with the original abridged version.)

Ledoux, S. F. (2013). Human multiple operant research equipment. *Journal of Behaviorology, 16* (2), 3–9. This paper describes some twenty–first century equipment that behaviorological experimenters could use in continuing the research programs described in Chapter 6 of Ledoux, 2014. A version also appears as Appendix 5 in Ledoux, 2015b.

Ledoux, S. F. (2014). *Running Out of Time—Introducing Behaviorology to Help Solve Global Problems.* Ottawa, CANADA: BehaveTech Publishing.

Ledoux, S. F. (2015a). Appendix 3 Addendum: Curricular courses and resources after 25 years (1990–2015). In S. F. Ledoux. *Origins and Components of Behaviorology—Third Edition* (pp. 314–326). Ottawa, CANADA: BehaveTech Publishing. (This paper provides resources for behaviorology curricula.)

Ledoux, S. F. (2015b). *Origins and Components of Behaviorology—Third Edition.* Ottawa, CANADA: BehaveTech Publishing.

Maurice, C., Green, G., & Luce, S. (Eds.). (1996). *Behavioral Intervention for Young Children with Autism.* Austin, TX: Pro–Ed.

Michael, J. L. (1982). Distinguishing between discriminative and motivational functions of stimuli. *Journal of the Experimental Analysis of Behavior, 37*, 149–155.

Peterson, N. & Ledoux, S. F. (2014). *An Introduction to Verbal Behavior—Second Edition.* Canton, NY: ABCs.

Seo, H. & Lee, D. (2009). Behavioral and neural changes after gains and losses of conditioned reinforcers. *Neuroscience, 29,* 3627–3641.

Sidman, M. (1960/1988). *Tactics of Scientific Research.* New York: Basic Books. Authors Cooperative, in Boston, MA, republished this book in 1988.

Sidman, M. (1994). *Equivalence Relations and Behavior: A Research Story.* Boston, MA: Authors Cooperative.

Sidman, M. (2001). *Coercion and its Fallout—Revised Edition.* Boston, MA: Authors Cooperative.

Skinner, B. F. (1948). *Walden Two.* New York: Macmillan. Reissued in 1976. (See the comments in the Bibliography.)

Skinner, B. F. (1957a). *Verbal Behavior.* New York: Appleton–Century–Crofts. The B. F. Skinner Foundation (www.bfskinner.org) in Cambridge, MA, republished this book in 1992.

Skinner, B. F. (1957b). The experimental analysis of behavior. *American Scientist, 45* (4), 343–371. Excerpts appeared in *American Scientist, 100* (1), 54–59, along with Ledoux, 2012a.

Skinner, B. F. (1963). Behaviorism at fifty. *Science, 140,* 951–958.

Skinner, B. F. (1974). *About Behaviorism.* New York: Knopf.

Watkins, C. L. (1997). *Project Follow Through: A Case Study of Contingencies Influencing Instructional Practices of the Educational Establishment.* Cambridge, MA: Cambridge Center for Behavioral Studies.

Watson, J. B. (1913). Psychology as the behaviorist views it. *Psychological Review, 20,* 158–177.

Wyatt, W. J. (1997). *The Millennium Man.* Hurricane, WV: Third Millennium Press. (See the comments in the Bibliography.)

Chapter 2
Where We are Going in This Book and Why:

Our starting assumptions—what we take for granted—plus a brief history, and an outline of *Part I …*

Living on the edge of a small town, I write this as a bright February sun illuminates nearby farm fields that are already devoid of snow. Deer pass through the unfenced yards between well–spaced homes in the twilight hours of morning and evening, tasting everything on their way to or from an undeveloped area at the center of the next block. One or another dog is always pulling an owner down the street. The resplendent view and clean air invite one to take a walk even without a pet, but a short distance in the stiff breeze, with the temperature below freezing, quickly compels a return to the warm side of the window with a cup of hot cocoa in hand. This is no idle scene; after most chapters, you will be able to return to this location and count the increasing number of our topics that make cameo appearances in this event snapshot.

This chapter provides some historical and philosophical premises of the natural science that we will cover. In that process it begins to inform readers about the many lines of progress that occurred over the last 100 years in the natural science of human nature and human behavior that we now call behaviorology. Later chapters will delve deeply not only into many of the contingency relations that determine behavior, but also into the contingency engineering available to help humanity solve its problems, as well as into some scientific answers that this discipline begins to supply for some of our ancient questions. These are all sound reasons for proceeding.

Besides, everyone wants to know why behavior happens, which is another good reason for proceeding. Of all the multitude of books on store shelves and online lists, the *vast* majority repeat pre–scientific or non–scientific notions that support magical or mystical accounts for behavior. Of all the books from the last 100 years, only a handful provides accounts based on the natural science of behavior, and you are reading one of them.

Resources and the Nature of Science

Before we begin investigating our topics, however, we first need to resolve a concern that we will face regularly regarding the examples that we will use, particularly our human–behavior examples. In early chapters and with basic principles and processes, simple as well as non–human behavior occasionally provides the examples best illustrating a particular point, because behaviorology concerns all behavior. However, the realistic explanation of human behavior remains our primary focus. The difficulty with human behavior examples (and many simple and other–animal behavior examples as well) centers on their inevitable complexity. Virtually every realistic example contains numerous factors and effects, and many of these often interact with each other. While we report an example to illustrate one or another concept, these other factors and effects *continue to demand explanation as well.* To meet such demands, we would have to put the whole book into each paragraph that contains an ordinary (i.e., a complex) human behavior example, which is of course impossible.

Resource Options

Instead we will remain ever patient, going without answers until satisfactory ones take a turn in this telling. We thereby avoid oversimplified examples. These easily mislead by falsely implying that some principles and concepts apply only to limited areas of behavior such as human abnormal behavior or circus–animal training. Many examples, for instance, will include some aspect of language (i.e., verbal) behavior that is not itself the point, so that part of the account for the example is missing. We gradually build the complexity toward some coverage of advanced topics like verbal behavior, because each chapter adds to the foundation for later chapters. Proceeding then through the complete sequence of chapters builds your repertoire. This will enable you to supply the missing account parts. Then, like our opening event snapshot, you can return to examples later and provide the rest of the story.

Regarding examples with complexities beyond the scope of this starter behaviorology book, you can turn to more elaborate treatments. These contain more principles, concepts, applications, and examples. At appropriate points I will provide explicit references. But the point of this book is not to tell the total story; instead the point is to explore a variety of basic topics that may lead the reader to interest in more thorough book treatments. My (2014) textbook, *Running Out of Time—Introducing Behaviorology to Help Solve Global Problems,* gets the undoubtedly biased recommendation as the best next book after this one. Actually, current natural scientists and engineers should read the textbook, because I designed it particularly to support their needs regarding the team contribution of the natural science of behavior to solving global problems.

The Behavior Nature of Science

We begin our topics with the general notion of science. When most people hear the word "science," they immediately think of one or more of the traditional *natural* sciences (e.g., physics, chemistry, biology, astronomy, geology). As our coverage is more explicit than this general notion, most of the time we will use the complete term, "natural science." On those occasions when we simply say "science," we still mean natural science, unless something else is clearly specified. With that as a given, consider what is perhaps a most basic and yet profound characteristic of science, a characteristic that only began to demand attention in the last few decades, especially with the rise of the natural science of behavior. What is this characteristic?

Science provides the foundations for a broad range of both beneficial— and occasionally damaging—products for humanity. The news media keep scientific and engineering developments in the public eye, although sometimes they focus excess attention on occasional disputes over who reached some milestone first, or the luckily even more rare occurrences of data faking. Why do such things happen? All of these—developments and squabbles and faking—happen because they all share the fundamental characteristic of science that demands our attention: *Science is behavior.* Those things happening will become less surprising and more comprehensible as you become more familiar with behavior and the variables responsible for it.

Consider science as behavior more specifically. Science is the thoughts, emotions, feelings, and muscle movements—all of which are behaviors—of those whom we call scientists, which includes me and maybe you and also the applied scientists we call engineers. All these behaviors occur under control of the same sorts of "causes" (or more technically, "functional independent variables") that affect the behaviors of everyone else as well (including other animals). Recognizing the nature of science *as behavior* helps us examine some aspects and constraints not only of scientific endeavors but also of any and all human endeavors, from the most mundane to the truly grand.

As with all behaving organisms, scientists are organisms whose behavior is entirely natural. That means it is neither magical nor spontaneous. Instead it occurs due to real, measurable variables. Nature contains a wide range of variables that naturally affect behavior in ways that scientific behaviors can discover and apply for humanity's benefit. These scientific behaviors include both respecting naturalism (the name of the general philosophy of science of the natural sciences) and experimentally analyzing phenomena, as well as both disseminating findings and developing them into practical products or procedures. While touching first on the question of philosophy, and returning to it as needed, we will soon turn to principles and practices. This book itself exemplifies the dissemination that helps prompt practical developments.

Constraining Assumptions Enhance Science

Before analyzing principles and practices, and even before asking how to analyze phenomena experimentally, we first consider some of the assumptions behind these activities. This includes assumptions behind the natural sciences in general followed by assumptions behind the natural science of behaviorology in particular.

Some Assumptions of Natural Science in General

Taken together, the general assumptions behind the natural sciences go by the name of *naturalism*. Alternatives to naturalism, such as any variety of mysticism, arise off–the–cuff or from appeals to tradition or common notions. Naturalism, however, arose from several hundred years worth of continuously accumulating experimental evidence and successful engineering practices and products. Parts of our global culture are quite enamored of one or another flavor of mysticism or superstition, some religious and some secular (i.e., non–theological). All parts of the culture, however, depend heavily on natural science to deal with everything that is real. Some folks may quibble over whether the existence of the universe depends on the activity of some divine being or on the efforts of the flying spaghetti monster. Others see it as due to measurable physical processes. But we *all* turn to science as much as possible for food, clothing, shelter, medicine, sanitation, communication, and so forth.

Measurability. A major starting point (i.e., a major assumption) in the effort to deal with whatever is real is that, at least in theory, things that are real are measurable along various dimensions. These include distance, mass, size, and temperature. Measuring things in these ways shows their reality. If measurement of an event (or an item or a process) is not even theoretically possible, then the event (or item or process) is apparently not real. We say "apparently," because science *assumes* reality only for things that are measurable. A lack of measurability places the status of the event (or item or process) *outside* the realm of the natural, and thus outside the realm of science. Of course, improved measurement technology may change the status of such events.

Some claims to measuring some things, however, mislead us. Some folks claim to measure "intelligence" with IQ tests, so intelligence must be real. But IQ tests only directly measure certain behaviors as an assumed stand–in for so–called "intelligence." On this basis science can only accept that the behaviors are real. Besides, the subsequent misleading claim turns the supposedly measured intelligence into a characteristic of some unmeasurable inner agent that supposedly causes other behavior. A later chapter deals with these kinds of problems in detail.

Natural functional history. Another assumption is that measurable events result not from magic or spontaneity but from natural histories that accumulate from the continuous chains of functionally related events behind them. We can trace some of that recent history. This provides the basis for predicting subsequent events and, to the extent that we have access and can get our hands on the preceding events at some relevant level, we can exert control over the subsequent events. For example, say you walk into your kitchen and see a broken glass on the floor with milk splashed all over the mat in front of the sink. Any notion that the glass and milk magically got there spontaneously proves quite unhelpful if you are trying to figure out what happened as part of avoiding further occurrences. Your effort proves more fruitful if paw prints track across the counter above the spill. From such clues you can sort out the major parts of the event's likely natural functional history. On that basis you can organize a prevention strategy (while cleaning up the mess) that might prove predictably successful.

Such natural histories help us to relate events to each other as well as to account for them on multiple levels of analysis, as various disciplines coordinate their parts of the complete account of events. For example, you may have surmised from the evidence that the cat knocked the milk glass off the counter, which helps you predict and control whether or not that happens again. Meanwhile, a chemist can tell you why the mat stinks later (after your cleanup left a milky residue in it), and a physicist can explain what ion exchanges are happening at the atomic level that leave you wanting to trash the mat.

Determinism. Sometimes the speed at which events transpire, or their complex interactions, outpaces our attempts to keep track of their measurements, making them seem chaotic or even somehow random. Another basic assumption comes into play here, namely that the universe and all the events in it are presumed to be orderly and lawful in their relations to each other. That is, we presume that nature determines every event completely, that no event is arbitrary, spontaneous, or capricious. This does not mean that everything *is* orderly and lawful, which is something that we really cannot know for sure (the nature of knowledge being a very advanced topic for a much later chapter on reality). Instead we work under this assumption, because we have experienced far more success staying alive and dealing with the world when we have worked under it than we had ever experienced without it. The name for this assumption is *determinism*.

Furthermore, when speed or complexity makes events seem random or chaotic, we take the problem as *not* due to nature but as due to the limitations of our measurement repertoires and technologies. Such limitations inevitably leave us with a certain amount of ignorance in that we cannot successfully account completely for the apparent randomness of the events. This difficulty endures because, even as we get better at measuring, our measurement technologies always seem to lag a bit behind the discovery of ever increasing complexity in the events up for measurement. Indeed this turns out to be such a common occurrence that we have developed several strategies to manage our residual ignorance, such as probability or chaos theories or statistical analysis.

While moving from the various assumptions behind the sciences in general to those behind behaviorology, consider this related connection. A couple of centuries ago, some cultural groups argued over whether or not we should obey the natural laws governing physical events. Gradually the reality settled in that we really have no such option. More recently some similar groups argued about whether or not we should obey the natural laws governing behavior. Again the reality is settling in that we really have no option in this matter either. Behaviorology is the natural science that brings that reality to the forefront, enabling knowledge of behavior, and the variables of which it is a function, to help both individuals and humanity stay alive and deal with the world.

All members of all species, including the human species—past, present, and future—are simply (actually, "complexly" would be more apt) wondrous sets of naturally evolved and elaborating physical and chemical processes and systems. Arising through the Law of Cumulative Complexity (introduced in Chapter 1), these systems unfold structures (i.e., a physiology) that share characteristics evoking biologically oriented verbal responses that summarize this complexity with exclamations including "life!" and "it's alive!" Our interest in behaviorology resides, in part, in the way it clarifies through experimentation so many of the details, particularly at the analytical level of environment/behavior interactions. These regard how natural laws pertain to the effects that energy changes have on some of these systems such that the systems mediate the phenomena called behavior. But going there requires some coverage of the assumptions behind behaviorology.

Some Assumptions of Behaviorology

While *naturalism* refers to those general assumption for all natural sciences, individual science disciplines often maintain further assumptions related to their particular subject matters. The assumptions of the natural sciences apply to the behavior subject–matter discipline as well. However, this discipline also has some subject–matter–specific assumptions. The systematized collection of additional assumptions behind behaviorology, as the natural science of behavior, go by the name of *radical behaviorism,* and several books elucidating its details are available for your perusal (e.g., Skinner, 1974). Here we mention but a few of its salient characteristics.

Parsimony. One assumption of behaviorology, which is also shared with all other natural sciences, concerns the necessity to *work with the simplest yet adequate explanation of an event.* We call this assumption *parsimony,* and it helps prevent waste by dissuading researchers from falling for overly complex accounts too quickly, because complex accounts usually cost more, and are more difficult—perhaps even impossible—to test experimentally. On the other hand, new data can show inadequacies in a current account. One should then move to test only the next feasible increment of complexity. If it tests out as adequate, then the account with that increment becomes the current best explanation of an event. If it does not, then you test the next increment for adequacy, and so forth, until you can again account for the event adequately.

Parsimony then cuts one way or the other, because an account can be **un**parsimonious either by being an unnecessarily complicated (and possibly even untestable) account, or by being, or becoming, inadequate as an account. Of course, some accounts are by definition in violation of parsimony, such as mystical accounts, which offer events that are unreal as explanations thereby eliminating the possibility of testing. Behaviorology stands firm with all other natural sciences in refusing to allow metaphysical (i.e., non–natural) events to have any abiding presence in explanatory accounts. Be aware, however, that different disciplines have different needs for attention to parsimony.

Traditional natural sciences generally have little need for attention to parsimony; they take parsimony for granted, due both to their ingrained centuries–long history of practice abiding by it, and to the gradual rate at which they accumulate the range of phenomena that each one covers. For example, physicists in the 1700s barely had any notion of the vastness of the universe or the minuteness of the sub–atomic realm. These subject matter components accumulated gradually as research horizons ever expanded, usually due to improvements in measurement technologies (e.g., instruments). If you can measure something, then go for it, explore it, discover whatever you can about it. Of course, no inner agents are telling scientists to measure, explore, or discover. These activities are naturally occurring behaviors under contingency control. In later chapters we will consider the many variables of which such scientific behavior, along with all other behavior, is a function.

To anticipate the differences better, between traditional natural science disciplines and those certain disciplines needing greater attention to parsimony, consider that, later, physicists in the 1800s operated on the level of Newtonian mechanics. Their experience provided no need or option for anything like quantum mechanics or relativity. But their 1900s counterparts could not operate without quantum mechanics or Einstein's relativity. The jump from Newton to Einstein actually resulted from a precision inadequacy that developed for Newtonian mechanics. This was due to measurement–technology refinements leaving the Newtonian–formula predictions off by amounts greater than the error margins inherent in the newer measurement equipment. Einstein's equations produced more precise predictions than Newton's equations, and the newer equipment verified the Einstein–equation predictions, leaving them more adequate, and hence parsimonious, than Newtonian mechanics. The process continues today (and tomorrow) with various string theories trying to manage the mathematical discrepancies between quantum mechanics and relativity. For an easy–to–read example of related phenomena, see Stephen Hawking and Leonard Mlodinow's *The Grand Design* (2010). The point is that all this happened without much reference to parsimony, because physicists—typical of traditional natural scientists—would have trouble dealing with such developments in any way other than parsimoniously. Also, and perhaps more importantly, they never had to confront the *whole* range of their subject matter all at once, without tools to deal with it.

On the other hand, any discipline that deals with behavior (e.g., theology, psychology, or behaviorology) confronts the whole range of its subject matter right from the start. This ranges from eating to emoting to speaking to thinking, basically simple to impressively complex, all on comprehensive continua encompassing so–called normality to abnormality. Meanwhile society always demands complete accounts and effective applications regarding all these behaviors all at once, regardless of whether or not practitioners had workable tools to produce those "complete accounts and effective applications." Under contingencies (i.e., dependencies) like these, we should not be surprised when some disciplines studying behavior fall for overly complex accounts (e.g., minds, psyches, selves, souls, or some other types of putative behavior–initiating self agents). These unparsimonious accounts can sound very impressive to those desperately seeking answers about behavior's causes. Contingencies compelling acceptance of these answers then leaves the seekers satisfied, and even excessively defensive, about their new–found answers, even though these answers actually provide little by way of working practical applications or interventions. Thus we can see why these disciplines need to give explicit attention to parsimony and adhere to it. But some of them either deny this outright, or perhaps they cannot afford the changes that adhering to parsimony would bring.

Behaviorology became an independent discipline in part because it opposed the commitments to unparsimonious accounts that non–natural, behavior–oriented disciplines maintain. Instead it stands with the traditional natural sciences, and regards parsimony as a component of their shared naturalism.

Four radical behaviorism assumptions. Behaviorology has additional assumptions, specific to its subject matter, under its informing philosophy of science, radical behaviorism. Of the various assumptions that characterize the radical behaviorist philosophy of science, four stand out for the role they played in the emergence of the behaviorology discipline. Due to the coverage they receive throughout the pages of this book, we will only touch on these assumptions briefly, mostly for their stage–setting value. (Some of the papers in Ledoux, 2015, also deal with these matters.)

(a) The first assumption regards behavior as a *natural* phenomenon. Partly, this respects the continuity of events in space and time that, in natural sciences, accumulates as a researchable natural history. We see behavior, not as originating in the capricious actions of one or another internal mystical self agent, but as a naturally occurring phenomenon under discoverable laws of nature just like any other real phenomenon. This enables prediction and control of the behavior subject matter, which provides a firm foundation for developing beneficial behaviorological engineering applications and interventions.

(b) The second assumption takes *experimental* control of relevant variables as the most appropriate basis for analysis of environment–behavior relations. As with other natural sciences, this leads to the application of that control in culturally valuable ways. Other kinds of control also play roles in behaviorology, although in less extensive ways.

(c) The third assumption specifies that the same full range of variables affecting other people's behavior also affects the behavior—scientific or not—of scientists, because they also are behaving organisms. Among these variables, scientists' philosophy of science comprises one of the most pertinent. This is because such philosophy exerts a kind of quality control over other scientific activities. For example, preventing mystical assumptions from misinforming either the selection of research questions, or the interpretation of research results, is a role that scientists' philosophy of science plays. This, of course, carries implications for other natural sciences as well as behaviorology. For example, centuries ago no fully developed naturalism informed the activities of the forerunners of physicists; instead the religious philosophy of the day informed their activities, such that they spent (Wasted?) plenty of time trying to figure out how many angels can dance on the head of a pin.

(d) The fourth assumption, perhaps the most important but also the most complicated, recognizes most private events, such as thinking, seeing, observing, or emoting, as most productively construed as covert, neural behaviors that are involved in the same lawful relationships that involve overt, muscle behaviors. Experimental data and interpretations, some of which get described in later chapters, have accrued in support of this assumption. Private events are lawful in the same way that one would regard public events. The skin presents no special boundary to the laws of the universe, and the nature of events on both sides of the skin is the same, with events measured in the same natural–science ways. Sometimes we even need access to the covert events. Attaining that access usually involves welcome coordination with physiology colleagues.

However, behaviorologists cannot grant scientific status to private events that someone invents to be causes of a behavior, such as internal hypothetical constructs conjured up conveniently with just the right characteristics to explain that behavior. Nor can we take real private events as the initiating causes of behavior, because we analytically pursue any causal chain to other, outside events for the sake of control in the subject matter. Instead, we see behavior, on the overt level, as neurologically based actions of the glands and muscles (both smooth and skeletal) while we see private events as covert behaviors observable and reportable only by the person, as a public–of–one, in whom other events compel their occurrence. These covert behaviors involve only neurological–level events; the behavior of "seeing in the absence of the thing seen" is one example. (This coverage is necessarily brief. In later chapters we will regularly return to such topics for increasingly greater depth. This detail–building repetition pattern is typical of education in the natural sciences.)

Single–subject designs. Other concerns inseparably intertwine with those characteristics of radical behaviorism, from sharing with other natural sciences the refusal to allow metaphysical events to enter explanatory accounts, and the necessity of parsimony in accounts of human behavior, to a preference for single–subject experimental designs rather than group statistical designs. While seeming more like a matter of experimental methodology than of philosophical assumptions, a vital philosophical connection supports the selection of research methodology. Single–subject designs and radical behaviorism have gone rather hand in hand throughout the history of natural behavior science, starting with the experimental preparations of Skinner's early laboratory at Harvard in the 1930s and continuing up through today's laboratory and applied research efforts. (For some methodology details, see Chapters 8 and 9 in Ledoux, 2014, because the topic tends to be a bit dry for the pages of a first book. For a revealing, comprehensive example of basic research that leads to a full research program, see Chapter 6 in that book.)

The most important reason for an interplay, between philosophical assumptions and research methods, stems from the assumptions about why variability occurs in experimental data. The term "variability" refers to the fact that repeated measurements of events seldom produce identical numbers (i.e., data). The measured amounts vary. If they vary a lot, we talk of high variability. If they only vary a little, we speak of low variability. Behaviorologists find that different, and unequal, assumptions about the source of the variability tend to come with different methodologies.

On the one hand, behaviorologists see some researchers, especially from non–natural disciplines (e.g., psychologists) assuming that observed variability arises from fanciful and spontaneous actions of some supposed inner agent that they have posited as causing a behavior of concern. The mystical status of that kind of variability automatically curtails direct access to the variability. Thus, one cannot reduce this variability. The most commonly accepted method to deal with variability that one cannot reduce involves using a group statistical design of one sort or another. In these methods, the mathematical manipulation of the data supposedly spreads out the variability across a large number of randomly selected subjects so that the results can suggest functional relations.

On the other hand, since behaviorologists, as natural scientists, cannot grant status to putative inner agents, those agents cannot be the source of observed variability. This means that variability stems from something else. Whatever that is, can group statistical designs deal with it? Can a different methodology deal with it better, perhaps by reducing it?

Looking back as far as Skinner's early work, behaviorologists identify a different source for variability, one related to how thoroughly, or not, we exert experimental control over the functional variables relevant to a behavior under study. Starting out simply, we began working with handfuls of subjects, three to six "per experiment." Actually three subjects would really be three experiments, as we consider the behavior of each subject individually, because behavior is primarily a phenomenon of individual organisms.

If our experimental arrangements actually controlled *every* variable relevant to the behavior of concern, *no* variability would occur. All the measurements would be the same, and the results would match predictions. But this seldom occurs, because we can rarely control all the variables. So predictions are always off by some amount; the measurements always vary by some amount. These amounts indicate the variability, which we thus see as arising from the effects *of the variables over which we did not exert experimental control.*

This kind of variability is not mysterious. It derives from the incompleteness of our control over functional variables. It reflects how much of the causal variables we are taking into account. The variability is larger when we control only a few of the relevant variables (leading to greater measured differences between predictions and outcomes). On the other hand, the variability is smaller when we control more of the relevant variables (leading to smaller measured differences between predictions and outcomes). This means we can reduce variability by taking more of the causal variables into account.

Realistically, economics must enter this picture, because taking variables into account is expensive, not just in terms of energy and some other resources, but in terms of funding. Generally, the greater the number of variables taken into account (i.e., measured and either held steady or manipulated) the higher the associated monetary expenses. As the importance of the experimental question increases, the more of these costs we must bear to take more of the relevant variables into account to answer the experimental question more thoroughly. This reduces variability and increases the success of prediction, control, interpretation, and application, which in some important ways justifies the increased funds society must authorize to cover the expenses. Conversely, for less important questions, with control being costly, we settle for controlling fewer variables and so must tolerate more variability along with associated reductions in the accuracy of prediction, control, and so on. After all, events are still lawful and orderly. We must, of course, be careful not to let costs determine the importance of the experimental questions.

Still, the point is that the source of variability does not reside in the magic of made up inner agents. The source resides in the causal variables that are not taken into account. That is, the source of variability resides in the amount of residual ignorance we must tolerate from affording a decreased amount of experimental control over all the variables relevant to the events under study.

Variability from that source then affects the selection of experimental methodology. We have no need to smooth out variability from discredited capricious inner agents across subjects in groups (which in any case tells us little about the individual behavior of those in the groups). Instead we need to adopt the methods that most effectively control, for a given amount of research funds, the most variables affecting a single subject (with three to six subjects studied, thereby providing a minimal level of replication, reliability, and generality for the results). Since the single–subject designs that we adopted early meet these requirements quite well, we continue to emphasize them in the natural science of behavior.

Anticipating what you will find in other books covering methodology, the two most common and effective types of single–subject designs are the "ABAB" design and "Multiple–Baseline" designs. In both types the subjects serve as their own control across different experimental phases, with each phase containing different conditions. We then accept the data as documenting experimental control of the relevant variables when the data show that behavior changes when, and *only* when, the experimental conditions change.

Mostly through these kinds of methodology, behaviorologists work to discern and apply the variables of which behavior is a function. These variables generally reside in an organism's species history, personal history, current situation and, particularly for people, the cultural setting. But again, what are these variables and what relates them to experimental analysis?

Variables in General Science and in Behaviorology

Variables in Natural Science

Returning to the question of how experimentally analyzing phenomena happens, let's review some basics. Let's start by repeating a point too easily taken for granted: Successful experimental analysis begins with accepting the constraints on natural science contained in the philosophy of naturalism. One of the most fundamental of these constraints is that natural scientists deal *only* with natural events as *independent variables* (ivs) and *dependent variables* (dvs). We exclude any untestable variables, including mystical, non–natural (i.e., anatural or supernatural or unnatural) variables, because they reside outside the range of real, at least theoretically testable, ivs and dvs. This exclusion has enabled many major successes, for at least the last 400 years, in expanding our knowledge and practices for dealing with the world.

Let's look more closely at ivs and dvs and their relationships, because they are at the core of how experimentally analyzing phenomena happens. ivs are the ones researchers change (manipulate) to see if different values have any effect, or different effects, on the dvs, which are the variables that they are studying. In other words, the effects of manipulating ivs are the changes that we measure in dvs. When a particular dv reliably changes the same way, and in the same amount, every time we change a particular iv, in a particular way or amount, we say that this dv changes as *a function of* that iv change.

For example, some electric stoves have ten buttons for each element, with each button tied to a different amount of electrical energy passing through and thereby heating an element. These buttons therefore constitute the values of the heat iv, and we manipulate the heat when we push different buttons. (The word *manipulation* gets some bad press, sometimes deserved, but here in science it has a quite neutral meaning.) By manipulating the heat, we control the speed at which the water in our teapot boils. That speed (actually, the rate of change in the water temperature) can be our dv. Contingencies involving guests getting impatient for tea compel pushing the button that provides the most energy to the stove element, making the water heat quickly. In contingency terms, the fastest water heating is contingent on (i.e., "depends on") pushing the highest energy button, which is contingent on (i.e., "depends on") guests waiting for tea. Alternatively, if conditions are such that we cannot use the hot water for several minutes anyway (perhaps because we are preparing the crumpets or scones that accompany the tea) then we push a button that provides less energy to the stove element, making the water heat more slowly.

As usual that example is more complex in a couple of ways. The whole situation actually intertwines both physical variables (amount of energy and speed of water heating) and behavioral variables (conditions controlling pushing one button rather than a different button, with appropriate results). And we excluded inner agents in the process, although at this point we still relied on our inner agent–supporting language to discuss the matter. All this anticipates our initial coverage of behavior IVs and DVs shortly. Also, the speed of water heating is a function of the amount of energy reaching the stove element, and the amount of energy is a function of which button is active. A reasonable summary of these relationships is that the fastest water–heating speed is *a function of* pushing the button that supplies the most energy.

An additional observation regards the direction that each variable takes. In that example, both variables go in the same direction; when stove heat goes up, the water temperature goes up, and vice versa. We call this a *direct* functional relationship. In another functional relationship, the variables go in opposite directions, which we then call an *inverse* functional relationship. For example, a *decrease* in pupillary size (a DV) is a function of an *increase* in illumination level (an IV), and vice versa. This is an *inverse* functional relationship.

We actually only observe the repeatable sequence of these events in time, that is, one seems always to follow the other ("seems," because observing every possible repetition—past, present, and future—is not possible). Enough such repetitions breeds a certain confidence that the sequences are not occurring by coincidence. We may then begin using the overused (and possibly misused) term "cause" to describe the sequence, saying that a certain IV causes a certain DV. We may even reach the point where the relationship is so well established, with such a high confidence level, that we begin calling it a "law" (e.g., gravity).

"Laws" are usually discovered by different disciplines, because different DVs and different scientific disciplines are intimately connected. DVs of a particular type make up the particular subject matter of a discipline. Generally, DVs involving energy constitute the basic subject matter of physics, while DVs involving matter constitute the basic subject matter of chemistry, and DVs involving life forms constitute the basic subject matter of biology. As we will soon see, DVs involving life functions constitute the basic subject matter of behaviorology. What are some of the DVs, and IVs, of behaviorology?

Variables in Behaviorology

Behaviorology is interested in the direct and inverse functional relations and laws that pertain to behavior, especially human behavior. With behavior in general, and changes in behavior in particular, as our DVs, let's consider just a general view of the range of IVs that control behavior. The classification, evaluation, and application of these IVs constitute the main topics of most of our subsequent chapters.

An easy, and so possibly helpful in the brief term, initial view of behavior and its "causes" takes this very simplified form:

$$A \rightarrow B \rightarrow C$$

We call this the "A–B–Cs of behaviorology." Here the "→" reads "functionally controls," while the "A" stands for antecedent events, the "B" stands for behaviors, and the "C" stands for consequences. Thus our formula reads "antecedent events functionally control behaviors, which functionally control consequences." As contingencies, the formula says the behavior is contingent on the antecedents and the consequences are contingent on the behavior. Later, as the complexity increases, we will also understand that the effectiveness of the antecedents is contingent on the consequences.

That formula indicates the most common kinds of variables surrounding behavior. A variety of stimulus variables comprise the antecedents of behavior, a variety of considerations surround behavior responses, and a variety of stimulus variables comprise the consequences, or "postcedents," of behavior. Stimuli are the vast range of real events in the internal and external behavior–controlling environments. More specifically they are various kinds, qualities, and intensities of energy changes at receptor cells. They occasion structural changes in the nervous system that result in the occurrence of behavior (or affect behavior in some other way) and we describe this interaction as the neural *mediation* of behavior under those environmental controls. All of this occurs entirely naturally. Internal and external environmental evocative or consequating energies affect the brain and nervous system, which thus neither originate nor initiate the occurrence of behavior. Instead, brains and nervous systems only mediate behavior as part of the physiology that is ever present and operating when behavior occurs. Essentially then, physiology is the locus at which environmental ivs and behavioral dvs interact. Similarly, the processes and outcomes that the terms in our behavior formula imply need no participation from mystical inner agents. Let's now take a brief look at some aspects of each term in our simplified behavior formula.

Antecedents. Several types of variables comprise the antecedents of behavior. Here we will mention the one that seems to be the most common, with the others coming up in later chapters. This most common antecedent stimulus is the "S^{Ev}." It stands for "evocative stimulus." These stimuli function to evoke behavior. Any setting contains many of these stimuli. Which ones affect you next depends on several other variables. These include the history of conditioning in your past that made some stimuli functional in relation to your behavior. The result is that, according to the currently operating functional relations, at least one of these stimuli *will* be affecting you. This takes the form of its presence *evoking* the relevant behavior. That means the stimulus functionally controls the occurrence of the response. If the stimulus provides the appropriate energy change at receptor cells, the physiology of the "carbon unit" that is you mediates the response (i.e., in the presence of the stimulus, the response occurs). For example, very likely previous conditioning has made flashing red lights evoke stopping. So if, as you drive around a corner, you come into the presence of a flashing red light, then you stop.

Or rather, excising the inner agent implied by the word "you," the presence of the flashing red light evokes the response of stopping, all entirely naturally. By the way, some folks object to saying "carbon units." Referring to them does not dehumanize humans, but respects the related scientific objectivity.

In a sense, simultaneously occurring S^{Ev}s compete with each other. What if, in the presence of that flashing red light, an officer directs you to turn left? This means you are in the presence of two obvious S^{Ev}s, a flashing red light and a directing officer. What happens? The officer's directions evoke the behavior of turning left! At least, that is what happens if your conditioning history has been in keeping with the laws of the land where directions from a present officer take precedence (i.e., carry more severe consequences for non–compliance) over automatically operating red lights. (See all the levels of complexity we can get even in a simple example? And, this early in the book, I cannot even come close to mentioning all these levels. Behaviorology makes possible our managing most of these multiple levels simultaneously. Marvelous.)

Again, usually you are in the presence of a multitude of evocative stimuli (S^{Ev}s) and which ones affect you, and in what sequence, may or may not be important. As I craft these examples here at my desk, the S^{Ev}s that surround me include the keys on the computer keyboard, the computer mouse, the computer screen, the papers on the desk containing my notes for this chapter, the flash drive on which I backup my work, the USB hub into which the flash drive plugs, the clock, the telephone, the tissue box, the lights, the stereo, the CD player (and the remote controls), even the dust bunny barely visible under the edge of the stereo, and so on and on and on! Which one will evoke my next response? Survey your own desk or current setting. I am sure you can list a similar number of S^{Ev}s and possibly still leave numerous ones unmentioned (as I did). Which one will evoke your next response? Right now the words in this book are again evoking your responses. However, what if you looked around and saw your clock showing that you have been enjoying reading for hours and are about to be late for a medical appointment? Now *that* contingency would indeed change behavior, from reading to traveling.

Behaviors. Let's turn now to "behaviors," the middle term of our most basic behavior–control formula. In practice, we differentiate between behavior and responses. While *behavior* (B) is the general term, we use the term *response* (R) for particular instances of a behavior. Responses occur under the control of currently operating variables; this makes every response a new, and thus different, response. This could be a problem since science usually deals with repeatable phenomena. But responses naturally fall into "response classes" on the basis of sharing the same stimulus consequences. Thus our experimental analysis operates on the level of response classes.

For example, in a well–managed classroom, students regularly raise their hands when the teacher asks a question. What is the question? It is, in part, an S^{Ev} evoking the behavior of hand raising. When the teacher acknowledges one of the students, the acknowledged student provides an answer. And when the answer is correct, the teacher provides intellectually and emotionally honest and appropriate comments that function as consequences with effects on similar subsequent behavior. For any particular student, many hand raises occur each day. None of them has identical characteristics with any of the others. They all differ topographically (e.g., up at different angles). But they all share the same kinds of consequences. Sharing the same kinds of consequences makes all those instances of hand raising into members of a single, repeatable response class that we can study and that we might simply call "hand raising."

Did you notice that even such common behavior is interactive? Such interactions describe what we call *interlocking contingencies*. The question is an S^{Ev} evoking the hand raising; the raised hand is an S^{Ev} evoking the acknowledgement; the acknowledgement is an S^{Ev} evoking answering; and an answer—correct or incorrect—is an S^{Ev} evoking appropriate consequential comments. We will address additional *behavior* considerations in later chapters.

Consequences. The third term in our formula receives much attention. It is not necessarily more important than the other two terms, but it is the part that helped differentiate operant behavior, to which this formula pertains, from the other kind of behavior, reflex behavior. (Both kinds get chapter–length treatments later.) Actually, the word *consequences* has a meaning far too specific to serve as the word in the third position of the formula, because that position depicts events that merely *follow* behavior in time. Standing for the final "c," it helped ease the introduction to the formula. (The "a–b–cs of behaviorology" is rather catchy.) But both it and the formula must develop further. Just as "antecedents" refers to events that precede behavior in time, we call those that follow behavior in time *postcedents*. In both cases we simply address solely the relative time relation. Consequences, however, are but one kind of postcedent.

Consequences divide into various types that we detail in a later chapter. Our concern just now is their status as "reinforcing" or "punishing" consequences, because this characteristic pertains to the central role of consequences. This central role involves making some responses occur more often (in which case the consequence earns the title "reinforcer") and making other responses occur less often (in which case the consequence earns the title "punisher"). For example, at certain specially designed intersections in Europe, drivers are photographed. Yes, that is common, but what happens next is less so. Those drivers who stop for the red light receive a free lottery ticket while those who run the light receive a payable traffic ticket. The data, for all those who drive that route, show that the number of drivers obeying the light increased and the number running the light decreased. Thus we can call the lottery tickets *reinforcers* and the traffic tickets *punishers*. However, we would prefer to see these stimuli occur immediately after the responses that produce them, and earn these titles at the level of the individual drivers rather than the crowd of drivers. We will see more individual examples in later chapters. We will also see that delays, such as from the mail delivery of these consequences, will compel an expansion in the details of our analysis. These real examples are always more involved than whatever particular principle we are using them to elucidate.

How does earning the reinforcer or punisher titles work? Recall that evocative stimuli (i.e., S^{Ev}s) evoke behavior by triggering changes in nervous–system structure that mediate the behavior occurrence (i.e., with ordinary motor behavior, the energy from S^Ev's makes sensory neuron bundles fire, which then make motor neuron bundles fire, which makes muscles contract, which *is* the behavior occurring). Reinforcers work also by triggering changes in the nervous system, but these changes are altered structures *that are more enduring* and thereby leave the body different. The result is that, after reinforcers occur, the occurrence of relevant evocative stimuli *more easily*—which we observe as more frequently—evoke the behavior again. And punishers work like reinforcers except that the enduring nervous–system structural changes leave the S^Ev's *less* effective, so that the behavior occurs less frequently.

Much more actually happens as *contingencies cause behavior* by operating through our physiology, which does not cause behavior. The details of the physiological events supporting the operation of both S^Ev's and consequences will come from cooperation with our colleagues in physiology.

All three terms together. We can elaborate these terms into a more clear statement of the fundamental formula for *initially* analyzing any behavior. The "A–B–Cs" formula readily becomes "S^Ev–R–S^R" (with "S^R" being a reinforcing stimulus). We call this formula *the three–term contingency* for obvious reasons. We can write many similar formulas, with three or more terms, to describe a range of alternative contingency forms. However, this particular one summarizes our *starting point* for analyzing any bit of behavior that becomes of interest or concern for us. We start by investigating the stimulus variables in the setting. Behavior never occurs in a vacuum, because numerous S^Ev's fill every setting (recall our "at our desk" example). Next—actually the order is not this strict—we examine the responses ("R" in the formula) for any characteristics that can affect outcomes. And we continue (as additional steps require further discussion later) by considering the many factors that affect the effective operation of the consequences. In due time we will find that our analysis of the contingencies in which behavior participates is deeper and far more extensive than our starting–point formula indicates (more material for later chapters).

Is "contingency" still an uncomfortable notion? Recall that a "contingency" may best be understood at this time as a *dependency:* The occurrence of the consequence *depends* on the occurrence of the response which itself *depends* on the occurrence of the evocative stimulus. Or, better, the consequence occurrence *is contingent on* the response occurrence, and the response occurrence *is contingent on* the evocative stimulus occurrence. For example, when out driving, as the car approaches some curves, those curves, as S^Ev's, evoke steering responses which staying safely on the roadway consequates. That consequence of staying on the road depends (is contingent) on proper steering responses which depend (are contingent) on the S^Ev directions of the curves.

Variables and Awareness

Here is something else that I find very important for thoroughly understanding human behavior. You should note (and possibly find fascinating and important as well) that *you need not be aware* of the curves, in that driving example, for them to evoke correct driving responses. Awareness is one of the neural behaviors of consciousness, and seems to operate as a "single channel" capability. If your single–channel neural–awareness responses are engaged, say, with daydreams of winning a lottery, the curves (as S^{Ev}s)—and continuing to stay on the road (as reinforcers)—will still effectively control your steering responses. (Up to a point, that is. "Don't try this at home" as they say—or alone, etc.) Ultimately you will observe (another neural behavior of consciousness) this not–needing–to–be–aware when, arriving at a red stop light further along the road, you find yourself wondering (yet another neural behavior of consciousness) how you got there, and where the last mile or two went. You neither crashed nor heard sirens wailing nor got pulled over, so your steering responses must have been adequate. However, those driving responses were occurring (under what we call *direct stimulus control*, developed in later chapters) right along with, that is, at the same time as, your daydreaming responses. The daydreaming responses prevented your being aware of the steering responses, because the needed neural channel, so to speak, was already occupied with those daydreaming responses. How often have you experienced something like that? It is not magic (although it seems wonderfully magical). It is just the behaviorological laws of nature at work, laws which operate without our needing to be aware of them or the variables involved. From where, however, did the behaviorology discipline come?

Some Behaviorology History

Emergence of Behaviorology

Behaviorologists separated their natural–science discipline from psychology after psychologists in general repeatedly refused numerous attempts over 50 years to change the non–natural discipline of psychology into a natural science. Psychology had separated from philosophy around the second half of the 1800s. Let's consider a simplified version of what happened. (Some folks will say that we tell a grossly oversimplified version, but it fits our space. Besides, readers can get a book on the early history of psychology and read for themselves the rich details in the story, at least as psychologists tell it.) Anyway, some philosophers back then had noticed the great strides of their physics, chemistry, and biology colleagues, strides that seemed to stem from the use of scientific methods. They figured that even as philosophers they might also make more strides if they too adopted these methods. Their fellow philosophers, however, essentially said, "No. Using scientific methods would not be philosophy; it would be a different discipline." So those who favored adopting scientific methods started a new discipline, psychology. However, they should also have adopted the philosophy of science of natural science. But they never did, not even the minimal constraint to work only with real, natural events as IVs and DVs. Instead they extended their philosophical outlook to embrace *accounting for behavior with a range of spontaneously operating internal agents* (i.e., agentialism). As a result psychology began as a non–natural discipline, and remains so today.

B. F. Skinner was an early exception to this operational pattern. In the early 1930s, he had contact with W. J. Crozier. Crozier was a leading professor in the biology department at Harvard University. From his contact with Crozier, he brought the philosophy of naturalism to his study of behavior, studying it as a natural phenomenon. Sticking to dealing only with real, natural variables, Skinner and his students and colleagues built up this natural science of behavior over the next several decades. All the while they not only worked in traditional psychology departments, where they were historically stuck, but they also attempted to get psychology to change into a natural science. They fostered this change by successfully producing, and accumulating masses of the standard kinds of formal evidence that normally, in natural–science disciplines, prompts that kind of change. But during this period, psychology kept a monopoly (aside from religion) on the human nature, human behavior subject matter. It even defined itself as "the study of behavior *and the mind,*" as it stuck to its secular (i.e., non–theological) version of mysticism.

As the early natural scientists of behavior accumulated advances during this period when psychology and their natural science of behavior shared their history, the name they used to describe their work changed several times. After "operant psychology" came "TEAB" (The Experimental Analysis of Behavior) and "behavior analysis." In the process, they founded various organizations and journals in support of their natural science. Ultimately, however, contingencies precluded getting psychology to become a natural science.

In 1987, some of these natural scientists of behavior looked again at the data from the attempts to change psychology. Being under contingencies that compelled data–based actions, they saw that they had succeeded in setting up numerous arrangements separate from their psychology colleagues, including separate courses, programs, journals, organizations, certification, and accreditation. They also saw that the numerous, continual, and overlapping change attempts, over about five decades, had gotten nowhere, even as the global problems of our world accelerated, demanding a share of attention from a natural science of behavior. Their traditional psychology colleagues seemed to be saying, "No. Accepting naturalism and rejecting inner–agent accounts, just to be a natural science, would not be psychology; it would be a different discipline." Under these circumstances, these natural scientists of behavior declared their independence from psychology, called their long established but newly separated discipline "behaviorology," and proceeded to found new professional organizations and journals.

So occurred the emergence of the natural science of behavior that we call behaviorology. It started around 1913, over 100 years ago, with John Watson's description of a behaviorist perspective. It developed with the variously named natural science of Skinner and his colleagues and their students for about 75 years after that. And it has been an *independently* organized natural science of behavior discipline, divorced from psychology, since 1987 (i.e., for about the last 30 years).

A complicating factor needs some attention. The behavior analysts took those and other independence–oriented steps while still closely associated with psychology. This allowed the psychology discipline to claim behavior analysis as part of itself. Decades ago some behavior analysts somewhat validated that claim when, apparently as one of the reasonable attempts to move psychology toward giving up mysticism in favor of natural science, they used the behavior–analysis label as the name for the journal of an official division of the American Psychological Association. Later the same division took "behavior analysis" as its own name. Focusing on (and perhaps distracted by) their successes with autism interventions, behavior analysts lost the name they had been using for their basic science. Raising insufficient or no clamor of objections, they left the mystical discipline of "behavior and the mind" to take over the "behavior analysis" label. Today's "Behavior Analysis," as "Applied Behavior Analysis," or ABA, is the engineering counterpart to the basic science we call behaviorology. (Currently many states have laws that differentiate between psychology and "behavior analysis," for concerns like licensing. But laws can change—either way—and seldom intersect academic disciplines anyway.)

The valid psychology claim to behavior analysis and its label leaves others, including natural scientists in general, worried about whether or not today's behavior analysts are as committed to natural science as their forebears were. The collegial relationships that some natural scientists of behavior have with some traditional natural scientists may reduce that worry a bit. But beyond such relationships, those natural scientists of behavior who remain "behavior analysts" invite avoidance, even scorn, from traditional natural scientists. They see these behavior analysts as refusing to go independent and leave the inherent secular mysticism of psychology behind, as refusing to become behaviorologists and use this independent label to name their separate basic natural science.

On the other hand, the contingency benefits of being part of psychology can include, among other things, an increase in job security and some safety from accountability. While problems from national economic swings can increase the value of such factors, a wide variety of contingencies (i.e., causal dependencies) can drive being part of psychology for different behavior analysts, including some—perhaps much—personal success in psychological work units. Even for those behavior analysts well trained in radical behaviorism and the natural science of behavior, these pro–psychology contingencies too often override the evocative and reinforcing potency of contingencies driving independence for natural behavior science. Could that include overriding the contingencies involving disciplinary integrity, and even the potentially greater benefits for humanity that could accrue from being able to work, with mutual respect, alongside traditional natural scientists to help solve global problems?

In the years 1984–1987, an extensive debate filled the published behavioral literature regarding, pro and con, the question of fully and officially separating the natural science and philosophy of behavior from psychology. Many discussants acknowledged numerous types of recognizable separation already present in varying degrees, including some academic departments and programs. But none of these early department separations occurred under full and formal declarations of independence, although some departments came close. For example, the "Department of Behavior Analysis"—named before the behaviorology label came into use—was fully separate from and independent of the Department of Psychology at the University of North Texas in Denton.

A more explicit independence move. The 1984–1987 debates culminated, in 1987, in a group of natural scientists of behavior meeting to reassess the situation and take action, as we mentioned earlier. They came to several conclusions. (a) If data from a half century of continuously attempting to change psychology into a natural science "from within"—by invoking standard, evidence–based methods that might take centuries and even then not work—showed failure to produce even slight movement in that direction, then changing psychology was not going to happen within a meaningful time span (e.g., before the opportunity passes in which to help humanity reduce global warming and so avoid its worst effects, a time frame of about 100 years as they rather optimistically understood it then). (b) Their natural science of behavior was not, and never actually had been, any kind of psychology as it had never accepted the basic psychological core of mystical agential origination of behavior. And (c) instead, their already well–established natural science would continue, at least in part, as a fully separate and independent discipline called behaviorology. This term first appeared in the late 1970s specifically to describe a natural science of behavior discipline completely separate from and independent of psychology. It is the only term, from among all proposed names, to have survived and grown in use.

For readers who prefer detailed histories, Lawrence Fraley and I—as participant–observers of the events—provided a 140–page, comprehensive, and thoroughly analyzed history of the emergence of behaviorology in our 1992 paper, "Origins, Status, and Mission of Behaviorology." (You can currently find it in Fraley & Ledoux, 2015.)

Audience, Styles, and Part I Topics

The Audience for this Book

Global culture members comprise the principal audience for this book, especially those who see the value of the naturalism that makes possible meaningful contributions from science and technology. The global culture has everything to lose if humanity fails to solve its global problems in a timely manner. This failure could occur due to an historical discrepancy that has arisen in the global culture, a discrepancy in objectivity around the globe. This discrepancy involves most of the global culture relying on natural science regarding physical, chemical, and life subject matters while at the same time also relying on mysticism and superstition regarding the human nature and human behavior subject matter. Behaviorology counters this discrepancy and, as you have seen, we have already begun to address it in this book.

That audience group includes everyone, although not everyone can appreciate this material. Let's list some of this audience group:

❧ lovers of life (e.g., from those who regularly examine the contents of nutrition labels on food packages in their local grocery store, to those who need to hunt to put food on their table);

❧ lovers of the arts of sight (e.g., from those who appreciate European oil paintings or Oriental scroll paintings to those who appreciate Native American pottery, basketry, weavings, or jewelry [cheap promotion: see Ledoux, 2016]);

❧ lovers of the arts of sound (e.g., those who listen to popular or traditional music recordings or attend concerts);

❧ lovers of science (e.g., those who enjoy each other's company at a backyard telescope eyepiece observing the moon or the phases of Venus or other solar system objects or the vast objects beyond our local stellar neighborhood);

❧ lovers of sport (e.g., from those who watch sports games inside to those who go outside and enjoy each other's company while target shooting at their local rifle or pistol range);

❧ even lovers of relaxation (e.g., those who read about or study or meet to discuss past, present, or even future scientific discoveries, developments, implications, and applications).

My apologies if I left anyone off that list. My experience with many of those interesting activities biased their selection for inclusion, which carries the important implication that we on this planet are all far more alike than our conditioning may have left us believing or feeling. And that extends to our language usage, with implications for some concerns about writing style.

Style Considerations in this Book

Rather than a reference–laced reporting of research results, this book relies on a conceptual treatment of topics. This treatment includes numerous everyday examples, directly emphasizing meaningful understanding and appreciation of the fundamental principles, practices, and implications of the behaviorology discipline.

As we reach and cover the more complex, advanced levels, we will also employ our findings to explore and explain some of the conceptually intricate and intriguing questions that have challenged human intellectuality and emotionality since ancient times. At such points I suspect you will agree that we can now begin scientifically to answer some of humanity's ancient questions.

One word of warning: If your past conditioning has left you likely to skip to the back of a book for a preview of the outcome, then you probably face some unusual risk with this book. Such ancient questions are truly compelling. But our coverage here cannot succeed yet in answering them in one of the next few chapters, because success requires the background of the chapters between here and the back of the book. If you skip all, or even some, of these chapters and jump ahead, then a safe prediction is that you will be confused, and thus disappointed, due to missing the appropriate preparation. Instead, please be patient, and stay with us as we try engagingly to get through the intervening chapters to the really illuminating, ever more fascinating parts.

A particular style concern, however, compels us to preview a little of the ancient language question. Language is verbal behavior, and it is a function of the same kinds of contingency variables that control all other behavior. One major class of variables, controlling the kinds of phrasings we use in English, stems from the agential viewpoint that reasonably existed at the time of the origin of language. Then, primitive animism was the most parsimonious view. Primitive animism explained movements as the result of inner spirits animating both organic and inorganic objects (e.g., animals and rocks and wind and water and clouds). As languages evolved they often retained the certain economy of words that agentialism coincidentally enables without reference to its shrinking accuracy. Thus today our language is laced with agential references, with personal pronouns as likely the most common. To say "I" or "you" or "he" or "she" is automatically to imply an inner agent of one or another variety (e.g., a mind or psyche or self or soul or person). Thus, stylistically, we may try to engage phrasings that lack, or at least reduce, those pronouns. The result, however, while scientifically more accurate, may sound stilted, a result with which I hope you will be patient. Experience shows that increased exposure to phrasings that support science reduces the discomfort they otherwise cause.

A similar problem confronts us over the use of active voice (e.g., Ed earned money) and passive voice (e.g., Money was earned *by Ed)*. Active voice, due to its direct subject–predicate–object structure, enhances clarity and readability, which accounts for its preference among authors, editors, and publishers. But look more closely. Active–voice structure often implies agency. Whatever is in the subject slot comprises the agent of the action. Passive voice avoids this problem but at the high cost of reducing readability and clarity. In spite of these problems, some scientific disciplines expect their authors to rely on passive voice because the "who" is distracting. Stylistically, we will here continue to rely on active voice, with few pronoun subjects, and with regular reminders that the remaining implied agencies actually contradict scientific realities.

In time, the increased quality of science and its products that may derive from recognizing and dealing with agential issues, including personal pronouns and active voice, will likely affect the way we regularly speak and write (i.e., the language itself, our verbal behavior). We may come to exhibit more and more verbal behavior that, consistent with scientific reality, lacks inner–agent connotations, and we will even gradually become more and more comfortable, individually and culturally, with this development. Meanwhile, when particular topics lead us to need more explicit scientific accuracy, some of the phrasings in this book may seem distorted to you due to the strength, only gradually reducing, of our lifelong agency–based verbal conditioning. Perhaps a new grammar is on the rise.

An "Outline" of Part I

The sequence of our Part I topics focuses on increasingly complicated and testable functional relations between environmental (i.e., both "nature" and "nurture") and behavioral variables. These extend into the technological interventions of behaviorology that beneficially apply the contingency causes of behavior in arrangements that we call *contingency engineering*.

Thus an "outline" of Part I takes the form of a sequential list of the included topics. The list shows our direction through our basic principles and practices:

- What behavior is *NOT* (Chapters 3 and 4);
- What behavior *IS* (Chapter 5);
- Causes and effects in sciences (Chapter 6);
- Reflex behavior (Chapter 7);
- Operant behavior: Antecedent stimuli (Chapter 8); Postcedent stimuli (Chapter 9); Response class and differential reinforcement (Chapter 10); Shaping (Chapter 11); Chaining (Chapter 12); Fading (Chapter 13); Schedules of Reinforcement (Chapter 14).

References (with some annotations)

Fraley, L. E. & Ledoux, S. F. (1992). Origins, status, and mission of behaviorology. In S. F. Ledoux. (2015). *Origins and Components of Behaviorology—Third Edition* (pp. 33–169). Ottawa, CANADA: BehaveTech Publishing.

Hawking, S. & Mlodinow, L. (2010). *The Grand Design.* New York: Bantam.

Ledoux, S. F. (2014). *Running Out of Time—Introducing Behaviorology to Help Solve Global Problems.* Ottawa, CANADA: BehaveTech Publishing.

Ledoux, S. F. (2015). *Origins and Components of Behaviorology—Third Edition.* Ottawa, CANADA: BehaveTech Publishing.

Ledoux, S. F. (2016). *Beautiful Sights and Sensations—Small Collections of Native American and Other Arts.* Canton, NY: ABCs.

Skinner, B. F. (1974). *About Behaviorism.* New York: Knopf. With the legitimate time–frame excuse of appearing before the 1987 adoption of the term behaviorology, this book portrays radical behaviorism as the philosophy of "behavior analysis." Back then this could be accurate, because behavior analysts were then moving to become an independent natural science under the behavior–analysis label. Sadly the efforts under *that* label were ultimately unsuccessful, because psychology claimed the behavior–analysis label. Although we now use *behaviorology* to denote the independent natural science of behavior, nevertheless this book is a wonderfully thorough and readable treatment of radical behaviorism, the philosophy of science of behaviorology. The Bibliography entry contains more details.

Chapter 3
What Is NOT Behavior or Cause

Explanatory fictions that fail to explain ...

The magazine, *Consumer Reports,* regularly carries articles on the benefits of maintaining a healthy weight, even if that requires some special diet. All too commonly, people claim that to maintain a diet one must *restrain oneself* from all the chocolates (Heaven forbid!) and one must exert lots of *will power* to eat only the right foods in the right amounts. Are self restraint and will power really the causes of successful dieting behavior? Let's examine a range of so–called causes with similar characteristics. Are they explanations of behavior, or are they fictions that fail to explain?

Setting the Fiction Problem Stage

Let's start, though, with a bigger picture. Recall our discussion of natural functional histories and levels of analysis. If we start well beyond the special considerations required at the sub–atomic level of what physicists call quanta, then, as we traverse the continuum of analysis levels from physics to chemistry to biology, we approach our level, the behaviorology level. At this level assemblages from atoms to molecules to compounds reach the expanding point where the accumulating complexity of natural chemical interactions, among all of these, builds structures that we call single–celled organisms. These continue naturally to accumulate complexity, building multi–celled organisms. Some of these organisms contain sufficiently complex neural structures for natural responses to stimuli to occur. This is the level of behaviorological analysis.

Considering the preceding levels, starting with physiology, we find that the access to behavior–causing variables generally remains out of our reach. We have discovered, however, that at the behaviorological analysis level we have a good deal of access to a wide range of contingencies as behavior–causing variables. Access enables improvements. This provides a good reason to stick with the behaviorological analysis level. At this level, the interactions of these organisms (i.e., these complex chemical systems, including humans) with each other and with all the various aspects of their surroundings, produce behavior phenomena. The "how" of these phenomena resides in the details of physiological structure and function occurring as (i.e., mediating) neural and muscular responses. Meanwhile the "why" of these phenomena resides in the details of the continuously accumulating changes, of those physiological structures, to the stimulating energy exchanges between parts of the physiological structures and parts of the internal and external environment. As implicit in our Law of Cumulative Complexity, all of this raises a cacophony of reciprocal interactions that are nonetheless lawful, orderly and, *for this analysis level,* predictable and controllable. Furthermore, prediction and control enable interpretation and explanation of complex behaviors by way of experimentally demonstrated natural laws of behavior. Later chapters introduce many of these laws and their application in both interpretations and interventions.

A main focus of this chapter, however, rests on a range of fictitious alternatives to scientific explanations. Recognize first, though, that while behaviorology, like all natural sciences, lacks explanations for everything, *fictitious alternatives are not explanations at all.* Again, for such reasons, we stick to the behaviorological level of analysis.

Why the pushy emphasis on why we stick to the behaviorological analysis level? Consider a related part of our big picture. While arts and sciences are not in opposition, conditioning of the traditional cultural type (i.e., the kind we all share in our growing–up background) causes problems, as does later, theological or secular, non–science supporting educational conditioning. The problems these cause involve pushing many into the organized ranks of cultural superstition and mysticism. Unsurprisingly these ranks reject behaviorological explanations, finding them somehow degrading or blasphemous. The alternative accounts that they would support, however, are unnecessary, redundant, harmful, and not explanations at all. This warrants our attention.

For all of history until about 100 years ago, fictional explanations for behavior constituted the main, if not the only, game in town. In his courageous book, *The Demon–Haunted World,* Carl Sagan (1995) referred to science "as a candle in the dark." We must turn that candle into a floodlight exposing the whole variety of unhelpful accounts for behavior while illuminating the helpful accounts from natural behavior science, and thereby support the role of this science in helping solve local and global problems. Along with other authors, Sagan has exposed a multitude of false and otherwise invalid explanations for many natural phenomena. Here we deal with the false and otherwise invalid accounts, particularly of the fiction kind, that pertain to behavior.

Note, however, that our most fundamental objection to fictitious accounts is *not* that they are fictional, that they do not exist, although this forcefully compels our attention. The real objection is that such accounts are irrelevant to the scientific understanding, prediction, control, and interpretation of our subject matter, behavior. These accounts involve no variables that direct experimental manipulation can show to be functionally related to behavior.

Common Explanatory Fictions

Our treatment of fictitious explanations for behavior will not be exhaustive. More complicated classes of fictional accounts remain for analysis, some in a later chapter, or for others to cover. Here we will consider the more common classes of fictitious accounts regarding behavior.

Some Origins of Fictitious Explanations

The origins of most explanatory fictions reside in the verbal conditioning most cultures put their members through early in life. Such conditioning built up over many, many generations, from times when language was a new behavior for humans. The only accounts for nearly *everything* then, and for many thousands of years, were versions of primitive animism. Refinements occurred and expanded but the core superstition and mysticism remained. Science made several appearances over the last few thousand years, only to be mostly lost each time, until about 400 years ago. As science expanded, the fictional accounts for most phenomena have retreated. Today fictional accounts generally still thrive only with respect to human nature and human behavior, and they are gradually but surely giving way in these areas also.

Linguistic Fictional Explanations

Language actually plays multiple roles in the continuation of fictional explanations (i.e., accounts that are fictions). It is the vehicle by which most cultural conditioning of new members occurs. This conditioning includes both operant components (e.g., language itself and traditional, usually mystical, cultural lore) and respondent components (e.g., emotional support and alteration of operants). Language also contains inherent allusions to mystical inner–agent fictional explanations of behavior. What led to these allusions?

The usual energy–conservation contingencies on the development of language led to economies (i.e., verbal shortcuts) in language use. These economies reflected and supported whatever mystical accounting for events was prevalent at the time of that development, including inner person agents. As a result, in English, the most obvious of the language economies that are still with us, and that contain inner–agent allusions, comprise the set of personal pronouns (i.e., I, you, he, she, it, we, they, me, her, him, us, them, my, your, his, hers, its, our, their, etc.). Each of these quite short words economically reduces a subject, object, or possessive to just a few letters, compared to the several syllables or words that otherwise occur. Try talking without pronouns.

In spite of the energy conserving economy, however, those personal pronouns inherently imply and, especially in the singular form, even seem to refer to the existence of internal mystical behavior–directing self agents (e.g., I, he, she). Such verbal shortcuts need not cause problems. Conditioning could manage them, making them merely refer to bodies mediating evoked behaviors, rather than inner agents. We will, of course, continue using these energy–saving verbal shortcuts, but we must excise the inner–agent implications, starting with education in behaviorological repertoires. Until so excised, these personal pronouns probably comprise the most ubiquitous form of inner–agent explanatory fiction, but not the only linguistic one.

Other agential language patterns involve the passive voice and, to a much greater extent, active voice. We previously touched on these difficulties when discussing writing style consideration. Passive voice leaves the agential "doer" of the verb action either out completely or gives it only the status of an afterthought. For example, in a passive–voice construction, the "by" clause often can be skipped. (See?) This constitutes a kind of improvement over active voice in terms of fictional causal agents, because active voice puts the agent right up front. Indeed, active voice puts the agent right in your face if conditioning that favors science has sensitized you to these issues. (Of course, no inner agent "you" resides in a reader's physiology…—See the problem?) But many people find passive voice at least a bit awkward or confusing.

So who uses passive voice? As one example, many scientific circles prefer passive voice for experimental reports, ostensibly because the passive voice helps emphasize aspects of the experiment rather than aspects of the researcher who is not the focus of the reports. Also, science maintains the struggle against unhelpful, untestable mystical accounts, a struggle the passive voice supports through de–emphasizing the agential "doer" (which, recall, exists not). While this also would at least intuitively tend to encourage the use of passive voice, especially among scientists, the passive voice construction generally makes for rough reading, and thus poor writing style. The conundrum for authors then, especially behavior–science authors, involves crafting sentences that are well written, easy to read, *and* non–agential. One method involves placing, in the subject space, events or properties or characteristics or processes that lack biological living status (as in the next three sentences). This way any question of agency fails to arise. Very likely, all facets of the culture could benefit from some linguistic changes in what passes for good grammar and writing style. Such changes support all the scientific realities benefitting the culture.

Managing linguistic fictitious explanations. Many scientific and other authors would welcome linguistic changes that allow them to write without automatically implying inner agents. They currently cope by trying to avoid personal pronouns and by carefully composing active but inner–agent–less sentences, all without that much success (as the writing here may show only too well). The results can trend toward a cumbersome and impractical writing style. A new, scientifically consistent set of linguistic practices (e.g., a new grammar) would help. Linguistic change of this magnitude, however, generally accrues so gradually as to be unnoticeable across even several human generations.

Meanwhile another kind of change helps us substantially, right now, to counter not only the implied linguistic types of fictional explanation for behavior but also the many other types that we will cover as well. This change arrives with the increased behaviorological analysis. The conditioning that is automatically involved in any behavior–science educational experience, like reading this book, necessarily and naturally leads to the generation and maintenance of some responses that are consistent with, and appropriately reflect, the scientific realities of behavior causation. After such conditioning, the contingencies of these scientific realities effectively counter the fictitious explanatory accounts that prior traditional cultural conditioning supports. Thus we manage personal pronouns and the active voice by seeing them for what they really only are, and using them only that way, namely, as energy–saving economic language tools (i.e., verbal shortcuts). As a residual problem, those throughout the rest of the culture, who as yet lack access to conditioning regarding behaviorological science, continue to see, and even support, ghosts behind every personal pronoun and active voice usage. They also leave other fictitious accounts for behavior uncorrected.

Cause and effect reminder. Before we visit more of the common explanatory fictions, let's review our use of the terms "cause" and "effect" as this will facilitate our efforts. Throughout this book, we use these terms only as verbal shortcuts for the functional relation between the longer terms "independent variable" (cause) and "dependent variable" (effect). Of course, we only contact pairs of variables in which one changes systematically with changes in the other. Any functional relation we describe is not necessarily something we actually see. Instead, the functional relation arises implicitly. Our contact with the *consistent* pattern of the dependent variable always following the independent variable increases our confidence in the reliability of the pattern. While arising implicitly, such functional relations, which occur throughout science, are not mystical, because they involve traceable energy transfers between the variables, transfers that may require a shift in analytical level to trace. For example, our behaviorological–analysis level may find a functional relation between an uncomfortably high indoor room temperature and turning on the air conditioner. But tracing the relevant energy flow requires a shift to the physics analytical level. We can make those shifts easily enough when we need them to better understand all the functional relations that actually account for behavior, but we have no requirement for those shifts at the practical level of the environmental control of behavior (although we may occasionally employ them). However, being *unable* to make such shifts implies a breach in the functional history of an event, and that usually leads to discovering a fictional rather than a real explanation, because real accounts make no such breaches while fictional accounts make them inherently or automatically. So, let's move on and deal with some fictional accounts for behavior.

Fictional Explanations and Some Constraints

Of the many types of fictional explanations for behavior, our focus here pertains to some of the more common types. By name these include reification, turning adjectives into nouns, nominal fallacy, circular reasoning, gratuitous physiologizing, and teleology. Even though many of these categories overlap each other in various ways, we will consider them separately. However, before describing these explanatory fictions, here are some comments about description versus explanation, and a brief review of parsimony, both of which constrain fictional accounts and help us better understand them.

Parsimony. Attending to parsimony plays a big part in dealing with fictional accounts of behavior, because all fictional explanations violate parsimony in one way or another, and so, again, are not really explanations at all. Rather than covering these violations separately with each fictitious account in turn, we cover them all here by recalling that *parsimony means working with only the simplest yet adequate explanation for an event.* Thus an account can violate parsimony either by being unnecessarily complex and untestable at this point (e.g., a mystical or superstitious account) or by being simply too scientifically *in*adequate to fill an explanatory role (e.g., time). Some accounts, especially some fictitious accounts, violate parsimony both ways.

Description versus explanation. Description is not explanation, and vice versa. The two are quite different concepts. Description involves elaborating or detailing *one* of the variables in a relation, usually the dependent variable, the effect, the behavior. This variable cannot make an explanation. Worse, such descriptions often mean talking about the behavior in such a general way that its response status disappears. Without a response, you end up with zero variables, even though linguistically the description still sounds like a plausible explanatory variable. On the other hand, explanation involves *two* variables, one an independent variable and one a dependent variable. Both are, at least theoretically, detectable and measurable (i.e., real) and together they exist as a functional relationship. The functional relationship then *is* the explanation. (It may or may not be valid; that is a separate question, usually decided experimentally.) With only one variable (or none), descriptions cannot be explanations, which require two real variables in a functional relationship.

For example, say that we see the responses of grabbing several cookies without hesitation whenever someone first offers them, and grabbing other children's toys whenever they show up with toys. Someone might *describe* these behaviors as impatient or showing impatience. That gives us some description. But have they explained the responses we saw? Indeed, where are the responses we saw? Impatient and showing impatience are so general that we have lost the responses (which can make "impatient," and "impatience," available for use as one type of fictional explanation involving adjectives and nouns, but that is a further reality we will get to shortly.) However, seeing just those responses provides insufficient information for an explanation. The responses are still only one variable, so even describing them leaves us short on explanation. Are there other variables we saw but failed to include in our analysis? Yes, indeed. Beyond the grabbing responses, we saw the cookies, and we saw the toys. Each of these provides a second variable for our analysis of the cause of the grabbing responses. (I bet you can predict where this is going.) While we would need to test them, one explanation is that eating the cookies causes future cookie grabbing, and keeping the toys causes future toy grabbing. Both of these provide explanations without descriptions confounding us. We could make a further analysis by noting that the cookie grabbing only occurs in the presence of cookies, and the toy grabbing only occurs in the presence of toys.

The cookies and toys could be evoking the grabbing responses. This would again provide us with two types of variables, IVS (the cookies and the toys) and DVS (the grabbing responses). Together these give us explanations (still needing testing) without descriptions confounding us.

Then again, say that we also see adults making inappropriate but polite comments after most grabs for cookies or toys, along the lines of, "That was quick," or "How cute." This provides plenty of adult attention as more consequences to the child's responses, which also provides yet another variable for our analysis of the cause (Causes?) of these grabbing responses. Both cookie grabbing and toy grabbing could be occurring at least partly as a function of adult attention. While we would also need to test this account, at least it too is an explanation without descriptions confounding us.

Alternatively, someone might *describe* these behaviors as greedy or showing greed (instead of impatient or impatience). Then they might try *explaining* the grabbing by claiming that the grabbing occurs, because of the "being greedy" or "having greed." The correct analysis, which you can make, involves variables quite similar to those we used to correct the impatient or impatience explanation claim. We must all avoid descriptions masquerading as explanations, especially since some subject matters rely heavily on descriptions, often as explanations (e.g., some approaches to child development) and the sophistication of some descriptions–as–explanations can make spotting them and dealing with them quite difficult tasks. For now, though, let's move on and consider other fictional accounts for behavior.

Fictional Explanations: Reification

The first fictional explanation we consider is *reification,* which comes from the verb "reify." Both words mean taking the kind of abstraction that involves "something lacking physical quantities or qualities," and treating it as if it were real, with a concrete, material existence. While initially an abstraction may begin as a mere conceptual device to help explore a topic, often metaphorically, the mere repetition of the abstraction often leads to its use as something real, as if it indeed had physical status. But it lacks any such status, and so its use as an explanation is scientifically inadequate. Scientific explanations must involve events with physical status, detectable and measurable.

For example, consider the abstraction that we call the "mind." In the process of secularizing the soul (a mini god powered by a maxi god) people began by using the "mind" to explore the possibilities metaphorically. The body behaved *as if* the mind was telling it what to do. Soon, however, they were using the "mind" as the secular but still non–physical controller of the body. Losing the metaphor left us saying, "The body behaves the way that the mind tells it to behave." This standard kind of comment reifies the mind.

Such talk began thousands of years ago, long before behavior science was available to counter the contingencies that drove ignoring the need for events to have physical status when developing explanations about events. Over this time frame, right up to our own, people—ordinary and professional—repeated the term so often that they began uncritically to consider it as, and use it to stand for, a real thing inside you. But, as a continuing abstraction, it still lacks any physical status. So its use as an explanation is an example of reification. As such it is scientifically inadequate because, again, scientific explanations must involve events with physical status, detectable and measurable.

For another example, consider the abstract concept of "personality." People used this word originally as a verbal shortcut to *describe* a consistent set or pattern of behaviors typical of a particular individual. However, it soon got used as the *explanation* for the behavior pattern, along with additional confusing layers. People repeated the term over and over, using it with other adjectives, each describing one or another *type* of personality, such as an introverted personality or an aggressive personality. Rather than being helpful, though, this compounds the problem. The adjectives began to describe the kinds of behaviors that the personality supposedly compels from the individual (e.g., "he slugged the other guy because of the aggression that his aggressive personality compels"). Furthermore, the adjectives make the noun, personality, seem all the more real, which it is not! "Personality" still lacks any physical status, along with any parts or processes attributed to it. So its use as an explanation is an example of reification. As such, personality is a scientifically inadequate explanation. What other reifications can you report?

Fictional Explanations: Converting Adjectives to Nouns

You may have noticed a common theme running through many explanatory fictions so far. They often involve *turning adjectives into nouns.* The adjectives may or may not be helpful, but the nouns simply become one or another explanatory fiction. Consider some typical adjective/noun pairs taken from our examples, noting that the pair members need not maintain the same form, although that is common. The adjective "impatient" becomes the noun "impatience." The adjective "greedy" becomes "greed." And "aggressive" becomes "aggression." Here are some other examples. "Intelligent" becomes "intelligence." "Paranoid" becomes "paranoia." The adjectives "hostile" and "antagonistic" become the nouns "hostility" and "antagonism."

In every one of those cases, turning the adjective into a noun leaves the impression that we now have some*thing* that can be a cause for some behavior. Actually, however, we have added nothing new to our analysis. Instead, we have a bigger problem, as we are now likely to stop analyzing, because we think we have a cause and need no further analyzing. But all we have is yet another fictional explanation.

That problem is so common to all types of fictional explanations that it bears repeating. As soon as you think you have a cause, you tend to stop looking. But if all you have is one or another type of fictitious cause, then you still lack what you need if your job is to intervene regarding the behavior for which you are seeking a cause. You still need a real, accessible cause. Worse, lacking a real cause can lead to harm if, without it, no effective intervention is forthcoming. This makes all explanatory fictions at least potentially harmful.

Converting adjectives into nouns characterizes a large portion of fictional explanations, because our English language so easily allows these conversions. And our traditional cultural conditioning takes the appropriate search for causal connections and perverts it into an acceptable search for mystical causes inside us. This makes these conversions particularly likely to occur. Yet all such causes are still explanatory fictions. We must be particularly careful to consider only causal variables that have the requisite physical status if our scientific attempts to understand, predict, control, and interpret behavior effectively are to have any beneficial effects. After all, *those* are the point of all scientific behavior.

Fictional Explanations: Nominal Fallacy

We use the term *nominal fallacy* for another common fictional explanation for behavior. This is a particular version of changing adjectives into nouns, because "nominal" means naming things. Linguistically this leaves us with nouns. After we observe a behavior, we often describe it. Then, our past linguistic and traditional cultural conditioning leaves us first pinning a name (i.e., a noun) onto our observation and, second, easily taking that name as the cause of the behavior we observed. Nominal fallacy, then, really just means giving a name to something and taking that name as the thing's cause.

Here is an example. You observe a skillful performance. Then you speak of the performer, not referring to the "body that performed" but to a mysterious (i.e., mystical) inner agent inside that body, as having talent. You might first describe the performance with the adjective "talented," before changing it into the noun "talent," but for many who conjure fictitious explanations, contingencies often compel skipping the description step. Finally, you explain the skillful performance in terms of that possessed talent saying, "She acted so well because she enjoys a wealth of talent." Yet you have only named your observation, and we still have only that one variable, so we cannot yet have a cause, which makes "talent" a nominal–fallacy type of fictional explanation.

Here is another example. You observe a young teenage boy snapping constantly at everyone around him. Then you speak of him as having a lot of hostility. Finally, you explain the snapping in terms of the hostility saying, "He snaps at everyone because he suffers much hostility." Yet you have only named your observation, and we still have only that one variable, so we cannot yet have a cause, which makes "hostility" a nominal–fallacy type of fictional explanation. (Fear not the repetitious phrasing. It can assist your understanding.)

Here is a final example, although you could easily come up with plenty more. You observe a sixth grade girl achieving good grades in high school level science and math. Then you speak of her as endowed with great intelligence. Finally, you explain the achievement in terms of that intelligence saying, "She achieves such good grades well above her level, because she commands so much intelligence." Yet you have only named your observation, and we still have only that one variable, so we cannot yet have a cause, which makes "intelligence" a nominal–fallacy type of fictional explanation.

Fictional Explanations: Circular Reasoning

We use the term *circular reasoning* for yet another common fictional explanation for behavior. Traditional cultural conditioning can compel inferring a "cause" of some sort from an observed behavior. Such a causal statement can sound linguistically satisfying. But this supposed cause gives us no new information about the behavior. It has no separate status as a variable. This gives the supposed cause the status of a fictitious explanation. We call it circular reasoning, because the cause appears as an inference *from the very behavior that it is supposed to explain.* For this reason some folks prefer to call it an *inferential circularity.*

Consider this example. A non–behaviorological practitioner or equally scientifically uninformed parent might be faced with a student's earning poor math grades. We have a fictional account if past cultural or educational conditioning evokes statements like "a 'mental block' within the student causes the poor math behavior." Some folks explicitly call it a *math* block. Either way, this statement gives us no new information about the poor math behavior, and it has no separate status as a variable. Instead its circularity becomes clear to those who ask the relevant questions.

"Why does the child do poorly in math?"
"Because he has a mental block to math."
"How can you tell [or, How do you know] that he has a mental block?"
"Because he does poorly in math."
"But, why does he do poorly in math?"
"Because he has a mental block."…

And so on, around we go in a circle, getting nowhere! The circular cause gets inferred from the behavior that needs an explanation. Invent some fake physiology, and the absurdity increases, as we see next.

Fictional Explanations: Gratuitous Physiologizing

Here is a somewhat common fictional explanation for behavior. It fascinates my students, at least after they become comfortable correctly pronouncing it. We call it *gratuitous physiologizing.* This involves inventing phony physiological accounts for behavior (the "physiologizing" part), and using them, because the physiological–analysis level makes the phony part sound more scientifically credible (the "gratuitous" part). But the physiology is merely made up. It is often circular. It also often fits under nominal fallacy. And it can involve changing adjectives into nouns. The mistaken physiological credibility makes these inner fictional causes look like variables separate from the behavior they are to explain, supposedly giving us two variables. Closer inspection, however, reveals the lack of scientific status. The "cause" is not real. It is merely another type of fictional explanation for behavior.

For example, consider again a practitioner faced with a student earning poor math grades. Some types of experience might lead him or her to claim that this problem with math is due to a "minimal brain dysfunction" (MBD) within the student. Why is the brain dysfunction "minimal?" Because the practitioner has had the student examined by a proper medical doctor, a neurologist, who cannot find any dysfunction of the student's brain; nothing is wrong. But something *must* be wrong, because the student is poor at math, and something must surely be causing this problem with math. Since nothing else seems responsible (not that much serious looking took place, apparently) the brain surely must be dysfunctional. If we cannot find the dysfunction, then it simply must be "minimal." As this argument goes, if it was not minimal, then we would be able to find it, and would have found it.

The problem is that the presumption of a brain dysfunction is gratuitous in the first place. It is unreal, invalid, unwarranted, all as in just invented. An initial assumption warranted by the situation, and worth exploring and fixing—as a behaviorologist would—is that something about the student's math–related environment is "dysfunctional." For some examples, consider these. Is the lighting adequate? Is the homework at the appropriate level? Is classroom assistance available? Is encouragement or help available at home? Would changing any of these help? Have contingencies yet evoked trying?

That the brain dysfunction is not real but gratuitous also becomes obvious from further scrutiny. Ask the circular–reasoning exposure questions:

"Why is the child poor at math?"

"Because he has an MBD."

"How can you tell that he has an MBD?"

"Because he is poor at math."

"But, why is he poor at math?"

"Because he has an MBD"… (and so on…).

Some school systems have actually used this "diagnosis." I find that a bit scary. Worse still, practitioners, especially those with the kind of education that would allow or encourage gratuitous physiologizing, have no intervention strategies, stemming directly from their mentalistic analyses, that are appropriate for dealing with MBDs (or, for that matter, with ineffective math–focused behaviors, or with dysfunctional math–related environments). They can only fall back on *intuitive* practices. That is, if these scientifically uninformed practitioners experience any success helping the math–poor student, that success must arise intuitively through practices that coincidently are congruent with the natural laws governing behavior. Behaviorology enables such practices by explicit design rather than by mere coincidence.

Real physiological events are sometimes directly investigated (appropriately, if by physiologists). But this still has little bearing on the emergence of an environmental–change technology that affects behavior, math behavior in this ongoing example, as opposed to a more medical technology such as some form of drug therapy. Drug therapy would be quite inappropriate for this example, and quite often inappropriate for other examples as well, because less invasive behaviorological–level interventions would work if tried. Too often, instead, supposed physiological events are only hypothesized, invented, or theorized, which continues a well–criticized pattern of non–explanation that we call gratuitous physiologizing.

Fictional Explanations: Teleology

A not–so–common, yet subtle as well as confusing, fictional explanation for behavior is one that we call *teleology,* which refers to the study of "future causes." These are causes that supposedly occur in the future with respect to the effect that they are causing. But the future, by definition, has not yet happened, so no future event can cause a present event. This is not to say that the present has no effect on the future. This is only to say that an event in the future cannot cause an event in the present, because the future event has not yet happened and so cannot have the physical status required of scientific causes. We refer to these future causes as teleological causes, and they constitute yet another type of explanatory fiction that cannot explain anything.

You often find teleological causes imbedded with a phrase like, "in order to." For instance, "He cleaned up his apartment to [or, in order to] keep his parents happy when they arrive for a visit." Or, "She cooked a special meal in order to please her special friend." While not covering causes for the behavior, these examples actually describe good contingencies between some behavior *and some consequence.* Some parental happiness *is* contingent upon a clean apartment. Some pleasure of a special friend *is* contingent upon a special meal. But what makes the consequence work involves the effects of *past,* and usually unacknowledged, contingencies.

Again, when you see or hear, "…to…" or "…in order to…," you very likely have a teleological fictitious account occurring. The example I use in my classes is one dear to my students' hearts. They study a lot and, as a result, earn good grades (well, most of them anyway; after all, *by design* most behaviorology courses apply, in the teaching of the course, the same science that the course teaches to the students). Those good grades show a valid connection between present and future, between behavior and consequence. Good grades are indeed contingent on studying hard, and studying hard produces good grades. But just what causes students to study? When I ask them that, their initial answers inevitably include some variation of, "Why, to get good grades, of course!" But getting good grades later *cannot* be the cause of studying hard now, because the grades are in the future. They have not yet happened. (And will not happen, unless the students study hard. Confusing, yes?) A claim that "getting good grades at the end of the term causes careful and thorough study now" is teleological, as is claiming that, "I study hard in order to get good grades." Both statements exemplify teleological fictitious explanations of behavior.

Before discussing with my students some of the actual causes of their studying efforts, I provide them a more complete account for why good final grades at the end of the term *cannot* cause their study behavior during the term. Faculty submit grades by taking their grade sheet to the administration building where the registrar's staff key the grades into the most secure computer on campus. From there the staff provide the grades to the students some days later. Using the fall semester for this example, consider that the campus entrance road has a fork where one way takes you past the campus maintenance center while the other way takes you past the administration building. (Both ways go around a loop road to various classroom buildings.) Now, the term ends in December, often with a few feet of snow on the ground and some ice on the road. A tanker truck arrives on campus to deliver a load of fuel to the maintenance center, but it takes the wrong fork and, slipping on the ice, crashes into the administration building while I am there turning in my grades. I am gone, the grades are gone, the computer is gone, even the records of the students' enrollment in the course are gone. They never get the grades! Yet all semester long, on a nearly daily basis, they worked at their studies earning the good grades that never got delivered. Why? The cause could not be getting the good grades, since the grades were in the future, and never occurred anyway. Such teleology is a fictional explanation.

When the students are done giggling over my fate, we discuss some of the possible real causes of their studying efforts. These pertain mainly to past causal variables, because many present ones, concerning mostly evocative stimuli particular to each student, involve complexities that we cover later in their course (just as we cover these later in this book). The past variables are ones that affect essentially everyone who ever attended elementary school. By the time these students enter college, most of them have extensive conditioning histories, spread over 12 or more years, wherein academic work of varying qualities produced a variety of consequences, many of which made such academic work either increase in rate, or at least maintain the then current rate (i.e., these consequences earned the label "reinforcing"). These ranged from stars and stickers for attempts and improvements in cursive letter forms and simple math facts and problems, in early grades, to compliments and letter–grade marks on completed homework assignments and tests—and marking period and final grades—at high school levels. That history leaves students' physically changed such that most assignments now successfully evoke—as real, testable causes—the appropriate study efforts that should produce—at the end of the term—good final grades.

The scientific inadequacy of teleological causes has no effect on the very real contingent connections between present and future events. Good grades are still contingent on good study. That is, because past good study produced good grades that occurred after the study, current assignments evoke present good study that produces good grades that may be delivered in the future. That is what causes present good study. The future good grades are not causing the present good study; saying so, as in "Students study in order to get good grades," rates as a teleological fictitious explanation. The same applies to saying

"I study in order to get good grades," only this also rates as mystically agential as no "I" inside the body "does" the studying.

Accessible Control and Fictional Accounts

Again, we sometimes talk as though our biggest objection to explanatory fictions is that they do not exist. However, while certainly a problem for them, this is not our chief objection. Rather, our most strenuous objection is that they cannot play any role in the prediction and control and understanding and interpretation of the behavior that they supposedly cause. They lack manipulable, independent–variable status. We have no access to change them in ways that bring about a change in the behavior they supposedly cause. One can claim that changing something else, such as an environmental variable, changes the fictitious variable, which then changes behavior in the sense of A causing B, and B causing C. But then the relation is similar to relations in mathematics: A causes C, and the middle term is at least unnecessary.

Another problem for fictitious variables, perhaps the worst problem for society, is that they leave the analyzers who invoke them comfortable. The analyzers seem to have found the sought–after cause. In reality, however, they have added nothing to their analysis. But, thinking that they have found the cause, they stop searching for real causal variables. The result is that accessible causal variables that we might change to improve (control) the behavior in question remain unanalyzed. We seldom continue looking when a convenient answer is available unless separate, practical contingencies for such behaviors are in force. The search essentially stops, because the fictitious mentalistic or cognitive analysis provides no compelling reasons to continue.

The dangers of inner causes may be less important when the behavior being explained presents no problems, such as excelling in math being explained by "intelligence" (inherited or not). But those dangers can be crippling in the opposite case, as we have seen. Yet one can still achieve a type of control. For example, the job specification of the practitioner with a math–poor student may supply the *practical* contingencies that require effective environment–controlling technologies that improve the behavior. The specification may require him or her to document successful help for that student. If he or she finds that student parked in front of a television set for five hours each day to the exclusion of study on school assignments, she or he may intuitively change the student's environment by pulling the plug on the set. If the ultimate result of that action is that grades improve, then that *functional* control results from the environmental change. That functional control does not result from the practitioner's mentalistically or cognitively focused explanatory fiction but in spite of it. His or her analysis may inadvertently coincide with the successful intervention, but it is not functionally related to that intervention. And this fact remains unaltered even when the practitioner tries to tie her or his fictions to the successful intervention by insisting that unplugging the television must have diminished the mental blocks or MBDs.

Not all problems stem from explanatory fictions. Sometimes the "behaviors" under consideration are not actually behaviors, as we see next.

What Else is *Not* Behavior

Some phenomena seem like behavior, usually because they involve some type of motion. However, they are not behavior, mostly because nerves are not energizing the motion. Here are some examples pertaining to some structural changes, and to traits, all of which sometimes masquerade as behavior.

Structural Growth or Decay are not Behavior

Various chemical processes that accumulate deposits as a growth mechanism are not behavior. The apparent elongating movement of finger nails, or toe nails, or claws, or hair, or fur, are chemical–deposition growths, not behaviors. Also, changes in height and weight are not behaviors, even though some of the variables that cause height and weight involve behaviors (e.g., exercise and eating). Similarly, bad breath is not behavior, even though inadequate dental–hygiene behaviors can cause it.

Those are only some of many simple non–behavior possibilities that I hope never fool you. More complex ones, like traits, constitute a greater challenge.

Traits are not Behavior

We may generally think of traits as a specific case of adjectives turned into nouns. They are not behavior. Regarding an event that you never witnessed, someone who witnessed the event turns a useful adjective that describes the event into a noun when reporting the event to you. If we take these nouns uncritically, they all too easily become a trait that someone could claim accounts for the behaviors that occurred in the unwitnessed event.

The adjectives may have originally described actual behaviors, but the nouns become traits, which are fictitious causes of behavior. The event reporter only invented the traits from the adjectives describing the original event, so the traits have no independent variable status, and so cannot cause behavior. For instance a friend tells you that a mutual acquaintance revealed a lot of obnoxiousness and hostility after listening with great impatience to an administrator at a meeting. Yet, except perhaps for "listening," your friend has not mentioned a single actual behavior. You are left to guess what the behaviors of the acquaintance actually were, as well as what the administrator said. Furthermore, the acquaintance's behaviors, whatever they were, did not occur as a result of the acquaintance supposedly harboring obnoxiousness, hostility, and impatience, as if these denoted traits of some inner self agent inside the acquaintance. Such mystical inner agents are, like traits themselves, irrelevant to a natural–science account of behavior, so "having" these traits, is of little import. In a later chapter, we will again consider traits, using love, which is *not* a trait, as an example.

Conclusion

Remember *Consumer Reports* and our example about whether self restraint and will power were the real causes of successful dieting behavior? Let's summarize the answer that you may have already supplied. We can immediately acknowledge self restraint and will power as fictitious causes of dieting, because they share common characteristics with several of the other explanatory fictions at the focus of this chapter. Even the analysts at *Consumer Reports* knew this intuitively (i.e., due to the non–verbal contingencies of being able to recommend practical steps that could indeed lead to success). In an article in the February 2009 issue of *Consumer Reports*, analysts "were able to identify six key behaviors that correlated the most strongly with having a healthy body mass index (BMI), a measure of weight that takes height into account" (p. 27). The six key behaviors were (a) watch portions, (b) limit fat, (c) eat fruits and vegetables, (d) eat whole grains rather than refined grains, (f) eat at home, and (g) exercise, exercise, exercise! Just reading the list of these behaviors helps bring the reader under some of the contingencies relating these behaviors to the reinforcing outcomes to which the phrase "successful dieting" alludes.

Dieting is predictable and controllable, and behaviorological principles apply to generating and maintaining this behavior. However, in a pattern typical of fictional accounts, the fictitious causes that we call self restraint and will power are not only unnecessary and redundant but also can cause the harm of continued ill–health, because they contribute nothing to dieting. Any dieting success actually accrues to a range of real independent variables. Behaviorological practitioners have applied these for decades (see Stuart & Davis, 1972). Again, while behaviorology, like all natural sciences, lacks explanations for everything, *fictitious alternatives are not explanations at all.*

Hopefully, our discussion in this chapter of some of the common types of scientifically *in*adequate explanations for behavior leads to a skeptical sensitivity to such accounts, evoking their rejection. The factual explanations and interpretations of later chapters are, and should be, just as subject to such skeptical sensitivity.❧

References

Sagan, C. (1995). *The Demon Haunted World—Science as a Candle in the Dark.* New York: Random House.
Stuart, R. B. & Davis, B. (1972). *Slim Chance in a Fat World: Behavioral Control of Obesity.* Champaign, IL: Research Press.❧

Chapter 4
What Is NOT Behavior
or Cause—II:

Further confusions and fallacies ...

$\mathcal{A}$s complexity in behavior increases, we begin to confront further confusions and fallacies that can arise in contingency analysis. Hold on! After covering so many fictitious causes of behavior in the previous chapter, can more exist? Yes, and this chapter covers only some of the remaining ones. The greater complexity of these remaining ones makes falling for them too easy. Becoming familiar with them helps avoid falling for them. Would "time" fit this category? Can the passage of time cause behavior? The answer can be convoluted. Let's consider it after tackling some less confusing difficulties.

We begin with a discussion about setting aside some sources that encourage complex analytical confusions and fallacies. Then we consider (a) the problem of analyzing the non–occurrence of behavior, (b) boredom, as an example of fictional constructs irrelevant to the variables controlling behavior, (c) bodily states, which are stimuli, not behavior, and (d) the difficulties with analyses involving stimuli that are remote in time. So let's turn to sources that support complex confusions and fallacies.

Some Sources to Set Aside

Behaviorology takes behavior as its subject matter. Theology and psychology also claim behavior as a subject matter. We touched on these disciplines in an earlier chapter due to the role they play in traditional cultural conditioning (i.e., the kind of early–life conditioning that leads to pre–scientific or unscientific responding, unless altered by later scientific education). Theology deals with behavior through non–scientific assumptions covering religious concepts. Psychology deals with behavior by substituting non–scientific secular (i.e., non–theological) concepts for religious concepts. As a result these disciplines both continue as major sources of many mystical explanations for behavior. They reject the assumptions that enable science to serve as a successful benefactor for humanity. And they consume resources that could otherwise increase the benefits for humanity from all the natural sciences. For these and other reasons, we should set these disciplines aside, and deal scientifically with behavior. Even if theology and psychology have contributions to make, science can probably better make many of these contributions.

Some of the other reasons to set these disciplines aside include their agential assumptions and some legal restrictions (covered soon). And as sources of fictitious explanations and analytical fallacies, these disciplines pose the danger of (if they are not already guilty of) bringing actual harm to individuals and the culture. Luckily, or perhaps out of personal ignorance, I know of no *individual* practitioner from either of these disciplines for whom such danger or guilt is an intellectually or emotionally satisfying outcome. On the contrary, those that I know personally would adamantly insist that they and their disciplines are only trying to help. However, they are contingency–controlled behaving organisms like everyone else (including behaviorologists and all other natural scientists). More importantly the contingencies that they are under continue compelling behavior in support both of their disciplines and of their practices, such as those agential assumptions and legal restrictions, that are proving harmful. Let's look a little more closely.

"Set Aside," Because of Agential and Legal Problems

The pre–scientific, fundamental agential concepts of psychology and theology pervade and mislead the current global culture. These agential concepts only allow practitioners from these disciplines successes when their intuitive practices coincide with scientific laws. These agential theories cannot *directly* produce the kind of best practices that a natural science can produce. Behavior–related practices of science come as direct implications and applications of the principles and concepts that stem from the scientific study of the natural laws governing behavior. Meanwhile, however, support for the agential concepts interferes with the swift dissemination of the natural behavior–science principles and concepts whose designed applications could benefit humanity far beyond the effects of those intuitive practices (e.g., until natural science got involved, no effective treatment for autism existed). Contingencies involving variables like the cultural status of these disciplines—gained perhaps somewhat deservedly while they were the only games in town—now compel interference with the successes of natural scientists of behavior. For this reason also, should we not set these disciplines aside?

Restrictions due to laws also cause problems. The supposed higher laws that theology claims reside outside the purview of science. However, the legally codified restrictions that support psychologists originally protected them and their clients by preventing the operation of charlatans. Now, however, these legal restriction often prevent behaviorological practitioners from applying the developments that natural scientists and engineers have achieved regarding behavior (e.g., see Maurice, 1993). Thus, scientific solutions for a wide range of individual and cultural problems receive far less attention than is their due.

Ironically, the available successes of these scientific, behaviorological practitioners can make psychology look a bit like it should face charges of charlatanism. For example, consider the negative recommendations regarding so many psychology–grounded therapies for treating autism. These appeared in the report of a multi–year project, completed in 1999, by the New York State Department of Health. The project evaluated the scientific research literature on the numerous types of available autism treatments so as to make recommendations based on scientific evidence of safety and efficacy. The final report (NYS Department of Health, 1999) either did not much recommend most of these interventions, or actually recommended *against* them (because they were harmful?) saying that they were "not to be used as an intervention" for young children with autism. Nevertheless, some psychologists and other therapists retain these so–called treatments and therapies. Yet any continued offering of treatments and therapies with poor recommendations, or with recommendations against them, at least raises ethics questions, and at most justifies charging psychology and these practitioners with charlatanism. For comparison, and to complete the record, the only fully recommended practices, in the NYS Department of Health report, were behaviorological practices, about which the report said, "It is recommended that principles of applied behavior analysis (ABA) and behavior intervention strategies be included as important elements in any intervention program of young children with autism" (NYS Department of Health, 1999, pp. 33–51).

"Set Aside," Because of Harm

For all the reasons we consider here, and perhaps more, many psychological treatments and therapies may, in the long run, be more harmful than helpful. And the harm done to clients and the culture may be direct or indirect.

Direct harm. Direct harm occurs when a client's condition worsens as a result of implementing a practice. For example, such a worsening can occur when a practitioner only engages in talk sessions lacking a scientific therapy plan while having a depressed client merely otherwise ingest anti–depressant drugs (which may also constitute a misuse of the benefits that pharmaceuticals can provide). Such practices provide little help, because they fail to address, scientifically, the independent variables responsible for "depression." As a result, over a period of time, the poor client either fails to get better or gets worse, both of which are forms of harm. Also, clutching at the truism that "these things take time" makes an inadequate excuse for being unprepared because one's psychological training emphasized untouchable agential accounts. This emphasis leads to down–playing the importance of addressing the measurable independent variables of problems as part of solving them.

Variables responsible for depression include the common reduction in the *General Level of Reinforcement* (GLR; see Cautela, 1994). A reduced GLR refers to a loss of reinforcers that can result from some kinds of environmental change. Behaviorological practices, such as those that raise the GLR, could provide some success at addressing these independent variables. With psychologists rejecting the natural science that covers the GLR, their being ***un***knowledgeable about the appropriateness of increasing the client's GLR provides little excuse for any harm that accrues from not increasing the GLR. This is a kind of indirect harm.

Indirect harm. Indirect harm accrues to clients when they cannot get help because cultural institutions, such as the law, bar the professionals who can help them from doing so. These same legalities often allow clients to see only those practitioners who have been culturally/legally designated (e.g., licensed) to help them. This holds, even if the training of these professionals derives from a discipline committed, not to practices grounded in a natural science of behavior, but to agential accounts of behavior, which remain irrelevant.

As an example, for years New York State's laws stood in the way of clients receiving the interventions that the state's own Health Department recommends for autism. These laws limited the scope of practice for behaviorological professionals so that they could not easily provide these interventions. These behaviorological professionals have the education and experience in the relevant field—one that behaviorological science informs—that enables them to earn an appropriate credential for implementing effective interventions. But the laws recognized only "psychology" credentials. Similar laws in some other states—supported over the decades by psychological guilds and lobbies—also prevent behaviorological professionals from providing interventions that work for needy clients. Such circumstances lead to harm for these clients. Given such reasons, why should we not set problematic disciplines aside?

Some laws even require behaviorological professionals to receive supervision from licensed psychologists. Yet the vast majority of psychologists have little or no training in the relevant natural science of behavior. Happily, many states take legal stands against such absurdities, as part of a movement towards fair, data–based and science–grounded legal treatment of helping professions.

A relevant point that we made in an earlier chapter bears further development. The theology/psychology disciplines, and the science of behaviorology, are neither equal nor culturally parallel. Over the last several hundred years, the distinction between theology and science has become quite well established. Similarly, over the last hundred years, the comprehensive research literature that thousands of natural scientists of behavior have produced, around the world, has equally established the distinction between the psychology and behaviorology disciplines. Specifically, *behaviorology is neither a part of, nor related in any meaningful way to, psychology of any kind.* And any contrary claim that psychologists or their discipline might make about containing behaviorology in any way would stand as a lie equal to some of those infamous whoppers that certain politicians told during the last century.

For all those reasons, we need to set agential disciplines aside. Consider a more general view. No one can objectively demonstrate the "mystical," whether theological or secular, even though it claims to offer certain explanations for everything. Science, on the other hand, remains inherently uncertain, and accepts that it can never explain all things, even as it strives to discover, understand, predict, control, interpret—in a word, explain—more and more. And natural scientists prefer to go without an explanation, while looking for a scientifically acceptable one, rather than accepting or conjuring some mystical account. The "mystical" insists that it accounts for everything, usually through a range of agential entities or processes, and that it is always right. Science, on the other hand, admits only to *natural* explanations of natural events. Scientific explanations derive from the best available experimental methods and evidence, and reside on the informing philosophy of naturalism. And science admits only provisionally to its explanations in the sense that science always remains open to new or better data that can alter or improve the previous best explanations.

In addition, the agential entities in the secular mysticism of psychology merely present a scaled–down version of the agential–entity power of theological mysticisms. Supposedly, for example, our culturally common heavenly maxi–god can move mountains. However, inner–agent mini gods (e.g., souls, minds, psyches, or selves) can only move body parts (e.g., arms and legs). Such power differences among mystical agents may strike us as curious. Natural scientists, however, require experimental results that enable beneficial engineering applications, while they watch for, and avoid, analytical confusions and fallacies. Let's now consider some of these.

Other Analytical Fallacies

We have covered some traditional cultural perspectives on behavior that derive from, and continue to support, pre–scientific, usually agential, analyses of behavior. We inevitably grow up under the contingencies of traditional culture (i.e., all those aspects of our culture that evoke and consequate—control—our behavior) including its superstitious and mystical components. These contingencies thus shape our early behavior, including repertoires of superstition and mysticism

For some people those contingencies maintain such repertoires throughout life. For other people additional contingencies, particularly those of science education, mitigate some of the negative effects of early cultural shaping by allowing contingencies that can condition more accurate and helpful intellectual and emotional responses regarding behavior. Still, that early–imposed traditional cultural legacy continuously compels some theoretical and practical fictions and fallacies for everyone.

Avoid Analyzing the Non–Occurrence of Behavior

Have you ever heard, or said, "Is there no end to the bothersome behaviors of others?" Since examples abound, we will let a short list stand for them all. For instance, does your mate fail to treat the toothpaste tube the way you prefer? Does a neighbor's lawn left unattended for weeks or months disturb you? Does a dog owner ignoring a dog's leaving dung on your nice lawn leave you dealing with a severe negative emotional reaction? (More fun ways to say this are available, but they are not printable.) All these particular examples refer to behaviors that are *not* occurring.

While we will examine some contingencies for one of those relatively simple examples here, far more important, and complex, examples certainly exist. Some of them concern the problems that stem from (a) someone not wearing a seat belt, or (b) someone not recycling (but just tossing recyclables into the trash), or (c) someone not attending to driving (but phoning or texting instead). In addition to the non–occurring behaviors that cause analysis problems, some of these examples contain the occurring behaviors upon which contingencies should focus.

Those kinds of concerns often evoke the early steps of an intervention to end or change the offending behavior. If we bring behaviorology to bear at this point, these early steps will likely include trying to clarify the current contingencies, and plan the intervention contingencies. In analyzing this kind of activity, one thinks in general terms, so we sometimes refer to behaviors rather than to more specific individual responses. Also, when we describe contingencies here, sometimes we combine emotional responses and feelings (which are actually effects of emotional responses), and sometimes we include the more public events available to observers. In all cases, however, we need to avoid including *non–occurring* behaviors in the contingencies that we describe, because that causes analysis problems.

Let's take some of those early steps with the contingencies in the example of an owner's dog leaving dung on your lawn. Let's start with some contingencies on the lawn owner. The odor/smell of the dung, and the sight of someone—as owners are responsible for picking up their dog's dung—leaving it on your lawn, are generally aversive stimuli. (*Odor* refers to energy streams from particular molecular gas pressures to which nasal membranes are sensitive, while *smell* refers to the response that neural structures mediate when such energy streams contact nasal membranes.)

Those aversive stimuli *both* elicit various negative emotional reactions, *and* evoke behavior that removes the dung from the lawn. Removing the dung reduces the negative (i.e., aversive) emotional reactions, and this reduction functions as a reinforcer (often called *relief*) that makes the kind of behavior that produced the reduction occur more quickly or more often when such aversive stimuli occur in the future. Meanwhile, the emotional reactions tend to exaggerate the forms of behavior that, in this case, remove the dung from the lawn (e.g., *stomping* out, shoveling *more deeply than necessary* to collect the dung, and *strongly* flinging it into the trash bin).

Remember that an observer of the lawn owner's reaction cannot directly see the emotions (for instance, the chemical dumps that the lawn owner feels as anger); only the lawn owner is privy to the emotions and some of the responses that they evoke. Still, since emotional arousal exaggerates the mediation of later responses for a short time, the observer reliably responds to (i.e., infers) the emotions from the wide range of events in the setting (i.e., the context) including, and perhaps especially, from hearing the string of hearty obscenities that seeing the dung on the lawn evokes from the lawn owner.

What about contingencies on the dog owner? Recall that in many jurisdictions, leaving your dog's dung on a neighbor's lawn is illegal. The general history that we all observe of the punishments that the legal system provides for illegal acts can help condition aversive feelings that we call *guilt*, and any subsequently evoked behavior tending toward illegality elicits more guilt feelings. In some cases parents or peers, including neighbors, provide the punishment for activities, which leads to the conditioning of similar aversive feelings that we call *shame*. Aversive feelings that we call a *sense of sin* occur when we engage in activities that religious authorities punish. Such aversive feelings are often a part of currently operating contingencies. Even the stimuli evoking thinking behavior regarding punishable responses come to elicit the emotions that lead to feelings of guilt, shame, and sin. So, let's write one such contingency that affects the dog owner.

Before we write that contingency, let's first briefly comment on some mechanics of writing contingencies (with details in later chapters). Recall that contingencies are causal dependencies. These often occur as a string of interconnected evocative stimuli (S^{Ev}s), behaviors, and consequences. Each of these relates to the others as independent variables and dependent variables in functional relations. We use arrows to show the functional relations. Here is a guilt contingency that affects the dog owner:

S^{Ev}		Behavior		Consequence		[Result]
Dung on lawn (aversive)	→	Walk away (with dog following on leash)	→	Diminishing feelings of guilt as distance increases (relief)	→	[Walk away more often or rapidly in the future after the dog leaves dung on a lawn]

Note that we have included some *predicted results* in some contingencies, although results are not strictly a part of contingency diagrams. Next, let's try planning some intervention contingencies. In an effort to describe a contemplated intervention that might improve the dog owner's behavior, and get the dung off the lawn, we might write this contingency for a dog owner:

S^{Ev}		Behavior		Consequence		[Result?]
Sight of dung on a lawn	→?	**Not** deal with the dung	→?	Feelings of guilt or shame remain	→?	[Bags the dung for proper disposal?]

But wait! That middle term (the behavior) in this contingency is a behavior *that does not occur*. It is a *non–behavior*. But stimuli only evoke behavior; stimuli only stimulate nervous–system structures the operation of which is the mediating *of a behavior,* **not** of a non–behavior. Furthermore, non–behaviors cannot produce results (i.e., consequences). With a non–behavior, no behavior actually occurs, and so none produces energy streams that then affect the environment in ways that produce other stimuli (other energy streams) that further affect the nervous system as consequences. Thus, *this contingency shows function arrows that do not function.* With no stimuli evoking the non–behavior, no behavior produces a consequence, so no result can occur. *All these arrows should appear crossed out* (i.e., —x—>), because none of these functions are possible. So, as a potential intervention, this "contingency" is useless. It does, however, show some of the kinds of reasons that should keep us from writing contingencies for non–behaviors.

Instead, we must write contingencies for behaviors *that occur,* and then implement the contingencies to establish those behaviors. In this dog–dung–dirtied–lawn case, one such intervention involves a couple of contingencies with stimuli evoking different dog–owner and lawn–owner behaviors. Here is what we might design, beginning with two related versions of one contingency on the lawn owner (who might be you):

S^{Ev}	Behavior	Consequence	[Result]
Approach of dog and dog owner	→ Hand good grocery bags to dog owner and request their use	→ Dog owner puts dung into a bag, leaving lawn clear of dung	→ [Lawn owner hands out bags when they are needed]

Here is a related version of that contingency on the lawn owner:

S^{Ev}	Behavior	Consequence	[Result]
Approach of dog and dog owner	→ Hand good grocery bags to dog owner and request their use	→ Dog owner puts dung into a bag, leaving lawn clear of dung	→ [Lawn owner feels relief from aversive stimuli of dung on lawn]

A further contingency on the lawn owner might even involve the dog owner also saying, as a consequence, "Thank you" for the bags, which could produce the same results that these two contingency versions list. (Try writing it out.) Note that only the lawn owner is privy to the result in the second version. Similarly, here is one of two contingencies on the dog owner, with only the dog owner privy to the result:

S^{Ev}	Behavior	Consequence	[Result]
Dung on lawn → with bag in hand	Put dung in bag (leaving lawn clear of dung)	→ Dog owner feels relief from guilt or shame	→ [Puts dung in bags regularly (and even carries bags for disposal of dung)]

And here is another contingency on the dog owner:

S^{Ev}	Behavior	Consequence	[Result]
Dung on lawn → with bag in hand	Put dung in bag (leaving lawn clear of dung)	→ Lawn owner says, "Thank you"	→ [Dog owner puts dung in bags regularly (and even carries bags...)]

Many interventions are possible, for that example as well as the others. In every case, remain wary of *non–behaviors,* and keep them out of contingencies. Meanwhile let's turn our attention to fictional constructs *in* contingencies, with "boredom" as an example.

Fictional Constructs Like Boredom

Fictional constructs represent another category of explanatory fictions. Here we want to consider the result when fictional constructs appear in our analysis of contingencies. In broad terms, past conditioning of unscientific behaviors make fictional constructs seem more convincing than the explanatory fictions previously covered. This makes fictional constructs more likely to interfere with analyzing contingencies effectively, especially regarding interventions.

Consider the common behavior explanation that some change in behavior occurred out of "boredom." We often hear this in the complaint form, "I'm bored," although no inner agent, "I," exists to "be" bored. The complaint could be, but usually is not, a verbal shortcut. Two words could stand in for over a dozen. "I'm bored" could replace a report that "the amount of reinforcement for current behavior has become too small to maintain the behavior," so it is extinguishing (i.e., undergoing a decrease in occurrences that may reach zero). However, when some contingency with more reinforcement compels another behavior to occur instead, scientifically uninformed people remain unaware of this contingency. Instead, they often say that the new behavior occurred *due to* their boredom. Inevitably, though, when we try to analyze the contingencies that such boredom statements seem to describe, we find the result in error. We find the contingency that we write invalid. We find the account a fiction.

While that also happens with the full range of explanatory fictions we covered earlier, let's elaborate boredom a bit. The feeling of boredom can be quite real. Reductions in reinforcement commonly elicit aversive emotional reactions producing feelings that we call "boredom." But the emotion/feeling is not the cause of a change from one behavior to another; instead both the emotion/feeling and the behavior change are a function of other variables.

Let's watch the problem unfold as we analyze a situation and write contingencies about what is happening. Let's suppose that you are a data entry clerk at a big business where you and your supervisor are both rabid basketball fans. You even cheer for the same team (which may be how you keep your job). If we accept your being bored with the repetitive nature of your work, then we might write contingencies treating the fictional construct, *boredom,* as if it were real (i.e., a type of reification). Then we can use boredom for the cause when we want to explain why, when your boss checks up on you, she finds you looking out of your cubicle at the television she has set up to watch the final game of the playoffs at which your team finally landed a spot. Our first contingency features boredom in its more common position, which is as the evocative stimulus; it is quite invalid and so lacks functional–relation arrows:

X INVALID CONTINGENCY:

S^{Ev}	Behavior	Consequence	[Result]
Boredom $-X\rightarrow$	Look at TV $-X\rightarrow$	Exciting $-X\rightarrow$ events happening at a fast pace on the TV screen	[Continue looking at TV]

Alternatively, we might be tempted to put the boredom as the consequence, as in this also invalid contingency:

X INVALID CONTINGENCY:

S^{Ev}	Behavior	**Consequence**	[Result]
Full $\rightarrow$ in–basket of un–entered data	Type in $-X\rightarrow$ some data	Get $-X\rightarrow$ bored	[Stop typing and switch from data entering to TV watching]

Both contingencies, however, are wrong, agential (Why is that?) and so on. At best, "boredom" simply implies that the evocative controls on your behavior are undergoing a transfer from one type of stimulus to another. A common source for such transfers resides in the different consequences for the two different behaviors. In this case the two behaviors are data typing and watching the television; these involve two different, actually competing contingencies. Here is the one on data typing, which conveys what your job entails:

S^{Ev}		Behavior		Consequence		[Result]
Full in–basket of un–entered data	→	Type in data	→	Rows of data appear appropriately on the screen	→	[Continue to type in data]

But here is the competing contingency, on television viewing:

S^{Ev}		Behavior		Consequence		[Result]
Final playoff game on television	→	Look at television	→	Exciting events happening at a fast pace on the television screen	→	[Continue looking at television]

Note, however, that the type and amount of reinforcement differs substantially for the consequences in these two contingencies. The narrow columns and rows of repetitive data slowly scrolling on the screen, from the appearance of the typed characters, provides only a constant stream of similarly unspectacular stimuli. The reinforcement pertains more to getting the batch of data entered successfully and keeping your job. However, the constantly changing scenes of the game in progress on the television, occasionally and irregularly punctuated with surprising and exciting developments, provide fast moving, always different, and sometimes spectacular events, all of which is reinforcing. (Later we will see these differerences as differences in reinforcement schedules.) The difference in these reinforcers reduces the power of the un–entered data to induce typing (which is its evocative function) and raises the power of the playoff game to induce watching (which is its evocative function). As a result, the boss observes you typing less and watching the game more. The reinforcement difference determines this outcome, not boredom, so boredom should not appear in the contingencies. The occasional increase in the intensity of the television reinforcers, like the roars of the arena crowd, further supports this effect, as those often abrupt intensity increases easily win any contest with the un–entered data for control of your orienting and attending behaviors.

Traditional agential cultural conditioning can also lead to a notion that boredom, as in "being bored," is a sort of bodily state. But that would only mean that it has the problems of bodily states as well as the problems of fictional constructs. The time has come to consider bodily states.

Bodily States Function as Stimuli, not Behaviors

Like fictional constructs, states of the body also cause problems when they sneak into our analysis of contingent relations, even though they are real events. A bodily state is not a behavior. Instead it serves as a stimulus. So we exclude bodily states from the behavior position in contingencies. Bodily states function as antecedent stimuli, and some might function as postcedent stimuli. Any feelings or bodily sensations associated with a bodily state are behaviors, usually of the neural kind that, also as real events, might also serve stimulus functions. Drug–induced physiological changes constitute a typical bodily–state example. The changes usually last long enough to constitute a bodily state to which responding occurs differently from responses to the state of the body without the drug present. Similarly, the difference between other bodily states and their counterparts also evokes different responses (e.g., how we describe them) to these different bodily states. As examples, let's consider (a) emotional arousal and relative calm, and (b) awake and asleep, starting with sleep.

Sleep. The term *sleep* refers to a bodily state on a kind of continuum, and it typically comes after the bodily state of tiredness (i.e., fatigue). Sleep pertains to a body that has temporarily lost the capacity to mediate many, perhaps most, kinds of behavior. Tiredness pertains to a body that is losing capacity to mediate behavior, perhaps due to excessive energy expenditures (e.g., a full day of hard work). Neither sleep nor tiredness are behaviors. Tiredness, however, can function as an aversive stimulus that evokes the behavior of lying down or going to bed. This produces a reinforcing consequence, for lying down or going to bed, of relaxing relief from the aversive tiredness.

We should not confuse sleep as a bodily state, which thus cannot be a behavior, with the various behaviors that can occur *during* sleep, such as turning over, breathing, and dreaming. However, when a body is sleeping, that sleep can function as a stimulus, perhaps evoking an observer's behavior such as the command to others to "Let him sleep." In that contingency sleep occupies the position of evocative stimulus, the first term in the contingency.

We face difficulties, however, when placing sleep in the consequence position in a contingency. Here are a couple of ways in which that cannot work. Some may see sleep as a reinforcer for not working long into the morning. But even before considering the consequence status of sleep, I expect that you said, "Whoa! That is a non–behavior; lacking an energy stream, it cannot function in contingencies. Start again." (I hope you said that, because you would be right!) Instead, some may see sleep as the reinforcer for being tired, even exhausted. But again, before considering sleep as a consequence, I expect that you said, "Hang on. That's not right. Those sound like more bodily states, which also cannot be behaviors. So again, start again!" (Right, again!)

A behavior that actually precedes the state of sleep could be something like getting into bed, a behavior that a clock showing 11:45 P.M. could evoke. But is sleep really a consequential reinforcing stimulus? Can we trace an energy stream flowing from the state of sleep to nervous–system structures responsible for mediating climbing into bed, and changing those structures such that getting into bed more readily occurs in the presence of the 11:45 P.M. evocative stimulus? Stimulus–evoked changes in neural structures mediate the behaviors of lying down or getting into bed, but does sleep change these structures in more permanent ways? Until our physiology colleagues confirm otherwise (which is possible, because contingency claims are falsifiable) we must answer "No" to these questions. Furthermore the *time* involved in sleeping, and "going to sleep," makes manipulating them as independent variables rather difficult. So far this prevents us either from testing sleep or from observing sleep produce an increase in a behavior that it follows. Without such observations, calling sleep a reinforcer seems at present premature. We may regard one's sleep as able to function as an evocative stimulus (for others), but it cannot function (so far as we presently know) as a consequential stimulus for the sleeper.

Emotional arousal. Some environmental stimulus events, inside or outside the body, evoke—actually we usually use the term *elicit* in the context of such reflexes, so—elicit chemical dumps into the bloodstream. That is, the energy streams from these stimulus events compel changes in nervous–system structures, and these changes mediate the glandular release of certain chemicals. We call that chemical *release* emotional behavior. The subsequent *presence* of those chemicals in the bloodstream is not behavior. However, this presence produces both the responses that we call feelings, and a general change in physiology that we call a bodily state of emotional arousal. When *other* stimuli occur during the aroused state (i.e., this general change in physiology) the behaviors that these other stimuli evoke or consequate occur in exaggerated or more intense forms compared to the behaviors they evoke or consequate when the body is not emotionally aroused. The aroused body is not behavior but it mediates behavior in exaggerated ways. Recall our example of aroused running after a bear elicits startle responses, compared with the normal running that a jogging schedule evokes.

Being real, however, the emotionally aroused body can function as an evocative stimulus in the first term of a three–term contingency. For example as a stimulus it can evoke changes in the behaviors of others that produce changes in distance from the aroused body. If the exaggerated responses that an aroused body mediates are aversive (e.g., dangerous) to others, then the exaggerated responses, as stimuli, evoke the responses of others that increase the distance between the aroused body and others. But if the exaggerated responses that the aroused body mediates are attractive (e.g., sexual) then they evoke responses that decrease the distance between the aroused body and others.

On the other hand, an emotionally aroused body, as a stimulus, fits poorly as the third term in a contingency, for reasons similar to those that applied to sleep. The arousal, as a general physiological change that we call a bodily state, alters—often exaggerates—the capacity of the body to mediate other behaviors, but that arousal seems unable to change the mediating structures in ways justifying calling arousal a reinforcer. New data may change this.

Furthermore we should not confuse the *state* of the body that we describe as emotionally aroused, which is not a behavior, with various behaviors that can happen *to* an aroused body, such as other responses that stimuli elicit and which we can distinguish from the aroused bodily state. Stimuli elicit the glandular release of body–altering chemicals. With those chemicals in the blood stream, the now changed body temporarily mediates subsequent behaviors differently from the behavior mediation of the pre–release state. We note such bodily changes explicitly when writing contingencies, because the presence or absence of such emotional states readily accounts for the differences in behavioral effects during comparisons of two or more contingencies. Some of this applies to drug–induced physiological changes as well, to which we now turn.

Drug–induced physiological change. A wide variety of medical, recreational and illegal drugs induce a variety of physiological changes in the body. Some of these changes are similar to the effects of some elicited glandular chemical releases except that elicitation does not introduce the chemicals. Instead swallowing, or injecting, or absorbing (e.g., from chemical patches) and so on, introduces the body–altering chemicals to the body. When any resulting state of the body with the drug present evokes a different label, compared with the state of the body without the drug present, we are dealing with different bodily states. Of these two, the one that grabs the most attention is the drug–induced bodily state.

A contingency–related discussion of drug–induced bodily states shares similarities with our discussion of emotionally aroused bodily states. A drug–induced bodily state, being real, can function as an evocative stimulus, in the first term of a three–term contingency. For example, as a stimulus it can evoke the verbal behavior of others, such as, "He is too drunk to drive!"

However, unlike sleep, a drug–induced bodily state might serve as a consequence, in the third term of a three–term contingency. To the extent that the drug state functions to alter the nervous–system structures that mediate the evoked ingestion of the drugs (by whatever method) the drug state could serve as a reinforcer or punisher. Actually, the *induction* of the drug state, the *immediate* physiological effect of taking the drug rather than an ongoing residual state, might serve these functions better than the state itself. If a drug state changes nervous–system structures so that these structures *more* readily mediate drug ingestion in the presence of the stimuli that evoke drug ingestion, then we could call this drug state a reinforcer. Similarly, if the drug state changes nervous–system structures so that these structures *less* readily mediate drug ingestion in the presence of the stimuli that evoke drug ingestion (e.g., cause a "bad trip") then we can call the drug state a punisher.

The difficulties with bodily states, like the problems with fictional constructs, concerns their inclusion in explicit contingencies. Some antecedent and postcedent stimuli, however, extend away in time from the occurrence of the behavior of concern, and this causes other difficulties for contingency analysis, difficulties to which we now turn.

Avoid Remote–in–Time Antecedents and Postcedents

In analyzing a behavior of concern, especially when designing an intervention to help a person in need, we often must take into account stimuli that remain remote in time. This includes not only some antecedent stimuli that occur back in the historical chain of functional events that led to the behavior, but also some postcedent events that occur long after the behavior of concern. However, while we are talking here about *real* events, both antecedent and postcedent, their remoteness in time from the current behavior necessitates their exclusion from the contingencies we write in our analysis of the *currently* functioning controls on the behavior of concern. We emphasize currently functioning controls, because this emphasis enables discovering the currently accessible independent variables that an intervention might change in ways that improve the behavior.

Put another way, we analyze contingencies for the antecedent and postcedent variables functioning *now*. Thus, we look for the antecedent stimuli that are *at present* functioning to evoke the behavior, while we also look for the postcedent stimuli that are *at present* functioning as immediate reinforcers or punishers for the behavior, especially postcedent stimuli that the behavior of concern is producing. If we write contingencies that include stimuli that are more remote in time (i.e., not current, present, or immediate) then this inclusion could lead to reduced effectiveness and poor outcomes, by masking the more immediate independent variables.

For contingencies that have components removed in time away from the behavior of concern, a term that has seen some use is the term *defective contingencies*. Defective contingencies may contain real events and describe actual relations, such as the relation regarding effective study behavior now producing a good grade later. But they do not describe the present variables related to a behavior of concern. Thus, we exclude these *defective* contingencies, when we speak of *contingencies of reinforcement*. For example, we exclude the relation between studying now and the later, delayed occurrence of a good grade not only due to the danger of teleology but also because the good–grade consequence occurs too far removed in time after the behavior to function in the manner of a stimulus that earns the title *reinforcer*. That is, every energy stream from the delayed appearance of a good grade happens long after the earlier study behavior, and so the appearance of that grade cannot affect the nervous–system structure as it mediated the study behavior. The grade appearance cannot make these structures mediate study behavior more readily in the presence of the relevant evocative stimuli. The grade appearance cannot *reinforce* the long ago behavior that produced it (although it may reinforce any behavior that *immediately* precedes its appearance).

When analyzing the contingencies currently controlling a behavior of concern, we first focus mostly on the *immediately* present antecedent and postcedent variables. These current variables, of course, extend a continuous sequence of antecedent and postcedent contingencies. And the sequence constitutes the natural functional history of the events of concern. Such sequences help us see patterns over time, and so aid prediction. A focus on present contingencies also avoids the distractions and other problems of antecedents and postcedents that are remote in time.

Conclusion

Having now tackled some other difficulties in this chapter, perhaps we are ready to consider that question that we posed about time. Can the passage of time cause behavior?

Here is a brief version of the answer. *Time* lacks independent variable status *with respect to behavior,* because it cannot be manipulated on the level at which we analyze behavior (e.g., at the level of physics analysis, manipulating time requires energies producing relativistic speeds, an energy/speed relation unavailable on the level at which we analyze behavior). Of course, the events that define time are capable of independent–variable status, and we must consider the ones related to behavior and take them into account. We will have occasion to return to this topic with some elaboration in a later chapter.

References

Cautela, J. R. (1994). General level of reinforcement II: Further elaborations. *Behaviorology, 2* (1), 1–16.

Maurice, C. (1993). *Let Me Hear Your Voice—A Family's Triumph Over Autism.* New York: Ballantine Books.

NYS Department of Health—Early Intervention Program. (1999). *Clinical Practice Guideline: Autism / Pervasive Developmental Disorders, Assessment and Intervention for Young Children (Age 0–3 Years) Quick Reference Guide.* Albany, NY: NYS Department of Health (Publication No. 4216; available online as a PDF).

Chapter 5
What Behavior IS:

Defining behavior, and some kinds of behavior …

*T*his book describes the contingency causes of behavior and provides some initial context and description of the discipline we call behaviorology. As a natural science, this discipline explains behavior as resulting from contingencies between behavior and its environments, both external and internal, which includes the *role* of physiology and genetics. Then the discipline applies this knowledge as contingency engineering that benefits individuals and humanity. Having talked about what science is, and what behavior is not, the time has come to define both behaviorology and behavior.

While we start with behaviorology, let's first mention an example of behavior that samples just how complex it can get. Imagine this scenario, which we will label later. You are pushing your left foot down while simultaneously turning a wheel about 70 degrees counterclockwise with your left hand. After holding that position briefly, you return the wheel to its original position and hold it there as you let up on your left foot. All this time you also pull back a little on the wheel—also with your left hand—while simultaneously turning your head back and forth, scanning your field of view. Meanwhile your right hand simultaneously pushes in a small plunger by a couple of inches. Now, no "you," as a mystical agent, initiates, does, or tells the body to behave those responses. They occur due to contingencies that make behavior happen. Studying these contingencies involves behaviorology, so let's now define it.

Definition of Behaviorology

We can state several definitions for behaviorology, some simple and others increasingly comprehensive. The simple ones may cause confusion as they are similar to the definitions that other disciplines use. The comprehensive ones may also cause confusion, due to their complexity (see Ledoux, 2014.)

Simply put, behaviorology is the scientific study of behavior. This sounds too simple, or perhaps this is just too similar to the definitions that other disciplines use. Let's try again. *Behaviorology is the natural science—and its related engineering technology—of the interactions between behavior and variables in the internal and external controlling environment.* This is a good basic definition of behaviorology, although some leaders in the discipline say simply that behaviorology is the natural science of environment–behavior relations. For readers demanding full disclosure, let's get more complex. But instead of a dozen–line–long comprehensive definition, let's use a set of parts that you can digest separately, *although together they make a full definition:*

  ⁊ *Behaviorology* is a natural–science discipline with experimental, analytical, philosophical, technological, conceptual, and theoretical components;

  ⁊ *Behaviorology* is a natural science among the life sciences;

  ⁊ *Behaviorology* emphasizes the causal mechanism of selection;

  ⁊ *Behaviorology* discovers, interprets, and applies the single and multiple variables that are in functional relations with the simple and complex, overt and covert behaviors of individual organisms (especially people) during their lifetime (and beyond, with respect to cultural practices); and

  ⁊ *Behaviorology* takes into account the socio–cultural and physical variables from the internal and external environments as well as variables from the biological history of the species.

  One can expand each of those parts, and even the parts of the parts, with extensive detail. But then, that kind of expansion describes much of our plan for the rest of the book. Meanwhile, *taken together, those parts define and clarify what behaviorology is and does.* What about its subject matter, behavior?

Considerations about Behavior

In part this book shares with you the perspective of behaviorologists about the most parsimonious views regarding *behavior and its explanations.* We view both of these as natural and lawful, even though they can be extremely complicated (although, my meteorology friends tell me, not as complicated as the weather).

  That status as *natural and lawful* leaves alternative mystical or superstitious accounts not only unneeded but, remember, mystical or superstitious alternatives are not explanations at all. A completely natural account, involving only real, physical events regarding *behavior and its explanations* is both possible and appropriate. So it remains our focus.

The Definition of Behavior

  We cannot easily separate behavior, organism, and environment. Without an organism with the necessary physiology, we have no behavior, and behavior always occurs in a setting or situation or context (i.e., an environment). This part of the environment exists outside the organism's skin while another part of that environment exists inside the organism's skin. Physical matter comprises all these three—organism, external environment, and internal environment— and they interactively exchange energies, with behavior arising from and contributing to the interactions.

  Thus, a definition could state that *behavior is muscle contractions that occur* **when stimuli make** *motor neurons fire,* because muscles need firing nerves to contract (i.e., "innervated" muscular movements). Yet this is too simplistic. After all, in previous chapters we have already touched on more sophisticated characteristics of behavior, and we will soon consider some of these in more detail, so let's try a more thorough definition. *Behavior is either innervated* *muscular movements or other neural activity involving energy exchanges within the organism, and their effects, that other energy exchanges within the organism, or beyond it in the external environment, elicit or evoke, or consequate.*

Whoa! At this point, that definition remains far too technical. Maybe we can try it after a dozen more chapters. For now, though, a definition between those previous two will work. *Behavior comprises countable events as either innervated muscular movements, or other neural activity,* **all produced or altered by stimulation.** While this defines a general notion of behavior, we use the term *response* for an instance of behavior, and we analyze changes in behavior in terms of response classes… But we get ahead of ourselves. Having considered what behavior is, let's consider a conceptual test regarding behavior.

The "Dead Body Test." Given that we confront a vast range of possible examples of what might be behavior, we would welcome wondrous tests that always clearly separate behavior from non–behavior. No such tests exist. However, the "Dead Body Test" provides a backwards starting point. The Dead Body Test says this: *"If a dead body can exhibit it, then it is not behavior."* As a reverse example, a dead body cannot "stand up," so standing up is a behavior.

Sometimes we say that a live body *exhibited* a behavior. By this we mean the natural account that a *stimulated* physiology mediated the behavior. For us "exhibit" certainly does not imply the mystical account that some inner agent spontaneously initiated the behavior. Nevertheless, the Dead Body Test works for either type of account, natural or mystical. On the one hand, superstitious folks accept that a soul or mind or self or person or other inner agent no longer inhabits the dead body, so whatever that dead body exhibits cannot be behavior. On the other hand, natural–science folks accept that to mediate behavior, the physiology must be alive, so whatever that dead body exhibits cannot be behavior.

We emphasize the latter, the natural–science, physiological behavior mediation. Behavior occurs as a function of energy changes (i.e., stimuli) in the internal and external environments that affect living physiological structures in ways that we see as mediating behavior. Yet we can have great difficulty separating this mediation either from the neural structural changes involved, or from the behavior occurrence. In some important senses, the mediation and the behavior are the same thing, with the behavior part more easily accessible (i.e., often we can see it). Actually the neural structural changes and the behavior occurrence have already been separated for study along disciplinary lines between physiology and behaviorology. Now we must combine these two again to study the mediation itself. We will consider it more in later chapters, but in preparation we must address many other points first. Let's start with a little more about those environments.

The External and Internal Environments of Behavior

While we presume the physiology of organisms, because another capable discipline covers that subject matter, *environment* provides the functional part of our behavior subject matter. Some disciplines have little need to take the environment as anything other than everything "around" us. Yet, for everyone and all disciplines, the environment has even more to it than this.

To grasp what more the environment has, recognize that you are "out there" as a part of my environment, and I am "out there" as a part of your environment. We call this part, for each of us, the *external* environment, the "totality of everything around" us, but it is not *everything*. If we were merely magical beings, we could stop here, as being "out there" changes nothing for magical beings. But we are not magical beings. Whatever we are, we still are. What we are, or are not, does not change merely because our past conditioning compels us to think that we are something else, something supposedly more interesting, at least for discussion in cocktail–party conversations.

Instead, reality provides more interesting things to discuss. So what are we? We are *physical* beings ("carbon–based units," according to a variety of sources), not magical beings. As such, science already makes clear that our skin constitutes no boundary of any sort to the laws of the universe. *Part of that universe is inside our skin,* which makes that inside part also a part of the overall environment. This part we call the *internal* environment. Now, I suspect that your insides make a more important part of *your* environment than they make of *my* environment, and vice versa. Still, physical things related to behavior go on inside the skin of each of us. For that reason we usually speak separately of the external environment and the internal environment. Compared with the internal environment, the external environment certainly allows easier access to its parts and processes for probes regarding functional relations. The internal environment, however, is not *in*accessible. It is merely *less* accessible. The stimuli producing behavior may exist on either side of the skin, as can any environmental effects resulting from a behavior.

For example, in the external environment, the light switch on the wall, at the top of some dark basement stairs, evokes flipping the switch. Then, as a consequence of that switch flipping, the stairway becomes visible, enabling safer transit down it, and back up.

As an internal environment example, consider the behavior of seeing the flashing of a dashboard idiot light in the shape of a gas pump. Like all behaviors, that seeing is a real event. It is a real, internal event that can serve (according to your past conditioning) the stimulus function of evoking one or the other or both of two response sequences (i.e., chains). It evokes an external response chain of looking for a service station and speeding to it and filling your tank. It also evokes an internal response chain of further waves of neurons firing that is the mediating of various behaviors such as seeing images of billowing air pollution contributing to more tornado filled storm clouds. As a consequence of those internal imaging responses (depending on your past conditioning) feelings of increased trepidation about, and worry over, global warming arise. Both the images and the feelings are real, but they occur internally, directly accessible only to a *public of one,* you, the body in which they occurred. Given the right past conditioning, these real events could also affect your subsequent overt behavior, including driving to the station without speeding, behavior that otherwise the idiot light evokes.

In general we presume that all real events that could potentially or actually control behavior constitute the environment. So when we behaviorologists say "environment," that term encompasses *all* of the environment including both the internal and external parts. Sometimes we stress the actual control of responding by using the phrase *behavior–controlling environment* (external and internal). Now, all this responding to external or internal environmental stimuli requires energy. From where does that energy come?

The Energy for Behavior

Every interaction between any part of the total environment and the body or behavior involves traceable energy. For any environmental stimulus to produce (i.e., evoke or elicit) behavior, that stimulus must transfer energy to the nervous system. Likewise, for any behavior to affect the environment, the behavior must transfer energy to the environment. And, for any environmental stimulus produced by an energy transfer *from* the body to result in more or less behavior on future occasions, the stimulus must transfer energy *back* to the nervous system. This changes the body structurally, making the body different so that it is the different body that receives new evocative environmental energy on future occasions and so mediates behavior differently on those occasions.

Most of these energy changes, however, especially those occurring only in the nervous system, are sufficiently potent only to *trigger* their effects; they are not potent enough to *fuel* those effects. Motor neuron firings take far less energy than muscle contractions take. The fuel for muscle–contraction effects comes from the reserves that the body builds and maintains as a part of the internal economy of the organism related to ongoing survival, to which behavior is an important contributing factor (e.g., eating and drinking). For example, you attend an author's talk at your local bookstore, and the speaker asks a question to which you have a relevant answer. The sound energy of the question is sufficient to *trigger* the release of enough of the potential energy that your body stores *to fuel the extension of your arm* into the air above your head, which the question evokes. Also, the sound energy of the question is sufficient to *trigger* the release of enough stored potential energy *to fuel the chains of neural responses* that the question also evokes, and which we call your private verbal thinking responses regarding the answer and how it gets phrased. Your hand–raising may even trigger (i.e., evoke) the speaker's response of calling on you...

Behavior Complexity

A difficult and yet fun part of following considerations about behavior, like those regarding energy, is managing the complexity of behavior. Our Law of Cumulative Complexity, which first appeared in Chapter 1, applies to behavior as it interacts with the contingencies that drive it. Think of our coordinated hand–foot–head example at the start of the chapter. (Figured out what behavior that refers to yet?) As in that example, one major way to manage behavior complexity involves looking at the component responses.

Our principles, however, derive from experiments, and experiments require repeatability. Each response, though, qualifies as a new response that has a different topography from other responses, even if it differs only ever so slightly. That is, every response differs at least slightly from all other responses in physical measurements, such as its extent, magnitude, direction (i.e., vector) or intensity. But even while differing from each other, responses also share some things in common, which gives us response classes.

Response class. To account for behavior in the tangle of confusing complexities, our experiments work through the concept of *response class*. A bunch of things that share something in common makes a class of those things. A group or variety of responses that have in common the same effects on the environment constitutes a response class. For example when you wash dishes by hand, the responses regarding which items get washed first varies from occasion to occasion. But all these washing responses share the common consequence that the items end up clean. So we say that all these responses are members of the same response class, in this case the response class of "dish washing."

During experimentation we *measure the response class members* to enable analysis, such as counting the number of times members of a particular response class occur in a measured amount of time. This can give us the rate of the behavior. We continue measuring the rate while observing whether or not any changes in rate occur when one or another independent variable changes. From the results of such experiments, we can derive various principles and conceptual details about behavior along with a range of complex behavior interpretations and applied interventions.

As an example, consider a child's behavior of saying, "please" at the dinner table when requesting food or other table items. Each instance, each response, differs from the others. Some are sincere and respectful while others are bland and trite and still others snide and obnoxious. Any instance of these types is a member of the response class that we could call "saying please." Our early interest, of course, is to get these responses going. So, we ignore the differences; we just treat all the instances the same way by letting them all produce the same consequences of receiving the properly requested item. Later we should intervene to reduce the inappropriate, impolite response class members (by applying an intervention that we will call "differential reinforcement" in a later chapter). Without being, or needing to be, aware of it, families engage in these processes regularly and often successfully. With response classes solving the problems of "responses always being new responses," and "responses always being different responses," we can now begin to classify the behaviors that make up various response classes.

Classification of Behavior

People have traditionally divided behavior into a number of sometimes different and sometimes overlapping types. Some types relate directly to our natural–science interest in human behavior. We will consider several behavior types across two broad classifications: (a) descriptive classifications of behavior and (b) functional classifications of behavior. Descriptive classifications stress connections based on shared characteristics. Functional classifications, which get most of our attention, stress connections related to the types of independent variables that the behaviors share. In some cases we will only mention a behavior type in passing here, reserving it for appearance in a later chapter.

Descriptive Classifications

Here we cover motor behavior and emotional behavior. Other chapters cover neural behavior, verbal behavior, and the overt/covert behavior distinction. In many cases the same behavior may appear under more than one of these descriptive classifications.

Motor behavior. When asked about behavior, most people respond in terms of behavior that occurs when muscles contract, especially muscles in the face, neck, shoulders, arms, hands, legs, and feet. People generally take such public behaviors as the most common, if not the only, type of behavior. Such *Motor behavior,* however, concerns much more. It consists of behaviors involving mostly *innervated* skeletal muscle contractions visible as whole body or limb movements. It also covers movements at small scales that reduce the ease of detection as, for example, would occur if someone were softly whistling into the wind while facing away from you. While you could not easily see them whistling, you might not hear them either. Note that muscles never contract spontaneously or because some inner agent magically tells them to contract. They only contract because some stimulation has made neurons fire, and that firing has made the muscles contract. This not only means that the stimulation made the muscles contract, made the behavior occur, but it also means that we describe the muscle movements as *innervated.* Since all such qualifying muscle movements require innervation, we more accurately refer to this behavior type as *neuro–muscular behavior* (and simply muscular or motor behavior for short). The examples are so numerous that particular ones may seem trite. Consider a knee bending or a hand reaching. Both are neuro–muscular behaviors (i.e., motor behaviors). As a more extravagant example, playing an original piano version of a Liszt Hungarian Rhapsody is, at least to some, a more interesting if rather vibrant and strenuous example of neuro–muscular behavior.

Note that to the performer of that Hungarian Rhapsody, however, playing it involves several other types of behavior as well. After all, while we can all rather easily observe each other's neuro–muscular behavior, we can individually, even more easily, and often more completely, observe our own neuro–muscular behavior, as is the case with the piano player. This is because we are privy to (i.e., affected by) a range of behaviors and stimuli that others cannot so easily access. Examples include the stimuli related to balance responses (i.e., proprioceptive stimuli) and the stimuli related to limb movement and placement (i.e., kinesthetic stimuli) as well as a variety of emotional behaviors.

Emotional behavior. We speak of *emotional behavior* when the activity that the body mediates involves certain glands releasing various chemicals into the bloodstream. Physiologists may see this as two separate steps, the action of the glands and the release of chemicals. At our behaviorological level of analysis, these two seem inseparable. On this matter we accept guidance from our physiology colleagues (and on other matters we provide them guidance).

In any case, the chemical release occurs as part of contingencies involving the releases as an effect in a functional relationship. The release occurs neither spontaneously nor because some mystical inner agent directed its occurrence. Furthermore it is not the expression of a putative inner agent's existence, nor is it the main cause of other behavior. Instead the chemical release occurs because of the occurrence of various external and internal stimuli. Some of these stimuli produce the release, because the body is already genetically structured to produce the release when these stimuli occur (i.e., without needing any kind of conditioning first). Other stimuli, however, produce the release only after some conditioning has also occurred (i.e., these stimuli must first be paired with stimuli that already produce the chemical release in what we call *respondent* conditioning) as we will soon see.

The chemical release, which we call emotion, has some further effects that center on two functions. Most obviously, the released chemicals function to produce effects, in conjunction with other real variables, that are detectable as feelings of many different kinds. We often give very specific, context driven names to these different feelings. Perhaps more importantly the released chemicals function to change the body in temporary ways that often lead to arousal or response exaggeration. The body then temporarily mediates behavior differently than it would have without the chemical release.

For example, let's say you are enrolled in a course. You stayed up much of Sunday night carefully reviewing previously well–studied material, because your job schedule would not allow time later for such activity, and your professor scheduled an exam at the end of the week. Then, at the Monday evening class, you are dozing off from the lack of sleep when your professor announces a "pop" quiz. This announcement elicits an emotional chemical release. The first effect of this release, which typically dissipates after a few moments, is that, in this context, it gets detected as feelings of surprise, anger, and pleasure—surprise because the professor has never before given a pop quiz, anger at being jerked around, and pleasure because last night's study has made you exceptionally well prepared. The second effect of this release, which can last for much longer due to its effects on external events, is that it temporarily changes the body which then mediates behavior differently than it would have without the chemical release. In this case, it energizes the body, counteracting your fatigue from too little sleep for several valuable minutes, and so enabling your preparation to produce much higher grade results on the pop quiz (which could have who knows what further long term effects on the future).

As other body parts or processes remove the chemicals from the bloodstream, the body returns to its normal operating levels. The body then mediates behavior without the additional controls from an emotionally changed body.

Functional Classifications

Every one of the behavior types that we considered under descriptive classifications involves one or both of the two kinds of behavior that we now consider under *functional classifications*. While the descriptive classification relied essentially on either the type, or the location, of the behaving body part, the functional classification relies on the types of independent variables that determine the behavior. Basically, eliciting stimuli control *respondent behaviors* while evocative and consequential stimuli control *operant behaviors*. This chapter turns out to be rather short, because the rest of the book provides extensive coverage regarding each of these functionally classified behavior types, especially the operant behavior. Thus our coverage here will only set the stage for that later coverage.

Respondent behavior. Early in the 1900s, researchers discovered that some stimuli, once their energy reached a certain level, completely compel the occurrence of a particular response. The occurrence of other stimuli cannot prevent this from happening. We call the effect of these stimuli "elicitation" and say that the stimulus elicits the response. Speaking generally we use the term "reflex" to refer to *both* the stimulus and the response together (i.e., a reflex requires reference to both a stimulus and a response). More specifically we call the stimulus an eliciting stimulus, and we call the response an instance of respondent behavior. Stimuli eliciting respondent behaviors are intimately bound up with phobias as well as the occurrence of all emotions and feelings, and with many bodily processes (i.e., the internal economy of the organism; e.g., digestion), and so many other things.

Indeed, when discovered, researchers began to search for all the eliciting stimuli for every human behavior. However, they were unable to discover all those stimuli, because stimuli do not elicit all human behaviors. Instead, another kind of functionally classified behavior exists.

Operant behavior. In the 1930s researchers discovered another kind of behavior, a kind with stimuli that evoke it, rather than elicit it. They called it *operant behavior* because, when stimuli evoke it, the behavior *operates* on the environment, changing that environment in ways that produce other stimuli that we can generally call consequences.

The occurrence of those consequential stimuli results in the operant behavior that produced them occurring more often in the future. To be more accurate, through energy exchanges, the occurrence of these consequential stimuli makes the evocative stimuli become more successful in the future in evoking the behavior. Of course, the stimuli themselves are not changing. Instead, their energies interact with the nervous system, the body, changing it so that the neuron firings, that are the behavior happening, occur more easily when the evocative stimuli appear in the future.

The evocation of operant behavior must happen, and always happens, but it differs from eliciting a respondent behavior. The most obvious difference is that consequences play little role in respondent behavior. Operant behavior, however, is operant because of the consequences. And some types of consequences make the operant behavior occur less often. Our later chapters provide more explanatory details on all of these considerations.

Operant behaviors tend to get more press, so to speak, and you are quite familiar with them. As we saw earlier, operant behaviors include walking, talking, singing, dancing, loving, thinking, studying, working, fighting, planning, partying, publishing, and so much more. All these behaviors are among the typical and ongoing results of operant processes. And all the operant processes go on continuously, on a moment by moment basis, for our whole lives, from before we are born until we die. Few processes can claim such significance or implications for human survival. No wonder the causes of behavior are of such vital interest to so many people.

Conclusion

Have you figured out what the behavior was that we described at the start of the chapter? It consisted of a complicated, coordinated, simultaneous set of hand and feet and head responses all comprising a particular behavior. Our definitions and subsequent discussion may have helped you figure out what kind of behavior was involved. Just to verify, we had described some behavior that the circumstances of piloting a small aircraft evoke. All those continuous and overlapping responses in a coordinated fashion occurred as part of the single behavior that we technically call a steep left turn. (Flying is fun!)

Again, no "you," as some internal agent, initiates or makes the steep turn. The turn occurs, because of contingencies, in instructions and flying, that successfully drive the "steep turn" combination of responses. If these contingencies did not drive those responses, then a crash would likely have occurred. The next chapter begins looking at the science behind behavior controlling contingencies.

References

Ledoux, S. F. (2014). *Running Out of Time—Introducing Behaviorology to Help Solve Global Problem*. Ottawa, CANADA: BehaveTech Publishing.

Chapter 6
"Causes" and "Effects":

How does science make sense of events? The relations of "Independent Variables" and "Dependent Variables" …

$\mathcal{T}$he aging Shaman's well trained apprentice reported his worries to his master. "I have been making many observations about the crops for the last couple of years, and I am concerned about the harvest this year." But the Shaman reassured his apprentice, "The rains have been good, the signs read months ago, and the appropriate number of virgins sacrificed. So things should be fine this year." This virgin–sacrifice requirement weighed heavily on the apprentice. He felt overwhelmed by having to choose, when he becomes shaman, who would be sacrificed. Maybe he was unfit to be Shaman.

Still, he took his master all around the fields and showed him all the observations that he had made. This year's rains came too soon. The many fields on the slopes got too dry, encouraging insect infestation, which lowered yields. The many bottom fields got too wet, encouraging scale, which also lowered yields. On the other hand, the few fields that had been abandoned for a few seasons showed increased yields when worked again. Also, the few farmers who tended their own fields had higher yields than fields tended by servants for a landlord. The apprentice concluded, "Overall, the evidence points to a poor harvest this year, less than half the usual yields. Many may starve."

After some thought, the Shaman replied, "You have great powers of observation. That is why you made a good apprentice. But after I am gone, you will have to take action in such circumstances. What must happen to save our people?" What must happen, indeed! Think about the evidence. What would *you* recommend?

We cover the reply of the apprentice in due course. First though, we need to examine some parameters surrounding behavior causes. These parameters include the nature of control, and the nature of the environment, and characteristics of the functional relations involving behavior in contingencies.

Behavior in Functional Relations

Earlier we discussed independent variables, dependent variables, and functional relations, along with our philosophy of science assumption that behavior, as a natural phenomenon, remains subject to natural laws. As behaviorologists we try to discover these laws. While they explain *why* behavior occurs, natural laws at the physiological level of analysis explain *how* these laws work.

We stress the connections between the physiological and behaviorological levels not because our accounts cannot stand alone. After all, they have stood alone, and they lead to extensions of knowledge, skills, principles, and practices regarding behavior. Rather, here we stress these connections as part of counterconditioning the traditional cultural lore. This lore conditions people to accept various agential causes for behavior. These causes include the ones that theological mysticisms (mostly external agents) and secular mysticisms (mostly internal agents) claim as explanations for the behavior subject matter. The interconnections between physiological and behaviorological variables show all the mystical accounts as not just unhelpful but unnecessary.

Nevertheless, we seldom have timely access for changing physiological variables in relation to behavior, so we must see them as somewhat irrelevant for practical purposes. Our ventures into the physiological level usually entail interpretive analyses that consistently overlap our functional analyses. This helps us manage the limited access we have to the workings of the behaving body, especially the nervous system that innervates and thus mediates the behaviors we study. For example, based on the observed reliability between the occurrence of some environmental stimulus and the occurrence of a particular response, we interpretively follow an energy trace from that environmental stimulus first into the nervous system and then out again as mediation of that particular response. While the physics and physiology levels of analysis allow direct access to the energy trace, we usually lack direct access to the trace at our behavior level of analysis. In this circumstance we can use terms like *evoke* or *elicit* to bear the weight of the functional relation. The stimulus (i.e., the energy trace from the stimulus) evokes or elicits the response. Let's consider some other aspects of functional relations and behavior.

Natural Laws Control All Behavior

Here is a very short but vital point, both philosophically and scientifically. It is short because we have covered it several different ways already. But it is so vital that it deserves its own section to stress its importance. It is so short that the side heading nearly says it all. To reiterate, *natural laws control all behavior,* which means, scientifically, that all behaviors, as dependent variables, occur as a result of the occurrence of other natural events as independent variables. For behavior, the controls reside in the functional relations that the terms *respondent conditioning* and *operant conditioning* summarize.

Environment–Behavior Functional Relations

As a short definition, some behaviorologists describe behaviorology as the science of environment–behavior relations. This covers the independent and dependent variables of most interest to behaviorologists. It includes all the independent variables that define both the settings in which a behaving organism exists and the consequences of behavior. Environment–behavior relations also covers all the dependent variables that define the behaviors, the innervated and so mediated activities, that happen through the organism. When, for example, we observe a behavior dependent variable reliably changing each time a setting independent variable changes, we describe a functional relation between them. If our observation is accurate, then a change in a setting variable becomes the basis for predicting a change in the behavior variable. Such analyses provide the scientific alternative to mystical agential accounts.

Still, with so many setting variables existing in any local environment, we can have quite some difficulty tracking which one or ones provided the energy trace that produced a particular behavior. Indeed, if our past conditioning has not made looking for those energy traces likely, then we may not even take their presence as a meaningful possibility. This leaves us taking the behavior as appearing spontaneously, out of nothing, which leaves us susceptible to positing the behavior as the product of some spontaneous internal activity. For example, if the energy that triggers a response enters the body through the eye, then we are hard pressed to trace that energy from its source, because the setting may contain thousands of potential sources. With no antecedent energy traces to account for the behavior, nor even any conditioned tendency to look for them, we may accept or invent a mystical account with just the right characteristics to explain the observed behavior.

Compounding that age–old agential error, we may further attribute the behavior to some intrinsic variance in the operations of the inner agent. This is an all too common attributed source of behavior occurrence. Variance, however, stems not from the capriciousness of the inner agent but from the difficulty in searching for and tracing the energy transfers from the environment into the body. Scientifically these energy transfers determine (i.e., control) the behavior. Any observed variance results from our incomplete accounting for *all* the relevant energy traces, which is not surprising given their great number *and* the great difficulty of following them. We manage this variance with different methods, such as describing it in terms of probability. Nevertheless, at our level of analysis, the behaviors, and the energy traces for that matter, are completely determined, and behaviorology strives to sort it all out as much as possible, and to apply the findings to benefit humanity. Let's peek a little at how that works.

General Analysis of Behavioral Events

The general approach to accounting for behavioral events proceeds by first analyzing the *antecedent* events (i.e., those occurring *before* the behavior of concern) for functional relations with behavior, and then, if appropriate, analyzing the *postcedent* events (i.e., those occurring *after* the behavior of concern) for functional relations with behavior. Sometimes the antecedent–event analysis is enough, if it documents *respondent* functional relations. When respondent relations are either not present or not a complete account, then we proceed further with the postcedent event analysis. This analysis usually documents *operant* functional relations. Over time, the usual repetitions of either case provide the foundation for work on interventions.

An Analytical Example

Let's look at a "simple" example of analysis in some detail. We must consider the antecedent event, the behavior of concern, and the postcedent event.

Here is the stage. You have had in your cellar for some years now an uncommonly good bottle of port from your favorite South Australian winery. You hope to serve it someday to an uncommonly good Australian friend, although you have no clue when, if ever, she might arrive. You have not seen her for years past and you may not see her for years to come. And you worry that she might not like your special port and think less of you as a result (as she hails from Queensland and has definite wine preferences that differ from yours). Then one day quite unexpectedly she turns up on your doorstep…

Our first analytical job involves identifying ① the behavior of concern, ② the antecedent event, and ③ the postcedent (and possibly consequent) event. Then we must, if possible, relate ② and ③ functionally to ①. If they are functionally related, we will use more accurate terms for them. We analyze, because we want to know what the variables are and how they are related. So we ask three questions. What is the behavior of concern? What evoked the behavior of concern? And what consequence did the behavior of concern produce?

Perhaps somewhat arbitrarily we take ① to be serving the special port. After all (to ask some of the questions differently) you have kept the port a long time, so why would you open it now? What difference might opening it make?

Using visual aids often assists analysis. Our first question is, "What is the behavior of concern?" Let's diagram the answer:

Antecedent Event	Behavior of Concern	Postcedent Event
?	Serving special port	?
②	①	③

Our second question is, "What is the antecedent event ②?" Let's diagram this answer also, because it might show a functional antecedent event that could explain why you would, after so many years, break out the port. In this case the answer is your friend's surprise arrival:

Antecedent Event	Response* of Concern	Postcedent Event
Friend's surprise arrival	Serve special port	?
②	①	③

*We change the general term *behavior* to the more specific term *response*, because this is a single instance of a behavior.

Our third question is, "What is the postcedent event ③?" Let's diagram this also, because it might show a functional postcedent event that could explain recurrences of the behavior of concern, port serving. As investigation reveals, your friend finds your special port one of the best she has ever tasted. So she compliments you profusely and, when she leaves, she says, ever so sweetly, "I'll be back." While each of these is a postcedent event, we predict that the one most relevant to future serving of special port is the profuse compliments:

Antecedent Event	*Response of Concern*	*Postcedent Event*
Friend's surprise arrival ②	Serve special port ①	**Profuse compliments** ③

If we have access to the rate of your friend's return visits, we will discover if your receiving those profuse compliments increases the rate of your serving your special port. If we discover that pattern holding, then we can use more sophisticated terms, which connote the functional relations, to diagram a three–term contingency. It contains two two–term contingencies (stimulus evokes response and response produces reinforcement) each of which is a functional relationship (which the arrows indicate, pointing in the direction of the functional relation):

Evocative Stimulus	*Response of Concern*	**Reinforcing Consequence**
Friend's arrival →	Serve special port →	Profuse compliments

Reading that last diagram provides answers to all our analytical questions. Your friend's arrival evokes serving your special port, which produces compliments for you and, as an outcome or result, the occurrence of those compliments increases the effectiveness of your friend's arrival as an evocative stimulus for port serving. This we see when you again and again serve special port whenever she arrives. (How is your supply holding up?!)

Our questions this time only focused on the events *immediately* preceding and following the behavior of concern. Clearly, earlier events led up to these events just as various later events may also affect similar future events. As our analytical knowledge and skills develop, we will deal with more complex events.

Generically, we describe such two–term and three–term contingencies as *contingencies of reinforcement.* This term summarizes the causes of operant behavior, the functional relations of operant conditioning, even though not all contingencies feature reinforcement. For some researchers, the term even covers the causes of reflex behavior, the functional relations of respondent conditioning.

Meanwhile, questions from other readers about the brand of port, or how to reach your friend, are not germane (i.e., relevant) to this discussion. However, *alternative* accounts for behavior are germane to the general discussion.

Alternative Accounts for Behavior

We have touched already on some alternative accounts for behavior, mostly of the fictional and mystical agential varieties. Here we touch on a couple of non–fictional accounts that nevertheless prove inadequate as accounts for behavior, namely, the passage of time and genetic endowments.

Some Caveats About Time

Consider the ubiquity of clocks. Now couple this with the current relatively sparse cultural conditioning regarding the functional effectiveness of antecedent energy traces in producing behavior. The result, unsurprisingly, leaves many people rather casually attributing the occurrence of behavior to the mere passage of time. That is, their explanation–related conditioning has been deficient, leading to untestable claims. One such claim says that, as time passes, a behavior's turn arrives and so it occurs. Another claim simply says that the passage of time causes behavior to occur.

Time, however, cannot have that functional status, because time lacks the kind of existence that would allow its manipulation at our level of analysis, which occurs at far slower speeds than the relativistic speeds needed for manipulation. Yet manipulation would be necessary to test time as an independent variable. Instead, we see time as an intellectual construct that, while useful in some contexts, we define as a non–spatial, non–reversible dimension consisting of a sequence of events. That is, a sequence of events *defines* time by filling it. Thus time itself, lacking manipulable physical reality, cannot transfer energy to the body and so cannot serve as a functional antecedent for behavior.

However, any of the physical antecedent events in a sequence that defines time *can* provide traceable energy transfers to the body and so serve as functional antecedents for behavior. These can be as complex as the covering of the usual series of sub–topics at a convention presentation about a research experiment (i.e., introduction, methods, results, discussion) or as simple as a progression of heartbeats or the sweep of the second hand on an analog clock face. Traceable energy transfers to the body from any such physical event in a time–defining sequence can produce behavior. But those *events* are producing the behavior, not the mere passage of time.

For example, your physician directs you to take a dose of medication everyday at 6 PM after you get home from work. Each day you get home sometime between 5:30 and 5:50, due to traffic. Once home, though, part of your behavior includes watching the clock. At 6 PM you take the medication. Time passing provided no energy trace to the body to evoke this behavior. Instead, the configuration of the clock face (e.g., "6:00 PM" on a digital clock face) provided the energy trace to the body and thereby evoked this behavior.

Note that your "clock watching" likely itself is more complicated than merely sitting and staring at the clock face until 6 PM arrives. You might more accurately describe it with the verbal shortcut, "keeping track of the time," as other stimuli evoke and consequate a series of other behaviors, such as cooking or setting the table for dinner. Meanwhile, as but one of many possibilities, the "keeping track of the time" behavior may involve regular, repetitious glances toward the clock interspersed among these other behaviors. However, regardless of the complexity, the passing of time still provides no antecedent energy trace evoking the medication–ingestion behavior. Instead the clock face finally reaches the 6 PM configuration, which then provides the energy trace during the last of the glances and so causes medicine ingestion to occur.

Some Caveats About Genes

With time not an explanation for behavior, what about genes? When contingencies, usually educational contingencies, move people away from either theological or secular (e.g., psychological) mystical accounts for behavior, many people derive some comfort in scientifically connected genetic accounts of behavior. Even hopelessly simplistic genetic explanations, such as "she does that because she was born that way" or "he does that because it's in his genes," carry some aura of scientific authority through a generalization process from real science. Sophisticated genetic accounts carry more such authority, even if unjustified. Consider examples like, "he works well with children because he inherited a very calm demeanor from his father," or "scientists say they have located genes for alcoholism and smoking."

One problem with such accounts is that, even if genes could cause behavior, genetic interventions to deal with problematic behavior currently remain in the realm of science fiction. We cannot readily change our genes, even if we can turn some off or on. We still lack the technology to manipulate genes as independent variables on the time scales at which behavior occurs and changes. However, the greater problem with all genetic accounts is that genes cannot exert any direct control over behavior, because essentially the workings of genes *only produce physical body structures.* That is the main biological role of genes. The body structures that result from genes—recall our Law of Cumulative Complexity—then interact with antecedent environmental energy traces. This interaction changes the body structures in ways that produce (as in mediate) behavior. Only in that *indirect* way can we say that genes affect behavior.

Original body structure may mediate one type of behavior better or more easily than another. Genetic body structures are already capable of mediating behaviors though always and only under the energy traces of unconditioned eliciting and evocative stimuli. For example, some neural structures connected to the iris are directly genetically produced, and these structures already mediate the behaviors of pupillary dilation and constriction, but *only* upon the energy trace of a change in the level of illumination (i.e., the pupillary reflex).

In other cases genetic body structures get changed by respondent and operant conditioning processes—which we will touch upon in upcoming chapters—and then are capable of mediating other behaviors. Again though, the mediation always and only occurs under the energy traces of eliciting and evocative stimuli. Consider an example, one possibly personally familiar, of spicy tacos eliciting salivating. The genetically produced neural structures that already mediate that salivating behavior get changed by respondent conditioning processes when eating the tacos gets paired with a restaurant logo. As a result these neural structures also mediate salivating in the mere presence of the restaurant logo.

Furthermore, the body structures of some people's genetic endowments may more easily change, relative to the body structures of other people's genetic endowments, due to the functional flow of environmental energy traces. Thus they may mediate certain behaviors more easily. However, without environmental support, even a favorable genetic endowment cannot mediate inspiring behavior. Your genes may have made you tall and slender, but the environment determines whether you plaster ceilings without a ladder for a living, or play professional basketball.

Consider this example. A couple (literally, a married couple) of colleagues, invites you over for dinner. They have three children, ages 15, 10, and 5. (They are mathematical physicists, OK?) Now, good music has a highly reinforcing effect on your behavior (i.e., when you hear some, you will go back for more, at least within the constraints of politeness). Well, over the course of the evening, you discover to your amazed delight that the two younger children play Chopin polonaises fabulously. Perhaps we could forgive you for jumping to the merely emotionally informed conclusion that these kids have inherited an amazing virtuosity. After all, both parents have played the piano a bit since childhood, and still do. So, what is going on?

What is responsible for the children's extensive and capable repertoire? What is the oldest son's excuse for not playing (an unfair but typical phrasing that translates as "why does the older son not play")? Are the younger son and the youngest, the daughter, just more musically inclined? What does that mean? What roles, if any, do environment and genes play in this playing?

Those are all typical questions for this type of situation. Let's answer those we can. We know that genes do not cause the great piano playing, because genes can only build body structure. However, these children's genes may have built structures that interact particularly smoothly with the musical stimuli in environmental contingencies, such that they mediate musical behaviors more easily than structures that the genetic endowments of others build. And what were the relevant contingencies? Investigation shows that, when the youngest turned three, the father got a good university professorship. So the parents bought a good piano and arranged for two or three lessons with a capable teacher, and several practice sessions, each week for both younger children for the last few years. They also arranged positive (i.e., non–coercive) contingencies supportive of effective lessons and practice–time usage, with lots of positive parental attention. All this resulted in expanding and improving the children's playing in leaps and bounds, leading to the performances that you experienced. This shows the mutually beneficial interplay between (a) gene–built structures that can facilitate some behaviors and (b) the occurrence of those behaviors under the efficient control of relevant environmental stimuli. Also, note that the younger child's playing impresses us more than the older child's, even though they play equally well and have had similar lesson and practice experience. This bias stems simply from the younger child being younger, a bias we can note when attributing the playing not to age or genes or some ability trait, but rather to an opportune set of circumstances and contingencies.

Also, perhaps you think this example is a bit unrealistic, because the finger spread required to play the polonaises is rather wide. Well, the parents are quite tall and strongly built, and so are the children. Indeed their growth rates exceeded the norms for their ages, leaving them always larger, in a healthy sense, than their peers, such that even the youngest has just enough finger spread to meet the notation demands of the music. This shows another way in which gene structures can facilitate some behavior, even though the behavior actually occurs only under the control of relevant environmental stimuli.

By the way, why was the oldest child, also of strong, capable body, not playing? The reason was probably not that he lacked the right genes; while he may have lacked them, we really have no way at present to know. The real reasons he was not playing, however, were two. Both pertain to a lack of opportunity, which concerns the environment, not genes. For one, back when he was growing up, until he was 11 or 12, only the mother held a university professorship while the father was free–lancing, so lessons were not affordable. As for the second reason, due to that same lack of funds, the family lacked a piano for playing and lessons and practicing. So again, no practicing was possible even if lessons had been affordable. Ultimately, we can see a third reason as well. With piano lessons and practice impossible, contingencies compelled other behavioral investments. While his siblings excel at piano and scholastics (for similar reasons) he has excelled at soccer and scholastics, again for similar reasons.

Here is another point to consider. Since genes have real, physical status, they *can* affect behavior. They can serve as evocative stimuli for behavior with respect to genes themselves, particularly verbal behavior. For example, when a police laboratory technician completes the necessary tests and concludes, "The genes from the crime scene match the genes from the suspect," genes and their measurements, as real environmental stimuli, evoke those verbal behaviors.

All these objections to genes as causes of behavior essentially repeat the objections that we have to other explicitly physiological variables as *causes* of behavior (as opposed to *mediating* behavior). In particular, we cannot *directly* manipulate either genes or other physiological variables, as independent variables, to affect behavior. In the case of other physiological variables, we can *indirectly* change them through alterations of the environment such as applying contingencies, or drugs, that change the physiology which then mediates behavior differently. However, drugs do not represent very elegant, and often not even very successful, solutions for behavior problems.

Traits are not Behavior

When we previously considered traits, we described them not as behavior but as adjectives that others, who witnessed an event, turn into nouns when telling you about the event (which you never witnessed). If we take these nouns uncritically, they all too easily become a supposed and reified material or entity present in the body that someone could claim accounts for the behaviors that occurred in the unwitnessed event.

In brief the adjectives may have originally described actual behaviors, but the nouns, at best, only label some characteristics of the behaviors. At worst the nouns become false causes of the behaviors. We say false, because the event reporter only invented the traits from the adjectives describing the original event, so the nouns have no independently verifiable existence. Recall our example of the friend who tells you that a mutual acquaintance revealed a lot of hostility after listening to an administrator at a meeting. Yet, except for "listening," your friend never mentioned a single actual behavior, leaving you to guess what the behaviors of the acquaintance actually were. Besides, whatever the acquaintance's behaviors were, the cause was not due to the acquaintance harboring hostility, as if this denoted a characteristics or trait of some inner self agent in the acquaintance. Like traits themselves, such mystical inner agents remain irrelevant to a natural–science account of behavior.

Love is not a Trait

To further clarify that traits are not behavior, let's consider *love,* which is not a trait. When some people say, "He is such a loving father," or "She is such a loving wife," they think they are alluding to love as a trait, as a characteristic of some mystical agent inside him or her, a characteristic that someone might falsely take as a cause of his or her behavior. But love is not a trait. They are instead alluding to complicated aspects of the father's or wife's or anyone's behavior, in relation to others, by using that single word, *love,* as a sort of verbal shortcut that summarizes a deeply poetic, multi–faceted and scientifically explicable natural phenomenon. Past cultural conditioning, however, may disable appreciation of this interpretation in favor of one or another more traditional notion. Our treatment here begins to elaborate the details behind this single word, *love,* as a verbal shortcut.

Rather than being a trait, love is behavior under some complicated and interacting sets of contingencies. Indeed, love is a range of behaviors, a range of effects of various contingencies. *Love* is the term we use to summarize these complicated and usually reciprocal effects between or among people. These effects occur, because the stimuli that elicit the feelings we call love, and the stimuli that reinforce the operant behaviors that we also cover when we speak of love, are the same stimuli (whether unconditioned or conditioned, which are topics we will address in future chapters). The interactions of these stimuli and effects produce the reciprocating behavior patterns involved in love.

Traditional notions of love come in a wide variety, nearly all of which are fascinating, charming, romantic, and so on. They derive from natural phenomena (e.g., the kind, amount, and specifics of one's past conditioning history) and so, along with love, are amenable to a scientific description. Since many of these traditional notions erupt from various conditioned mystical self–agent accounts in the common lore, they will conflict with any scientific account. But our focus remains a scientific description of love.

In the past we have perhaps been under contingencies that left us fooled about what love is, and now we can begin to make elegant improvements. You see, love does not change character simply because we consider it scientifically. Whatever love is, that is what love was, is, and remains after our scientific description. This exercise is worthwhile, due to the potential implications for life and living that become possible from an improved understanding of love when we consider it scientifically, even if our consideration here only interpretively expands our understanding on the basis of the few general principles of behavior that we have touched on so far.

Only ever so rarely will we find an instance—and be glad for it—when contingencies induce benefits merely through astute, intuitive observation rather than as an outcome of scientific activity. For instance, the author of the works of Shakespeare (poems and plays and all) had no science of behavior as a writing guide, and probably shared with contemporaries some mystical notions about love. Yet this writer was still an astute observer and reporter of human behavior and the contingencies in which it occurs. He well applied those observations in accurate, interpretive and poetic portrayals of human behavioral possibilities, including our averages, excesses, deficits, loves, faults, and heroics. Over the last several thousand years, we can gladly count a few other rare, and well known, speakers, artists, and authors (e.g., Lao Zi, Buddha, Confucius, Jesus) whose works—natural products of other natural contingencies—have bestowed similarly astute and beneficial observations and applications on us, even though they too were without the benefits that science can add. Let's elaborate a little about how science handles "love."

Consistent with the rest of natural behavior science, we may interpret the word *love* as referring to complex interactions of respondent behaviors and operant behaviors, along with their eliciting and evocative and reinforcing stimuli, all with particular characteristics. Perhaps as much from biological evolution as cultural selection, we value these interactions the most when they occur between *two* opposite–gender, or even same–gender, bodies, although we also value them when they involve more people, and we even often value them when they involve only one person plus some class of objects or events. In his 1953 textbook, B. F. Skinner introduced the topic of love with reference to its vernacular usage where love can refer to the predictability of some particular behaviors occurring more than other behaviors. For example, we can safely predict that someone otherwise describable as in love has a greater "…tendency to aid, favor, be with, and caress" rather than "injure in any way" their partner or other object of their love (p. 162). Elsewhere in his book, Skinner points out that "love might be analyzed as the mutual tendency of two individuals to reinforce each other, where the reinforcement may or may not be sexual" (p. 310). These points speak to changes in operant behavior that respondent emotional chemical releases support (more topics for future chapters).

First, though, you should not take "in–love" as implying a bodily state, or a state of being, that is responsible for loving behavior. The temptation toward this implication stems from the general temptation toward mystical inner agents from our common, traditional cultural conditioning. This temptation also stems from the sometimes lengthy occasions when emotion–produced feelings linger, or when ongoing stimulus changes recharge such feelings.

Let's examine love more closely by slowly detailing many of its interconnected parts. Generally, love involves certain respondent emotions (with their accompanying feelings) that alter the ways (e.g., arousal) in which certain neural structural changes mediate lots of related operant behaviors in a positive feedback loop that can continually increase these behaviors and their related emotions and feelings and reinforcing effects.

Say… What?! Let's try to unpack that into a bunch of smaller but interrelated pieces. All consequences that earn the title *reinforcer* change neural structures such that the stimuli that evoked the behavior that produced the reinforcers become more functionally effective in evoking that behavior upon future occurrences, which is particularly effective with the *ongoing presence* of a partner. Furthermore reinforcing stimuli also condition other stimuli to serve reinforcing functions, because they reliably occur at the same time (a pairing process). In addition, reinforcing stimuli also often serve an eliciting function that in the past we have treated too lightly as a mere side effect of reinforcement. This function involves eliciting the release of chemicals into the blood stream, an elicited release that we label emotion and that, as real events, produces real detection responses that evoke the label "feelings," some of which are part of *love*.

Those feelings, as real events, also serve as consequences that alter the way neural structures mediate subsequent behaviors. Normally, any emotional arousal *only* temporarily changes the body, while the chemicals circulate, such that the body mediates behaviors (which other environmental events must still evoke) differently from the way it would mediate those behaviors if the emotional arousal had not occurred. However, when the feelings resulting from emotional responses consequate the behaviors that produced the stimuli eliciting those emotional responses, the resulting structural changes may endure beyond the moments when the chemicals course through the bloodstream. This may constitute an important method by which the occurrence of stimuli that we call reinforcers leave the body changed such that the body more easily mediates in the future the behavior that produced the stimuli. Future research will clarify to what extent this holds.

All the events in those comments are *real* events, all are at least theoretically measurable, *and all occur together in a complex mix in love,* clearly giving love the status of a real event. When we speak of love, we speak at least of the feelings that elicited emotional chemical releases produce. But we really speak of so much more than that when we speak of love.

Consider two people in love, first on the level of operant behavior, and then on the level of respondent behavior. Injuring the other in any way is not a part of their mutual evoked behaviors. Instead, for each, mutual conditioning has made the very presence of the other evoke responses describable as loving (i.e., they each aid, favor, stay with, and caress the other). And in love, any sign *that you reinforce the other* is reinforcing to you. (Technically, we should have said that any sign *that your own behavior reinforces the behavior of the other* reinforcers your behavior, because reinforcers *only* affect behaviors, not organisms. However, to simplify phrasing with this complex topic, we occasionally engage the verbal shortcut of "reinforcing people.")

A few such mutual reinforcers hold unconditioned status, such as attention and copulation. "Unconditioned" means that the stimuli function as reinforcers without any conditioning having occurred. However, most of the relevant reinforcers hold conditioned status—having occurred with the unconditioned reinforcers—including the sight, sound and feel of the other. All of these inevitably pair with the other during reinforcing activities. Indeed, such pairing is how the signs themselves, that one's behavior has reinforced the other, become reinforcers for the one! And if the other's behavior *also* functions as a reinforcer for one's behavior, then we often speak of romantic love.

Many of these reinforcers end up being contingent on (i.e., delivered to the other on) occasions when the other's behavior supplies reinforcers to the one. In this way, both the other and the one reciprocally provide (often so smoothly and quickly as to be simultaneous) these reinforcers to each other. Taken together, these operant components of love involve an ongoing, mutually beneficial, reinforcing interaction that emphasizes evidence of each having successfully reinforced the other, with each finding that success itself reinforcing. All this supports continuing the mutually reinforcing interactions.

For clarity, stimulus changes do not occur in the stimuli, which remain as they were. Rather the changes occur in the neural structures, which now mediate operant and respondent behaviors differently. Nevertheless the number and range of variables eliciting the emotions we detect as feelings of love multiply along with the operant reciprocal reinforcing events, all interactively present together, constituting what we call love.

Summary of love. Here is another author's summary about the behaviorological analysis of love (including two familiar "what one *does*" agential verbal shortcuts that you can translate):

> Love is a term describing one's feelings that are elicited when
> one's own behavior is reinforced by evidence of its reinforcing effect
> on another organism—that is, when the reinforcer for what one

does consists of any sign that what one does functions as a reinforcer of the other person's behavior. Romantic (i.e., mutual) love further requires that the other person's behavior be a reinforcer for your behavior. People "in love" with each other are reinforced when their own respective behaviors prove reinforcing to the other party. "I love you" means that my behavior toward you is reinforced by evidence that my behavior ranks high among *your* reinforcers. Also, we are especially likely to say "love" when the degree of respondent conditioning has been high, and the elicited emotional reactions are especially strong (Fraley, 2008, p. 929; emphasis in the original).

Conclusion

So science can help us understand even topics that traditionally people considered beyond science. Behavior in general constitutes one such topic, as does love in particular. So what about farming practices? With your educational background that includes some science, your recommendations about what can save the apprentice's people, from our story at the start of the chapter, surely avoided the conditioned pitfalls of prescientific knowledge. Such conditioned prescientific pitfalls, however, constitute the apprentice's controlling environment, which he (i.e., his behavior) cannot escape. As tears filled his eyes, the only possible answer escaped his lips. "We need more virgins. Is it too late?" (See Healey, 2015.)

   Think about where else in today's world we see similar outcomes. In general, for example, extensive, repeatable and precise evidence detailing the natural–science causes of behavior get ignored in favor of theological maxi–god accounts or psychological (agential) mini–god accounts. To some folks (How about you?) this sounds rather like the apprentice and the Shaman sticking to the notion that sacrificing virgins accounts for good harvests. More particularly, some proposed solutions to global problems also ignore the scientific realities about human behavior. Is that wise? Reserving that topic for our last chapter, let's turn to more details about the scientific accounts for human behavior. ❧

References (with some annotations)

Fraley, L. E. (2008). *General Behaviorology: The Natural Science of Human Behavior.* Canton, NY: ABCs.

Healey, G. J. (2015 August 25). Applied science. *St. Lawrence Plaindealer* (p. A2). Canton, NY. (This chapter adapted the apprentice's story from the longer version that appeared in this newspaper column.)

Skinner, B. F. (1953). *Science and Human Behavior.* New York: Macmillan. (The Free Press, New York, published a paperback edition in 1965.)

Chapter 7
Reflex Behavior,
Also Known As Respondent Behavior:

What comprises some "simple" but vital behaviors?
Some causal variables that mostly act
both alone and before behavior ...

A century ago, when a child first went to the dentist, few precautions were available to avoid coincidental events that could cause fear reactions regarding future dental visits. This was because little had been discovered scientifically then about the relevant behavior–environment causal relations. So what could go wrong? Let's consider an example. The nurse escorts the child from the antiseptically scented waiting room to the smaller room with the big, throne–like chair in the center. Entering with much anticipation and little trepidation, the child finds everything feeling fascinating and new, from the instrument cases containing so many shiny metal implements to the room's white walls and the dentist's and assistant's dazzling white coats and big smiles to the strange wires, lights, and tubes hanging from posts near the chair or from the ceiling. After settling into the chair, and with little by way of reassuring explanation or commentary about the room or equipment or procedures, or people, the child finds the dentist and assistant in his or her face, and inside her or his mouth. This all begins to seem a bit frightening (i.e., all these unfamiliar stimuli start eliciting feelings of fright). Then the drilling starts, with its surprising, shrill sound, and weird vibrating sensations in the mouth. And then comes the pain, which occurs along with all these other stimuli, all pairing with each other. When the dentist is done, *everything* about the visit, not just the pain causing drill, elicits fear responses. Thus any further visit requires the parents to drag the child kicking and screaming to the dentist.

Must a visit to the dentist follow such patterns and outcomes? What happens now? Let's first try to understand what has happened already.

Respondent, and Operant, Conditioning

The best way to introduce reflex/respondent behavior turns out to involve presenting it with comparisons and contrasts with the other functional type of behavior, operant behavior. Our coverage will follow that lead.

Each type of behavior that we considered previously, under *descriptive* classifications, involves one or both of the two kinds of behavior that we considered under *functional* classifications. While the descriptive classification relied essentially on either the type, or the location, of the behaving body part, the functional classification relied on the types of independent variables determining the behavior. Basically, eliciting stimuli control *respondent behaviors* while evocative and consequential stimuli control *operant behaviors.*

Respondent Behavior

We credit Ivan Pavlov, a Russian physiologist researching digestive processes, with discovering what we now call *respondent behavior* and *respondent conditioning* early in the 1900s. Pavlov discovered that some functional stimuli, once they reach a certain threshold (i.e., energy level), completely compel the occurrence of a particular response. Other supplemental stimuli cannot prevent this from happening. We call the effect of these functional stimuli "elicitation" and say that the stimulus *elicits* the response. Since we are dealing with bodies that must mediate (not originate or initiate) behavior, what is really happening?

What is really happening is that the body is structured such that the energies from the stimulus automatically make the body mediate the behavior. If the body structure is the structure that the genes produced, then we call both the stimulus, and the response that it elicits, *unconditioned*, because no conditioning history was needed for the stimulus to elicit the response. For example, if a companion surprises you by stuffing a small but quite spicy taco hors d'oeuvre into your mouth, you salivate. That little spicy taco is an *unconditioned stimulus* that elicits salivating as an *unconditioned response.*

Pavlov also discovered that other stimuli, ones that started out having no effect on the response under study (so we call them *neutral stimuli)* can come to have eliciting effects. Indeed, nearly any stimulus consistently paired with an unconditioned stimulus will start eliciting the response, but this requires other terms. Since we call pairing the stimuli *the conditioning process,* we call the no–longer–neutral stimulus the *conditioned stimulus.* We then use the term *conditioned response* for the response that the conditioned stimulus elicits. The energy transfers in the conditioning process change the body so that now the occurrence of the conditioned stimulus automatically makes the body mediate the behavior as a conditioned response. These terms, of course, are for our convenience; meanwhile, the body merely mediates. For example, if those tiny spicy tacos (an unconditioned stimulus) are only on the menu of a particular restaurant chain with a notable logo (originally a neutral stimulus) and you eat them regularly (which each time pairs the tacos with the logo), then quite soon the logo itself (now a conditioned stimulus) will elicit salivating (now as a conditioned response). Whenever you see the logo—on the street, on an advertising sign, on your cell phone—you salivate somewhat. Virtually everyone has had similar experiences; think of hamburgers and yellow arches.

The conditioned stimulus will continue to work this way so long as it and the unconditioned stimulus occur together at least occasionally. Happily (or perhaps not) if the restaurant drops those tacos from the menu so that they never again pair with the logo, then the logo will eventually stop eliciting salivating (unless, of course, the restaurant serves other mouth–watering delicacies that you keep eating). We use the term *extinction* for the process in which the response stops occurring after the stimulus pairings stop. Actually, we call it *respondent extinction,* because another type of extinction exists for which we traditionally reserve this term. In respondent extinction, the unconditioned stimulus and the conditioned stimulus no longer occur together with the result that the conditioned stimulus ceases to function as such and returns to being a neutral stimulus. Of course, the stimulus has not changed, rather the body has changed due to the changes in energy transfers.

Respondent behavior and respondent conditioning extend far beyond just tacos and slobbering. They are intimately bound up with the occurrence of all emotions and feelings, and with all sorts of physiological happenings that we refer to as the internal economy of organism, and with phobias, and with many other events and processes.

Indeed, after Pavlov discovered eliciting stimuli and their effects, many other researchers began to search for eliciting stimuli for every human behavior. However, they were unable to discover all those stimuli because, as they were— and we are—soon to discover, stimuli do not elicit all human behaviors. Instead, another kind of behavior exists, to which we now turn.

Operant Behavior

We credit B. F. Skinner, one of the earliest natural scientists of behavior, with discovering this other kind of behavior and conditioning in the 1930s. He called it *operant* because, when stimuli evoke this kind of behavior, the behavior *operates* on the environment, changing that environment in ways that we call consequences. While Skinner understood that *stimuli evoke operant behavior,* he found that using the word "emit" helped to increase the contrast between operant and respondent behaviors, which helped others to "understand" these behaviors (i.e., to respond correctly in discussions and research about them). So he described operants as behaviors an *organism emits* while describing respondents as behaviors that *stimuli elicit.* For operants, however, *evoke* more accurately describes what the stimuli do, while "emits" can imply, or leave room for, a putative and unnecessary inner agent to "do" the emitting (or to tell the body to do it). Skinner recognized this danger. But his career efforts to stop it proved inadequate. For the sake of more success, we avoid the term "emits." Instead we favor the full disclosure of *stimuli evoking operant behavior.*

The evocation of operant behavior must happen, and always happens, but it is not the same as eliciting a behavior (a point to which we will return shortly). The consequences of the operant behavior transfer energy back to the body where this energy changes the body physically. The energy changes some neural structures so that the body now *more readily* mediates similar behavior *when similar evocative stimuli occur again.* Since we have less access to the body than our physiology colleagues, because that is not our level of analysis, we address that energy effect from consequences as *two* observable changes (the details of which our physiology colleagues will someday elaborate): After the consequences occur, (a) we see the evocative stimuli function more effectively in the future (meaning that they more reliably evoke the behavior), and (b) we see the behavior increase in rate (e.g., more occurrences per time period) across future instances, which we sometimes describe as an increase in probability.

Rate, though, is just a potent and rigorous measurement method, and the stimuli are not what changes. What actually happens? If you really want to know, we will have to get a little geeky. (Luckily this happens only rarely.) What actually happens is that the body changes. Let's repeat some details as a review before elaborating: The energy from the consequences changes some neural structures so that the now different body *more readily* mediates similar behavior *when similar evocative stimuli occur again.* Importantly, note that this effect of the consequences defines the consequential stimuli as reinforcers.

Consequences, however, can have another effect. Sometimes the energy from the consequences changes some neural structures so that the body now *less* readily mediates similar behavior. This effect of consequences defines the consequential stimuli as punishers, which also have respondent effects that generally are the opposite of the respondent effects of reinforcers (more later).

All these processes go on continuously, on a moment by moment basis, for our whole lives, from before we are born until we die. Many—even some medical practitioners—consider the permanent cessation of operant processes, and death, as the same thing.

We cover many of the principles, processes, and procedures for these two types of behavior and conditioning in this and later chapters. These two types—respondent and operant—provide the general structure for dealing behaviorologically with human nature and human behavior. Here are a couple more distinctions that increase our understanding.

Elicit/Evoke Distinction & Respondent/Operant Distinction

The contingencies in respondent and operant behavior and conditioning form the foundation for effective understanding, prediction, and control of our behavior. Helping us in these endeavors is the *elicit/evoke* distinction.

While an elicitation is a kind of evocation, we restrict the term *elicit* to respondent behavior (which others call *reflex* behavior). We say "the stimulus elicits the response." That is, we use *elicit* when a stimulus energy transfer *triggers* the neural–structure mediation of fuel (i.e., food energy) supported responses that other stimulus energy transfers cannot stop or change, and to which consequences are essentially irrelevant. More simply, some stimuli *elicit* responses, such as the stimulus of an increase in ambient light compelling the response of constriction (i.e., a decrease) in pupil size. Other stimuli cannot stop the constriction and consequences show no effects on the constriction. For example, the sun's reappearance from behind a cloud *elicits* pupillary constriction; the occurrence of other stimulus energy transfers cannot stop the constriction, and any consequences make little difference in the constriction.

Let's now expand those points about "elicit" more completely (i.e., let's add the "qualifier" of unconditioned or conditioned status): We use *elicit* when a particular stimulus transfers energy to the body and that energy transfer *triggers* neural structures to mediate a fuel–supported response that *the occurrence of other stimulus energy transfers to the body cannot stop or otherwise change, and to which consequences are generally irrelevant*. The neural structures can either be ones that genes built (i.e., of unconditioned status) or ones that earlier energy transfers have changed (i.e., of conditioned status).

Here is what differentiates respondent behavior and conditioning from operant behavior and conditioning. With the former (i.e., respondent) *other stimuli cannot stop or otherwise change the mediation of the response, and consequences are essentially irrelevant,* while in the latter (i.e., operant) the occurrence of other stimulus energies *can interfere* with response mediation, resulting in a different response, and *consequences are vitally relevant.* Due to that differentiation, we use *evoke* (rather than elicit) with operant behavior.

We say "the stimulus *evokes* the response, which other stimuli then consequate." We use *evoke* when a stimulus energy transfer triggers the neural–structure mediation of fuel–supported responses that other stimulus energy transfers *can* stop or change, and to which consequences are essential. More simply, some stimuli *evoke* responses and, equally importantly, those responses produce consequences that alter the future occurrence of similar responses. That is, those responses produce consequences that change the neural structures such that they more readily (with reinforcers) or less readily (with punishers) mediate similar responses when similar evocative stimuli occur again.

Let's now expand that point about "evoke" more completely, while adding the "qualifier" of unconditioned or conditioned status: We use *evoke* when a particular stimulus transfers energy to the body and that energy transfer triggers neural structures to mediate a fuel–supported response that *the occurrence of other stimulus energy transfers* **can** *stop or otherwise change,* which interferes with response mediation and results in the occurrence of a different response, *and to which consequences are essential.* The neural structures can either be ones that genes built (i.e., of unconditioned status) or ones that earlier energy transfers have changed (i.e., of conditioned status).

As an example, particularly of how other evocative stimuli can interfere with the mediation of an *evoked* behavior, producing a different behavior, consider this scenario. While enjoying a post–lunch siesta in your dorm room, your alarm sounds, and the clock reads 2:50. This gives you 10 minutes to get to your 3 PM afternoon class, and you start walking toward the classroom. The clock stimulus has evoked going to class (where more reinforcers are likely more contingent on your preparation and your interaction than on your mere presence). But on the way to class you pass someone whose car is stuck in the mud and he gestures toward you for assistance. That gesture stimulus now evokes detouring to lend a hand. After helping, the sight of the classroom building might re–evoke going to class, although the amount of mud on your hands first evokes another detour, this time to the wash room. In any case, you can see how other evocative stimuli can interfere with the mediation of an evoked behavior, producing a different behavior. It also shows that the clock was not an eliciting stimulus as, in that case, none of the detours could occur. By the way, if the contingencies of your past conditioning have not left you ready to help "out of the goodness of your heart," then your current money–related contingencies (typical for college students) leaves the $20 bill, which the car owner waves at you (as a last resort to get your help) more effective in evoking your aid, and we still see you detour to help move the car out of the mud. Also, "out of the goodness of your heart" is not an adequate scientific account; in due time you will be able to provide a proper scientific alternative.

Distinction summary. Aside from operants and respondents that involve strictly neural firings, note these three differences between operants and respondents. (1) Operants usually involve innervated striated or striped or skeletal muscles, while respondents usually involve innervated smooth muscles and glands. (2) With operant behaviors, much conditioning involves reinforcers following an evoked response with the effects that (a) the evocative–stimulus function becomes more effective, and (b) the rate at which similar responses occur increases. With all respondent behaviors, conditioning involves the pairing of two stimuli with the effect that the new stimulus now also elicits the response. And (3) evocative stimuli produce operant responses that affect the environment, producing consequences, while eliciting stimuli produce respondent responses, such as emotions and feelings and other physiological changes involving the internal economy of the organism.

"Conditioning" as an Analytical Confusion

While we could have covered this topic in a previous chapter, as another analytical confusion, it helps more here as we transition from this chapter on reflex behavior to the series of chapters on operant behavior.

What we call conditioning happens through various contingencies between causal variables and behaviorial effects. In our analyses of behavior, however, we use the term *conditioning* in the names of two processes, respondent conditioning and operant conditioning. Beyond this general labeling, we also use the term *conditioning* in some other possibly confusing ways. Let's sort these out through quick reviews of the two other ways that we use *conditioning* in respondent conditioning and the two other ways that we use *conditioning* in operant conditioning, in addition to these names. We still connect our analyses with physics and physiology through energy streams and mediation.

Respondent "Conditioning"

In respondent conditioning the verbal response "*conditioning*" refers to both the *pairing* of stimulus events and the *effects* of the process of pairing stimulus events. Any of these events could actually be behaviors serving stimulus functions as real events. This usage applies to any and all pairings of such stimulus events regardless of whether or not the natural history of, or the occurrence of, the pairing involves the intervention behaviors of humans.

More specifically, in the respondent conditioning context, *conditioning* describes what happens as a result of two stimulus energy streams stimulating the nervous system at about the same time (i.e., a "pairing"). The most obvious case starts with a stimulus energy stream (from the stimulus that we call the *unconditioned stimulus*) that *already* elicits a nervous–system change that mediates, due to genetically produced structures, a response (that we call the *unconditioned response*). Think about the startle response that the puff of air, into one of your eyes, elicits during one of the tests that your eye doctor administers during a comprehensive eye exam.

Sometimes that unconditioned stimulus energy stream occurs at about the same time (a pairing) as another stimulus energy stream from a stimulus that is originally neutral (and which at this point we call the neutral stimulus) in that it fails to elicit that nervous–system change (i.e., it fails to elicit the change that the unconditioned stimulus elicits, the change that mediates the response). One or more such pairings result in an altered nervous–system structure (i.e., the *conditioning*, the pairing, alters nervous–system structure) such that subsequent occurrences of the previously neutral stimulus will also now elicit a nervous–system change that also mediates a response (very much like the unconditioned response). This new nervous–system change occurs due to *conditioning*–altered structures (i.e., due to the occurrence of the two energy streams, the pairing, having the effect of altering nervous–system structure). When this happens, we call the no–longer–neutral stimulus a *conditioned stimulus*.

The conditioned stimulus continues to elicit a nervous–system change that mediates a response, one that we now call the *conditioned response* (which is essentially the same response that the relevant unconditioned stimulus elicits) as long as the conditioning process (i.e., the pairing with the unconditioned stimulus) reoccurs at least occasionally. Think about the startle response that the cold chin rest elicits when you contact it *again* when the doctor asks you to settle down for the test of your other eye. The chin rest was originally a neutral stimulus with respect to startle responses, but then it was present in a pairing with the puff of air into your first tested eye, so now when your chin contacts the chin rest, it elicits a conditioned startle response.

Respondent conditioning summary. In respondent conditioning we actually use the term *conditioning* three ways. (a) We use *conditioning* as the name of the phenomenon itself, respondent conditioning. (b) We also use *conditioning* to refer both to the pairing process, and (c) to the outcome of that process. The process involves the pairing of an unconditioned stimulus and a neutral stimulus, while the outcome is the production of a new function for a previously neutral stimulus.

Operant "Conditioning"

With operant conditioning I will not repeat for you so many of the physiological details the way I did in our discussion of respondent conditioning. Nevertheless, recognize that the stimulus changes in operant conditioning also actually occur via energy streams affecting the physiological level rather than through changes in the stimuli themselves. After all, as in respondent conditioning, the stimuli in operant conditioning do not change. If you give your 18–month–old daughter a crisp $100 bill (an action which I cannot recommend, for several very good reasons) that bill will likely evoke responses like her putting it in her mouth. But if you give it to her when she is 18 *years* old, it will evoke very different responses. Yet it is the same $100 bill; *it has not changed.* But the physiology that is your daughter, especially the nervous system, has changed, due to 18 years of continuous, moment–by–moment conditioning (which continues…). Furthermore, you should carefully and explicitly try to review those physiological changes, at the same level we used in our respondent conditioning discussion, through your own thinking behavior before moving on to the next chapter section. (Completing such exercises builds your understanding—your verbal, intellectual repertoire—far faster than reading repetitive writing, which can even save several pages.)

In operant conditioning the term *conditioning* can refer to either of two different processes. The main usage of *conditioning* refers to the process of certain environmental independent variables changing behavior, regardless of whether or not the natural history of the environment–behavior relations involves intervention behaviors of humans. The other, simpler usage refers to the process of pairing different stimuli (again, regardless of whether or not the natural history of the pairing involves intervention behaviors of humans) that produces conditioned (i.e., secondary) reinforcers and punishers out of stimuli previously neutral with respect to the behavior of concern. Let's consider the simpler usage first.

"Conditioning" conditioned reinforcers and punishers. Recall that the terms *reinforcer* and *punisher* only apply to stimuli that follow behavior. As a result of following behavior, when relevant evocative stimuli reoccur, that kind of behavior occurs more readily, with reinforcers, or less readily, with punishers.

One process in operant conditioning parallels the respondent conditioning pairing process. *Conditioning* refers to the process of pairing two stimuli, a stimulus that *already* has a reinforcing (or punishing) effect on behavior (and that we therefore call the *unconditioned,* or *primary,* reinforcer or punisher) and a *neutral* stimulus that has no reinforcing or punishing effect on behavior. After enough pairings (i.e., after enough occasions when the unconditioned reinforcer or punisher occurs close in time with the neutral stimulus, and sometimes only one pairing is enough) the cumulative result of the pairings is a change *in the function of* the neutral stimulus. The previously neutral stimulus gains the functions of the unconditioned reinforcer (or punisher) with which it shared the pairing process. It now functions as a reinforcer (or punisher) and it continues to function this way so long as the *conditioning* process (i.e., the pairing with the unconditioned reinforcer or punisher) reoccurs at least occasionally. *After* we verify that this function change occurs with the previously neutral stimulus, due to the pairing—*conditioning*—process, we call the stimulus a *conditioned,* or *secondary,* reinforcer (or punisher). Of course, any of these stimuli can be the real event of a response, overt or covert, subsequently serving stimulus functions. Our Law of Cumulative Complexity applies.

"Conditioning" changes in behavior. In operant conditioning the main usage of *conditioning* refers to a process that leads to behavior changing. In this usage *conditioning* refers to the functional effect of certain environmental independent variables, in contingent relationships with behavior, happening in ways that make behavior change. These changes can involve the generation of new behavior and the maintenance of ongoing behavior. *Conditioning* then, in the operant sense, refers to the functional effect of stimuli, especially postcedent stimuli, in changing the nervous system such that it inevitably mediates either more of a previous behavior (leading us to speak of the postcedent stimuli as reinforcers) or less of a previous behavior (leading us to speak of the postcedent stimuli as punishers).

Recall that *postcedent* stimuli refer to stimuli that occur after a behavior (many of which the behavior physically produces). The energy streams from such stimuli induce changes in the nervous–system structure that mediates the kind of behavior that those stimuli have followed. These nervous–system structural changes are such that when energy streams from the stimuli that *evoke* the responses reach these changed structures on future occasions, the structures mediate the responses either more, or less, readily (leading us to speak of the related postcedent stimuli as reinforcers or punishers respectively). Remember, stimuli evoke *all* behavior, even though when they evoke respondent behavior, we instead speak of them as *eliciting* that respondent behavior.

Operant conditioning summary. In operant conditioning we actually use the term *conditioning* three ways. (a) We use *conditioning* as the name of the phenomenon itself, operant conditioning. (b) We use *conditioning* to describe the possibly repeated pairing of two stimuli, an unconditioned reinforcer (or punisher) and a neutral stimulus, with the result that the previously neutral stimulus now functions the same way as the paired reinforcer (or punisher). And (c) we use *conditioning* to refer to the change in evoked behavior, the increase (or decrease) in responding, that occurs when reinforcing (or punishing) stimuli follow responding. Remember that the term reinforcer applies when behavior occurs more often, and the term punisher applies when behavior occurs less often; in both cases we speak of *conditioning* having occurred.

Two–Term and Three–Term Contingencies

As part of introducing the causal role of two–term and three–term contingencies in making behavior happen, let's turn some words into symbols and graphics. First, though, a few words will be helpful regarding the difference between a normal verbal–episode description (spoken or written) and a symbolically written string of contingent relations. We describe the aspects of a behavioral episode as we would narrate a story. We use normal words and phrases to convey the parts, their relations to each other, and how the parts and relations change and develop and possibly even conclude. Understanding an episode, however, benefits from the clarity that comes from brief and concise symbolic representation of those parts and relations and developments. For this we use a simple symbolic notation system. To show the basic conventions of this system, let's develop a simple example.

Beginning with narration, when Jack spots Jill on the other side of several fresh–produce aisles at the grocery store, he politely but somewhat loudly calls out her name, and she smiles and waves to him. To analyze this narrative episode (i.e., to try to increase our understanding of what is happening and how the parts relate to each other) we must first determine whether the behavior that concerns us is Jack's or Jill's. Since we are usually interested in both, we start with the first behavior as our initial behavior of concern, and we portray the relevant events regarding this behavior as briefly and abstractly as possible under symbolic headings typical of contingency descriptions. Thus, in this case, we would write:

$\underline{S}^{Ev}$		Response		Consequence
Jill	$\rightarrow$	"Jill!"	$\rightarrow$	Jill smiles and waves

[This is a three–term contingency in which
Jill = the person, and "Jill!" = the shout of her name.]

Jill's appearance across the aisles is the antecedent stimulus controlling, by evoking, Jack's shout, "Jill!" Jack's shout is the response. And this response produces its consequence, which requires a phrase due to its several parts, namely Jill's smile and wave. Actually, Jack's shout is not the first behavior in this episode; we might have been interested in the behavior of his *seeing* Jill. And we might also be interested in what happens after Jill smiles and waves. All of these constitute possibilities for further examples and analysis.

Other conventions in this kind of symbolic notation include arrows, as well as other symbols that this example lacks. Arrows indicate functional, or contingent, relations and should only be used when such relations are present. If none are present, then no arrow appears between the variables. In cases where a previously operative functional relation no longer holds, then an "x" should appear on the arrow shaft (i.e., —x—>).

If we have not yet identified one of the parts of a contingency, or will not identify a part due to its unimportance or irrelevance to our concerns, we can still include that part by indicating its status with "———?———"; this symbol simply serves as a place–holder for us. It shows that some response or stimulus was an operative part of the depicted contingency. These conventions, and a few additional, rare ones with obvious meaning, appear in our consideration of two–term and three–term contingencies.

Two–Term Contingencies

Recall that in general we analyze contingencies to understand, predict, control, and interpret behavior, and control happens with or without stimulus input involving people, including the control that provides our feelings of happiness and freedom. More specifically, we analyze contingencies to answer two vital questions: First, we ask the basic question: *Why does the behavior occur?* The answer to this first question leads to identifying the independent variables (i.e., causes) of which the behavior, as the dependent variable, is a function. Second, we ask the practical question: *Is access to changing the causal variables available, or must we change analytical levels (e.g., from the internal to the external environment) to make access available?* The answer to this second question determines the possibility of therapeutic interventions, should these be or become necessary, because effective interventions *require* access to the causal variables (as a change in variables leads to different behaviors).

Note that our basic *"Why?"* question can have several different answers. The answer can be only an antecedent stimulus of the eliciting type (S^{EI}) as occurs before respondent behavior. Or, the answer can be just an antecedent stimulus of the evocative type (S^{Ev}). Or it can be both an antecedent stimulus and a consequential stimulus, as in operant behavior with, for example, antecedent evocative stimuli and postcedent reinforcing stimuli. The answer can even be more complex; the antecedents could also include a neutral stimulus (S^N), or even a function–altering stimulus (S^{FA}, about which we will say more).

A respondent contingency example. Since we will examine several operant contingency examples that have antecedent stimuli and consequential stimuli, here is a familiar respondent contingency example with only an antecedent stimulus. If you go into your bathroom in the morning and are facing the mirror when you flip the light switch, you will see (if you are awake enough) your pupils shrink in size. We call this the pupillary constriction reflex, and it is an *unconditioned* reflex, meaning that the response follows the stimulus without any previous pairing of stimuli—that is, without any conditioning—needing to occur first. That is, the stimulus—the increase in illumination—elicits the response—the decrease in pupillary size—due to affecting some neural structures that derive directly from our genetic endowment. Here is how we would analytically diagram this respondent contingency:

Eliciting Stimulus (S^{El})		Elicited Response
Increase in light level	$\rightarrow$	Decrease in pupil size

The presence of this reflex helps prevent blindness. Blindness from its absence could easily have led long ago to a body's becoming old sabertooth's lunch prior to puberty, showing us how contingencies of survival would, in our species history, select for the genes that build the neural structures that mediate this reflex. As a result you would be rather hard pressed to find any human being (read: human body) alive today lacking this reflex. This example shows that we will find some of the causes of human behavior in our species history. Other causes occur in each individual's own life history, often called one's "personal" history. These include *conditioned* respondent reflexes along with a variety of operant variables. Let's consider some of those operant variables.

Operant contingency examples. Ultimately we must answer both our basic *"Why?"* question and our practical *"Access?"* question by making repeated observations with, and changes in, *real* variables, although we may get along with only describing or discussing them when simply completing legitimate interpretative analyses. Of course, in a book like this one, such legitimate interpretative analyses are exactly the point.

So let's begin some interpretative analyses by recycling our Jack and Jill example, first analyzing its parts as a pair of two–term contingencies. Recall that when Jack saw Jill from across several fresh fruit and vegetable aisles at the grocery store, he rather loudly called out her name. Looking up, she saw Jack and smiled and waved to him. Let's analyze several things that are happening.

The first contingency involves the functional relationship between the stimulus, Jill, and Jack's response, "Jill!" This would lead us to write:

S^{Ev}		Response
Jill	$\rightarrow$	"Jill!"

By the way, what if Jack was *not* seeing Jill when he called out her name? What if he was gazing at the perfect watermelon for their patio party, an item for which they were both looking, and called her name so that he could point out this watermelon to her. Jill then smiles, waves and also says, "I'm over here." In this case we would write the first contingency this way:

S^{Ev}		Response
Best watermelon	$\rightarrow$	"Jill!"

This shows that not *all* contingency parts of an episode need to change just because one or another part changes.

The second contingency in our original episode involves the functional relationship between Jack's response, "Jill!" and the consequence, which is Jill's smiling and waving. This would lead us to write:

<u>Response</u>   <u>Consequence</u>

"Jill!" $\rightarrow$ Jill smiles and waves

In our alternate, watermelon episode, the second contingency would only differ slightly. We would write it this way:

<u>Response</u>   <u>Consequence</u>

"Jill!" $\rightarrow$ Jill smiles, and waves,
and says, "… over here."

Sticking with our original episode, if we put these two separate two–term contingencies together, we get our original three–term contingency. As before we would write this pair together this way:

$\underline{S}^{Ev}$ <u>Response</u>   <u>Consequence</u>

Jill $\rightarrow$ "Jill!" $\rightarrow$ Jill smiles and waves

Were you to write out the combined contingencies for the watermelon version of the episode, here is what you would write:

$\underline{S}^{Ev}$ <u>Response</u>   <u>Consequence</u>

Best $\rightarrow$ "Jill!" $\rightarrow$ Jill smiles, and waves,
watermelon   and says, "… over here."

Since these examples move us into three–term contingencies, let's make a more careful analysis with these diagrams to see better what is happening and how the variables relate to each other.

Three–term Contingencies

Contingency diagrams not only depict what is happening, but we can also expand them to indicate how the depicted functional relations operate. Recall that the variables we depict in these diagrams all participate in the exchange of energy traces between internal and external environments and behaviors via the natural nerve, and often muscle, processes that we call mediating behavior.

Diagramming strengthening. The diagrams can convey some of how these processes work without getting into the physiological details. As a temporary aid to help us work through such diagrams, let's add numerical markers (i.e., numbers in circles) to some of them. In general, these numbers follow the sequence of antecedents, behaviors (or responses, if we are being specific) and consequences. Here is a generic three–term contingency:

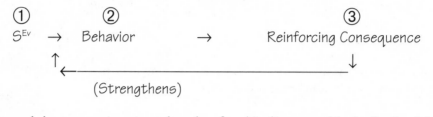

Beyond the conventions introduced so far, this diagram adds the feedback loop that shows the occurrence of the consequence affecting the functional relation between the stimulus and the response. The diagram shows that not only is ② a function of ① while ③ is a function of ②, but it also shows the strengthening effect of ③ on the relation between the S^{Ev} (①) and the behavior (②). We can show this effect by enhancing the first function arrow, giving us this diagram:

*The function of this kind of postcedent ③ is to strengthen the relation between ① and ②. However, other than physiologically, we can detect such an effect only by observing an increase in the occurrences of the behavior ② across future occurrences of the S^{Ev} ①.

Multiple antecedents. What if the number of variables increases, as might occur with more complex human behaviors? Generally, each analytic diagram contains only one behavior or response. With more of these, you need multiple diagrams. Dealing with multiple consequences occurs on a case by case basis, in one or more diagrams as appropriate. Multiple, and functionally related, antecedents, however, are rather common, because some stimuli are effective *only if other stimuli are functional first.* For instance, a stimulus that occurs alone may be neutral (i.e., an S^N) with respect to the behavior of concern, but this stimulus may become evocative if it occurs in the presence of some other particular stimulus. We would designate this other stimulus as a "function–altering" stimulus (an S^{FA}) because its presence alters the function of another stimulus. (You might sometimes see the term "enabling stimulus" [S^{En}] which other researchers use instead of the term *function–altering stimulus*.) Using just our diagrams, here is what can happen.

We start with a lack of functions (i.e., no arrows) since the antecedent stimulus is neutral (so nothing gets evoked):

$\underline{S}^{N}$ Behavior   Consequence

[a neutral stimulus]   [no effect on the behavior]   [so no consequence]

The appearance of a function–altering stimulus, however, brings about changes that can establish a series of functions:

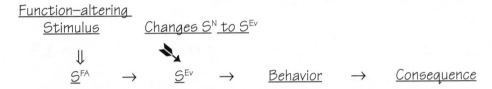

Function–altering
  Stimulus   Changes $\underline{S}^{N}$ to $\underline{S}^{Ev}$

       ⇓

   $\underline{S}^{FA}$ → $\underline{S}^{Ev}$ →   Behavior   →   Consequence

Diagramming combinations. Let's first combine the feedback loop with our Jack and Jill example, and then develop the function–altering diagrammatic conventions, adding the feedback loop again. Here is the first combination:

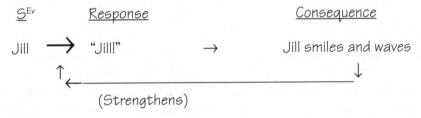

$\underline{S}^{Ev}$ Response   Consequence

Jill ⟶   "Jill!"   →   Jill smiles and waves

        (Strengthens)

To show the function–altering effect, we elaborate the watermelon version of our Jack and Jill episode. Jill evokes Jack's calling "Jill!" *only after* he spots the best watermelon. Jack need not look up and see Jill, because she has remained a present, though neutral, stimulus only a couple of aisles away:

First we have the neutral stimulus, which by definition has no functions:

$\underline{S}^{N}$ Response   Consequence

Jill   [no effect on behavior]   [so no consequence]

Then we add the function–altering stimulus and see the resultant changes:

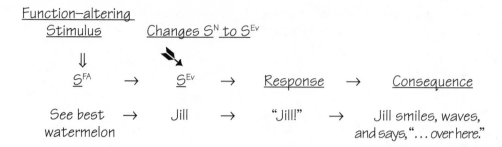

Function–altering
    Stimulus   Changes S^N to S^{Ev}

$$S^{FA} \quad \rightarrow \quad S^{Ev} \quad \rightarrow \quad \text{Response} \quad \rightarrow \quad \text{Consequence}$$

| See best watermelon | → | Jill | → | "Jill!" | → | Jill smiles, waves, and says, "…over here." |

Now let's add the reinforcing consequence feedback loop:

$$S^{FA} \quad \rightarrow \quad \underset{S^{Ev}}{\overset{S^N \text{ to } \downarrow}{}} \quad \rightarrow \quad \text{Response} \quad \rightarrow \quad \text{Consequence*}$$

| See best watermelon | → | Jill | → | "Jill!" | → | Jill smiles, waves, and says, "…over here." |

*Note that the reinforcing consequence strengthens **all** antecedent functional relations.

Our exposure to all these basic contingency–diagram conventions provides a foundation that will make more complex contingencies easier to manage when we come upon them. This will happen as we get deeper into the series of chapters on operant behavior, which help us deal more effectively with the world of behavioral phenomena. While we seldom actually engage such diagrams again in these pages, you can now write them whenever that can help your own understanding.♣

Conclusion

In our "first dentist visit" example from 100 years ago, the child was a victim of respondent conditioning, much of which occurred coincidentally. When a child first goes to the dentist today, dental school training has conditioned the dentist to take precautions not only to minimize unavoidable fear–producing stimuli, but to avoid most coincidental events that could also cause fear reactions regarding future dental visits. And today dentists have better equipment, materials, resources, and procedures than they had a century ago. Furthermore, when something causes these problems today, in spite of their precautions, dentists have adopted a range of standard behaviorological practices for successfully counterconditioning the problem reactions.✑

Chapter 8
Non–Reflex Behavior, Also Known As Operant Behavior:

What kind of behavior requires causal variables both before *and* after it? Some operant causal variables act *before* behavior (i.e., *antecedent* variables). …

$\mathcal{W}$e see ghosts. Scientifically, that statement sounds silly. Perhaps we should say that we think we see ghosts, and wish we had not or did not. Scientifically, however, these statements also sound a bit silly. What is happening? What stimuli make the behavior of "seeing ghosts" happen.

Before we go there, we need some background on the role of antecedent stimuli in making operant behavior occur. Having mentioned antecedents in earlier chapters, here we add some details to our coverage. Our topics begin with the concept of *context,* which covers all aspects of the setting in which behavior occurs. That quickly takes us into the broad area of stimulus control, where we cover three interrelated processes, (a) *evocation,* (b) *generalization,* which has two types, response generalization and stimulus generalization, and (c) *concept formation,* which includes the concept of stimulus class. All of these processes involve various contingencies that affect behavior. Lastly we consider some scientific/behaviorological interpretation regarding seeing ghosts. Right now, though, let's start with the *context* in which behavior occurs.

The Context of Behavior

Behavior always occurs in some context. Even in the vacuum of space, behavior occurs in a context, often in a capsule of some sort (e.g., a part of the International Space Station, a shuttle, or a lunar lander) or just a space suit. Without at least a space suit, behavior in space ceases quite quickly and permanently. The context is the local environment—another name for "space suit" is "environment suit"—and this environment always includes not only a part of what we call the *external* environment but also the behaving organism and its insides, the internal environment. After all, the laws of nature operate on both sides of the skin. Some of those laws of nature govern behavior, which we summarize in our basic behavior formula. This formula constitutes our *starting* point for analyzing the contingencies regarding any behavior:

Antecedent functional independent variables
⇓
Behavior dependent variables
⇓
Postcedent functional independent variables

We refer to the antecedent–variables part of the formula as the *context* in which behavior occurs. Various aspects of context play vital roles in the contingencies controlling behavior. And the context shares control through some interrelationships between antecedent and postcedent variables.

Often *many* contextual variables share control over a behavior, requiring that we juggle as many as possible in dealing with the behavior. Just as we have talked so far as if reinforcers follow every response (which is not the case, as our chapter on reinforcement schedules will explain) we have also talked as if only one antecedent stimulus is involved in evoking a response, which also is not the case, as this chapter explains. We accomplish this juggling by specifying *n–term contingencies* (i.e., contingencies containing as many more terms as we need to finish the analysis, instead of the mere three or four terms in our previous contingencies). These contingencies feature several contextual antecedent stimuli, each of which *alters the function* of the next stimulus until we reach the stimulus that actually evokes the behavior of concern. Equally importantly, *n–term contingencies* also keep the concept of context within science, by setting aside as irrelevant any inner–agent—or other mystical—accounts of context.

With more complex contingencies, the context actually consists of numerous, in–flux combinations of interacting antecedent variables. We manage this complexity by focusing on the generally small number of these functional variables that bear most of the responsibility for the behavior of concern. Meanwhile, to begin to see the roles that function–altering stimuli play, we must first consider some other aspects of *stimulus control.*

Stimulus Control

The term *stimulus control* refers to the broad area of antecedent control of behavior by stimuli. The two basic processes traditionally included under the stimulus–control term are *generalization* and *evocation.*

To understand stimulus–control processes, we need not involve the physiological level of analysis. As you recall, physiology addresses the question of *how* behavior happens while our contingency analysis addresses the question of *why* behavior happens. Nevertheless, reference to physiological events sometimes helps us more fully comprehend what is going on in the contingencies involving antecedent, and even postcedent, events.

When we discuss antecedent phenomena, like generalization and evocation, what kind of physiological events help us understand them better? Basically these antecedent phenomena involve effects that we describe as stemming from stimulus–produced behavior mediation either by genetically produced neural structures or by neural structures that have gradually accumulated structural changes through uncountable conditioning episodes across a human's or other animal's history right up to the present moment. Such episodes, of course, affect postcedent phenomena as well, including the full range of contingencies of reinforcement (e.g., contingencies which involve—separately or in combinations—processes such as reinforcement, punishment, extinction, evocation, and generalization).

Generalization

The phenomena that we describe as *generalization* concerns a pair of observations that parallel our two generalization types, stimulus generalization and response generalization. After conditioning produces a behavior in the presence of one stimulus, we observe that same behavior occurring to some extent in the presence of other similar stimuli; this is *stimulus generalization;* in it the stimuli differ. We also observe that, after a stimulus comes to control one response, that same stimulus also controls other similar responses to varying extents as well; this is *response generalization;* in it the responses differ. As an example of stimulus generalization, after conditioning produces the behavior of a professional astronomer photographing celestial objects with a large observatory telescope, we often observe the similar stimuli of backyard astronomical equipment also evoking her behavior of photographing celestial objects. And as an example of response generalization, after ballistics–formula stimuli come to control the behavior of an artillery unit commander in national defense forces, we might see those same ballistics–formula stimuli evoking her behavior of testing, for fun, the ballistic potentials of various rifle and pistol cartridge loads on the target range. Let's elaborate on each of our two types of generalization in turn.

Response generalization, & response class. Basically, *response generalization involves different responses.* More completely, *response generalization involves the same stimulus evoking different responses.* These different responses share two related and often inseparable characteristics. (a) For one characteristic, these responses are members of the same *response class.* A class of something involves a bunch of things that have something in common. This means that a response class is a group or set of responses that have something in common. What they have in common is the same effect on the environment (e.g., they have the same reinforcing consequence). (b) For the other characteristic, these responses vary in topography, sometimes substantially. This means that the responses vary from one another in physical form, for instance, in one or another of their dimensions, such as direction or speed or intensity.

Here is an example about door handles that shows both of these characteristics, response class membership and topography variation. Normally, the horizontal handle on your door evokes the simple response of grasping it with your whole hand and pushing it downward, because that response produces the reinforcing consequence of the door opening. Of course, other members of the "door opening" response class, although varying in topography, would also produce that consequence. Pushing with any one finger alone, or with a fist, or even with an elbow or a knee produces that consequence. All of these variations in response topography share the similarity that each would get the door open; the door–handle stimulus has an evocative effect on all of them.

The door handle's evocative effect, however, produces a *particular* response topography *under appropriate conditions.* When you arrive at your door toting four full grocery bags, the door handle only evokes the elbow or knee members of the door–opening response class. This happens even though the handle has never evoked such responses before, so reinforcers have never followed them before, in this context. However, as elements of other responses, other stimuli have evoked elbow–use and knee–use responses before, and reinforcers did follow on those occasions. We use the term generalization, response generalization, to describe the occurrence of the elbow–use or knee–use response on this occasion.

Similarly, if you were carrying, not grocery bags, but a big, open–topped box from which groceries were threatening to spill, the door handle would not evoke either the elbow or knee response, either of which could destabilize the box. In this case the handle would instead evoke a verbal response perhaps in this form: "I cannot get the door. Would you please get it?" If, as a result of this question response, the door opens due to your spouse, who was shopping with you, turning the handle, then your verbal response—so utterly different in topography from the other door–opening response class members—proves also to be a door–opening response–class member. Again, we use the term response generalization to describe the occurrence of this verbal response.

Now imagine the generalization process, as part of all our contingency processes (including those we have yet to consider) extending across the board to all types and levels of behavior phenomena. Not only is behavior complex, but we can also account for it with these processes. Humans need no magical processes in their accounts. What about *stimulus* generalization?

Stimulus generalization. Basically, while response generalization involves different responses, *stimulus generalization involves different stimuli.* More completely, while response generalization involves the same stimulus evoking different responses, *stimulus generalization involves different stimuli evoking the same type of response.* The rule of thumb, from the overall research results, is that, while the evocative stimulus involved in conditioning the response evokes the *most* responding, *the more similar a new evocative stimulus is*—in comparison to the evocative stimulus involved in conditioning the response—

the more responding that new stimulus evokes. Conversely, the less similar a new evocative stimulus is, the less responding that new stimulus evokes.

Here is an example that uses hypothetical data similar to the real data from relevant research studies. Let's say you condition responding to occur in the presence of a yellow light. That is, using the evocation–training procedure that we will talk more about shortly, you reinforce responses occurring when the yellow light is on, hence these responses continue, while you ignore any responses when the yellow light is off, hence these responses extinguish. Later, without any reinforcements, you present a test series of lights, each one of a different color, sometimes yellow, red, orange, green, or blue, in a randomized order. And you keep track of whether the light on each trial evokes the response or not. You then plot your data. To keep this example simple (and unrealistically perfect), our hypothetical data show that while the yellow light evoked the response on 99% of its presentations, green and orange each evoked the response on 70% of their presentations, and blue and red only evoked the response on 30% of their presentations.

Figure 8–1 plots those hypothetical data points. The pattern of plotted points traces what we call a *generalization gradient.* Such gradients from actual studies, often much like this one, show that the more similar a new evocative stimulus is to the original evocative stimulus (in this case, a yellow light) the more responding that new stimulus evokes (e.g., orange and green). Conversely, the less similar a new evocative stimulus is, the less responding the new stimulus evokes (e.g., red and blue; indeed, red and blue lights may function more like a light that is off, which would evoke few if any responses).

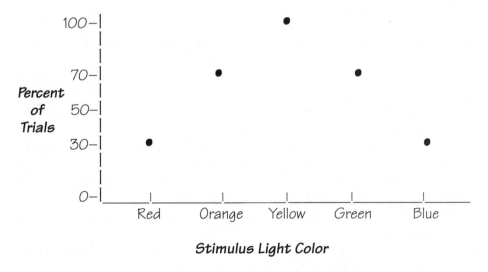

Figure 8–1. The percent of trials on which each stimulus–light color evoked a response (hypothetical).

Now let's turn to our next stimulus–control topic, the process of evocation. Under this topic we can discuss that training procedure we mentioned.

Evocation

While the process of generalization *enlarges* the set of evocative stimulus relations, the process of evocation *shrinks* the set of evocative stimulus relations. The evocation process often reduces the number of stimuli that evoke the response of concern while making the remaining evocative stimuli more effective. Let's consider to what these terms refer.

The stimulus–evocation process. We first mentioned this process when we increased the precision of our basic behavior formula from A–B–cs (antecedents—behaviors—consequences) to A–B–Ps (antecedents—behaviors—postcedents). This prepared us to move on to explicit three–term contingencies (e.g., $S^{Ev}\text{–>}R\text{–>}S^R$). The basic evocation process appears quite simple. Due to relevant past conditioning, the S^{Ev}s (*evocative* stimuli) evoke the response. That is evocation. The "relevant past conditioning" mostly involved stimuli following the responses that the S^{Ev}s evoked. Since we then observed these responses occurring more often, we called these postcedent stimuli *reinforcers*. We can describe that as the reinforcers strengthening the responses in the presence of the evocative stimuli, because that is what we see.

However, your improving behaviorological repertoire benefits from more precision, particularly about the relation between the evocative stimuli and the reinforcing stimuli, even if the details are less visible. In terms of the whole contingency, the reinforcers actually serve a function different from merely "strengthening responses." *The occurrence of reinforcers actually strengthens the relation between the evocative stimuli and the responses that these stimuli evoke.* That is, the energy traces from reinforcers bring about enduring changes in neural microstructures, the microstructures that later energy traces contact, energy traces from future occurrences of the evocative stimuli. As a result of the enduring microstructural changes from reinforcers, the evocative stimuli *more* effectively induce the neural mediation of the response, which thus occurs more often (the observation of which is, again, what leads us to call these postcedent stimuli *reinforcers)*. Also remember that neural mediation includes muscle movement, since muscles contract only when appropriate neurons fire.

All this works with punishers as well, which we say weaken the responses that they follow. However, *the occurrence of punishers actually weakens the relation between the evocative stimuli and the responses that these stimuli evoke.* Speaking more briefly, the punishers bring about enduring changes in neural microstructures, the microstructures that future occurrences of the evocative stimuli affect. As a result of *these* changes, the evocative stimuli *less* effectively induce the neural mediation of the response, which thus occurs less often (the observation of which leads us to call these postcedent stimuli *punishers).*

For example consider receiving dollars after mowing a lawn well (which induces more good lawn mowing) but losing dollars in a fine after jaywalking (which induces less jaywalking). Here is the effect of those consequences, one a reinforcer and the other a punisher, on the related evocative stimuli. Receiving dollars after mowing *strengthens* the function of the neighbor's tall grass as an evocative stimulus for the mowing behavior, while losing dollars after jaywalking *weakens* the function of a street clear of traffic as an evocative stimulus for the jaywalking behavior. Like all behavior, these behaviors are scientifically quite lawful and orderly, obeying contingency relations.

Let's look even more closely at evocative stimuli and reinforcer functioning. With a relevant conditioning history, evocative stimuli transfer energy that triggers changes in nervous system structures (on a moment by moment basis). These changes include the neural firings that *are* the mediation of behavior (neural or motor, and involving more substantial bodily energy resources). Meanwhile reinforcers also transfer energy that triggers changes in nervous system structures, but these changes result in altered structures *that are more enduring,* leaving the body different, such that the later occurrence of the relevant evocative stimuli more easily—but still involving substantial bodily energy resources—evokes the behavior again, which we observe as the behavior occurring more frequently.

Now you can take a turn. That account for reinforcers needs a little rephrasing to turn it into an account for punishers. Try it. Then we go on to consider other details about how evocative stimuli come to evoke behavior.

Evocation training. As we just saw, the effects of the energy traces that we call reinforcers *provide the conditioning history* that makes S^{Ev}s (i.e., evocative stimuli) evoke behavior. The reinforcing stimuli functionally feed energy back into the organism's nervous system, changing it so that the now different nervous system mediates behavior differently. When the S^{Ev} that had evoked the behavior before confronts the now changed organism again, the nervous system mediates the same appropriate behavior differently as in more readily or quickly. (Stimuli called punishers work like those called reinforcers except that the enduring nervous–system structural changes leave the S^{Ev}s *less* effective as evocative stimuli so that the behavior occurs less frequently.) How does this happen? How can we establish a stimulus as an evocative stimulus?

For starters, and sticking with publicly observable events, the environment—the context—always contains numerous potentially functional antecedent stimuli. If a response occurs and produces a reinforcer when a particular antecedent stimulus is present, and this happens fairly consistently, then we see the later appearance of this stimulus as producing—that is, as evoking—this kind of response (which usually produces the reinforcer). We then call this antecedent stimulus an evocative stimulus.

A simple definition of an evocative stimulus implies that function along with hinting how that function occurs: An evocative stimulus is a stimulus in whose presence a reinforcer follows the evoked response. (We say "in *whose* presence," because the stimulus can be, and often is, an organism, including humans.) So how does a functioning evocative stimulus come about?

Research contingencies produced a laboratory procedure for conditioning functional evocative stimuli. The procedure usually involves first conditioning some standard response with reinforcers but without any added antecedent stimuli. Then the procedure continues by presenting a series of trials with two kinds of antecedent stimuli. Presenting these trials *is* evocation training. On some trials, in the presence of a particular antecedent stimulus, reinforcement follows the response. In this case we call this antecedent stimulus an evocative stimulus, and the response occurs more reliably in its presence. On other trials, in the presence of a different antecedent stimulus, *no* reinforcement follows the response. In this case the response occurs less reliably in its presence. We say that the response extinguishes, and we call this different antecedent stimulus an "S^Δ." We have no formal name for this antecedent stimulus other than this symbol. We pronounce this symbol "ess–delta," because "Δ" is the fourth letter of the Greek alphabet, Delta. It corresponds to our English letter "D," which we used early on, in "S^D," for "discriminative stimulus," before some contingencies compelled the use of "S^{Ev}," contingencies that involved "S^D" evoking confusing and incorrect agential responses. (See Ledoux, 2014, Chapter 12, for details.)

Note that the behavior that extinguishes under the S^Δ is the same kind of behavior that continues under the S^{Ev}. With the S^{Ev} present, the behavior that reinforcement follows occurs more often. But with the S^Δ present, the behavior that nothing follows occurs less often. Without reinforcement, this behavior extinguishes. As an example, recall the stimuli in our recent hypothetical generalization–gradient experiment. Our starting steps are pertinent. We would have started by reinforcing (i.e., conditioning) a standard response with each of our half–dozen subjects. Perhaps the experiment conditions presses on a telegraph key by each of our college–student subjects, with points as reinforcers that get exchanged for money. With presses established, we would then have limited the reinforcer occurrences only to presses that occurred in the presence of a light that was illuminated yellow, which would be the S^{Ev}, the training stimulus. In this case, the presence of the non–illuminated light (i.e., the light in the "off" condition) would be the S^Δ. Responding would become well established in the presence of the "on" yellow light, but would extinguish in the presence of the "off" light. Evocation training would then be complete.

That example prompts some other points as well. For one, lots of errors occur in this kind of training (i.e., lots of responding occurs under S^Δ until such responding extinguishes). We will consider a method for *errorless evocation* when we tackle the "fading" procedure in a later chapter.

Another chapter introduces the procedure of backward chaining, which stems from an inherent characteristic of evocative stimuli. We call this characteristic the "dual function of a stimulus." It refers to each evocative stimulus also *becoming a conditioned reinforcing stimulus,* because it is present, in a pairing process, whenever the reinforcer occurs. Thus these antecedent stimuli come to serve two functions, that of an evocative stimulus for the next response and that of a reinforcing stimulus for a previous response.

Also, the example implies that various kinds of stimuli can work as training pairs for S^{Ev}s and S^{A}s. These antecedent stimuli can merely involve the presence *or absence* of the same stimulus. Similarly, these antecedent stimuli can be the same stimulus in either an "on" condition or an "off" condition. Or, we can use two different stimuli, such as a green light as the S^{Ev} and a red light as the S^{A}, or vice versa. Different considerations will determine which of these pairs we see operating in our observations, or use in our experiments (e.g., if our research subjects are color–blind, then using colored lights as antecedent stimuli is definitely inadvisable). Furthermore, the world outside the laboratory is filled with stimuli, some of which function as S^{Ev}s and some of which function as S^{A}s. You need not look far around to find them. Try it.

Are S^{Ev}s the only kind of functional antecedent stimulus? No; another common kind of functional antecedent stimulus that we consider here is the function–altering stimulus, the S^{FA}.

Function–Altering Stimuli

As we noted in a previous chapter, the role of evocative stimuli in controlling behavior expands substantially in the presence of *function–altering stimuli*. These are stimuli in the presence of which other stimuli function differently from the way they otherwise would function. Up to now our discussions have usually implied that only one antecedent stimulus is involved in evoking a response. While this does occur, more commonly a few stimuli, even several, participate in evoking a response, because some stimuli function effectively as evocatives only if other stimuli are functional first (or at the same time). For instance, stimulus B might be neutral with respect to our behavior of concern (i.e., if B occurs alone, nothing else happens; it is an S^{N}). However, if B occurs in the presence of stimulus A, then we may find that B is no longer neutral and instead now functions evocatively. We would designate stimulus A as a "function–altering" stimulus (an S^{FA}), because its presence alters the function of stimulus B, which we would now designate as an S^{Ev}. Several such S^{FA}s might occur, together or one at a time, before some stimulus evokes a response.

To deal with those situations, we expand our analysis from our minimal three–term contingencies to *n–term contingencies* (i.e., contingencies containing as many terms as we need to make a full analysis). These n–term contingencies feature several contextual antecedent stimuli, each of which functions to alter the function of another stimulus until we reach the stimulus that actually evokes the behavior. Let's look at two relatively simple examples.

The first example contains only one function–altering stimulus, and it alters the function of another stimulus *from the status of a neutral stimulus to the status of an evocative stimulus*. At work, a longtime co–worker asks you where you are taking your spouse to dinner to celebrate your anniversary that day. While I encourage you to consider the possible role of that coworker–provided stimulus with respect to certain covert (i.e., neural) responses—in the class we call remembering—regarding anniversaries, let's move on to a more practical matter. You have less than $20 in your wallet, and paying for dinner with a credit card could dampen the enjoyment of the experience, because you and your spouse had agreed to cut back on credit card use. Then, while rummaging in your pockets to check your exact change, you find a stipend check that you had placed there last week. It is for a rather small amount (and, as a verbal shortcut replacing a proper account—with no agency implied—we could say you forgot it). But the amount would more than cover a nice evening out.

How does the check become cash? You usually pass your bank on your way home from work. But usually the bank functions as a neutral stimulus with respect to stopping while travelling home. However, the check alters the function of your bank such that today the bank functions as an evocative stimulus for stopping and cashing the check. This, in turn, provides the cash you need for the night on the town with your spouse. Here is how we would diagram the example:

S^{FA}		$S^N \searrow$ S^{Ev}		R		S^r
Check in pocket	$\rightarrow$	Bank	$\rightarrow$	Cash check	$\rightarrow$	More funds for dinner

How about a slightly more complex example? This second example also contains only one function–altering stimulus, but this case shows another possibility. In this case the function–altering stimulus alters the function of an already evocative stimulus. The S^{FA} alters the function of the S^{Ev} from evoking one behavior to evoking another behavior. A desired video recording on a store shelf generally functions as an evocative stimulus for buying the video. But, for a body conditioned poorly in terms of legal behavior, the occurrence of a power outage while that body is at the store alters the function of the video on the shelf. For that body, with the power out, the video no longer functions as an evocative stimulus for buying it; instead the video functions as an evocative stimulus for stealing it (e.g., slipping it into a pocket and slipping out the door). Here is a diagram of the multi–term contingency in this example.

S^{FA}		$S^{Ev}_1 \searrow$ (for buying) S^{Ev}_2 (for stealing)		R		S^r
Power outage	$\rightarrow$	Video on shelf	$\rightarrow$	Stealing [rather than buying]	$\rightarrow$	Video in hand

As with most of our examples, this example is more complicated than I have presented it. Can you point out how? For instance, consider that the power outage, by distracting the clerks and disabling the security cameras, reduces the threat of punishing consequences for this stealing response.

Furthermore, note that this example also shows us two opposed meanings of the term *responsibility,* as in "people are responsible for their actions." The two meanings are the usual agential meaning and a more recently recognizable scientific meaning. Regarding the agential meaning, supposedly the human being involved is responsible for the bad behavior, because the bad behavior occurred upon the mystical inner–person agent telling the body to take the action; thus we should hold the person—inseparable from the body—responsible, and blame and punish him or her. However, scientifically accounting for the bad behavior in this example leaves no room for such agential claims about responsibility. Nevertheless, regarding the scientific meaning of responsibility, we can say that, as part of the general contingencies operating in society, the behaving body *is* responsible for the evoked action, but only in the sense that the behavior has consequences; the body is responsible in the sense that this body will experience the consequences of the evoked action (e.g., a fine, or jail time), and in this case those consequences are supposed to make such actions less likely in the future. Any notion of responsibility, then, inheres in the contingencies rather than in putative person agents.

Separately, we should ask a question. Will agential notions inform the substance of those consequences (the way they inform the typical consequences that our current scientifically uninformed—with respect to behavior—penal system manages)? Or will science inform them (as Fraley discussed at length in his book, *Behaviorological Rehabilitation and the Criminal Justice System,* 2013)? Sadly, at present our society remains more in the grip of the agential notions.

In yet other ways, this example is more complicated than I have presented it. It typifies one of my early warnings to you, namely that *all* human behavior examples are more complex, and have more to their full accounts, than a beginning book can cover and still be merely a primer.

More about multi–term contingencies. We should also cover more of the potential functional relations present in multi–term contingencies, more of the large number of multiple terms involved in the context of behavior, such as the number we address when we speak of *n–term* contingencies. For instance one S^{FA} alters the function of an S^{N}, making it function as an S^{FA} that alters the function of another S^{N}, making it function as an S^{FA} that alters the function of yet another S^{N}, and so on, until an S^{FA} alters the function of an S^{N}, making it function as an S^{Ev} that evokes a response. And as we saw, an S^{FA} can also alter the function of each of any number of S^{Ev}s from S^{Ev}s for behaviors of one kind to S^{Ev}s for behaviors of other kinds, as the last example's diagram showed for one S^{Ev}. And a single S^{FA} can also, among other things, alter several S^{N}s and S^{Ev}s together. And what about multiple S^{FA}s? Our Law of Cumulative Complexity can lead to these kinds of contingencies all around us. As real events they allow some success at understanding, predicting, controlling, and interpreting our behavior, something that agential or other mystical accounts disallow.

Here is a diagram for one way multiple S^{FA}s can affect multiple S^Ns. "R" stands for some specific response, and the diagram includes the arrows that show the strengthening effect of the reinforcer occurrence (S^r) on *all* of the earlier stimulus functions:

$$S^{FA}_1 \quad \rightarrow \quad {}^{S^N}\searrow S^{FA}_2 \quad \rightarrow \quad {}^{S^N}_n\searrow S^{FA}_n\cdots \quad \rightarrow \quad {}^{S^N}\searrow S^{Ev} \quad \rightarrow \quad R \quad \rightarrow \quad S^r$$

(Reinforcer strengthens **all** antecedent functional relations)

As with similar contingency diagrams, you would read that diagram this way: In the presence of the first S^{FA}, an S^N functions as another S^{FA} in whose presence an *n* number of neutral stimuli function as an *n* number of S^{FA}s until, in the presence of the last S^{FA}, an S^N functions as an S^{Ev} that evokes a response that a reinforcer follows. And the occurrence of the reinforcer strengthens *all* of the previous antecedent functions. Let's apply this conceptual description of contingencies to an everyday example.

Here is a diagram similar to the diagram for our earlier check–cashing basic example of a function–altering stimulus at work. This time the arrows show the strengthening effect of the S^r on *all* the earlier stimulus functions. What we see is that, as a result of the reinforcer occurrence, a future check in the pocket will function more effectively as an S^{FA} that makes an otherwise behaviorally neutral bank (S^N) function more effectively as an S^{Ev} in evoking the "stop and cash the check" responses:

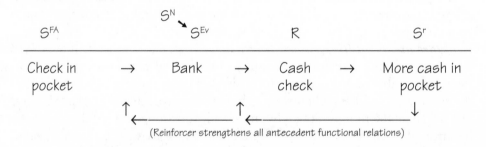

(Reinforcer strengthens all antecedent functional relations)

To expand the example, you can include the evocative role of the co–worker whose question asking (about your anniversary activity) evoked rummaging in your pocket, which turned up the stipend check. This helps us see the rather continuous flux of stimuli and responses that constitutes the contingencies that make our behavior happen.

Consciousness responses and direct stimulus control. In anticipation of the later chapter on details about consciousness, we should at least mention a connection to stimulus control. All the various behaviors of consciousness (e.g., awareness, recognition, comprehension, thinking, problem solving) are strictly covert neural behaviors that data (Fraley, 2008) suggest occur on a single channel, so to speak. That is, apparently one behavior at a time happens with each behavior–capable body part, although behavior can happen with several behavior–capable body parts simultaneously. For consciousness behaviors, the brain comprises the behavior–capable body part. For our familiar overt neuro–muscular behaviors, the behavior–capable body parts include mouth, hands, arms, legs, and so on. Sometimes neural consciousness behaviors add to the stimulus controls of our familiar overt muscular behaviors. For example the bitter cold outdoor temperature forecast makes you *think* about the sore throat you developed the last time you ventured forth in only a thin coat. This combines with the other evocative stimuli for dressing warmly, inducing you this time to wear a heavier, better insulated coat.

At other times neural functioning provides additional controls over overt behavior without the occurrence of consciousness behaviors, under a kind of stimulus control that we call *direct stimulus control*. With direct stimulus control, one need not be at all aware—a consciousness behavior—of function–altering stimuli and evocative stimuli, or even of postcedents like reinforcing stimuli, for any of these stimuli to affect behavior. When function–altering stimuli and evocative stimuli control overt behavior in the absence of additional stimulus controls from, as real events, neural consciousness behaviors, we speak of *direct stimulus control*.

Here is a quick example. Recall that like most people, you have likely found yourself (no inner agents implied) at some point driving along a road when an uncommon stimulus has evoked some heavy braking, such as several turkeys crossing the road in front of you as you reach the top of a hill. At that moment, maybe after a string of silent (or loud) expletives, your neural consciousness behavior of thinking involves wondering (another neural consciousness behavior) how you got there, that is, where the last couple of miles went. During those missing miles, your consciousness behavior involved a *neural–only* response chain that we might call daydreaming, perhaps about a recently attended movie (or some other such daydream–like response chain). Meanwhile *all your usual neuro–muscular driving responses* (e.g., steering) had been successfully occurring under the direct stimulus control of the various stimulus features of the road (e.g., the center line and gentle curves). After all, no accident occurred. Direct stimulus control greatly expands our understanding of the contingencies responsible for much human complex behavior.

Concept Formation

Stimulus control also relates intimately to the notion of a *concept*. Since we have covered some essential components of stimulus–control, we can consider the concept of concepts more closely. The term concept often refers to the basic set of characteristics that cover some topic, such as a kitchen, or a pet, or smaller or larger groups of these (e.g., cats or domesticated animals). In each of these and innumerable other cases, you might have difficulty listing only the necessary characteristics needed for defining the concept, because exceptions confuse the attempt. Yet while we have difficulty *defining* a properly conditioned concept, we still have little trouble "applying" it by correctly indicating whether some possible instance is an example of the concept or not (e.g., a chair with a missing leg is still a chair, although sitting on it is not recommended). This is not because concepts are magical and science is irrelevant to them. Rather it is because concepts have no existence apart from the contingencies covering the related stimulus properties and the responses they evoke and consequate.

In other words, each concept is *in* the contingencies. It is not the outcome of any inner agent gathering information upon which to conjure the concept. Being in the contingencies, concepts have a physical status that inner agents lack. A related reason resides in the processes through which contingencies produce concepts. What may amaze us is the discovery that a description of the concept–producing process in general is simple enough that some researchers have demonstrated the process with other animal subjects like pigeons.

While I generally refrain from non–human examples in this book, I include this example, because the pigeons were under contingencies regarding "applying" the concept of *people*. Let's view the process as an experimental procedure in which the contingencies bring about a concept of people, a procedure that I describe as if a professor had just completed the study. In general this procedure comes from several research articles (with references provided in Ledoux, 2014, p. 294). Some poetic license, within scientific constraints, makes this example as interesting as its title, which Whaley and Malott (1971, p. 179) called "The Pecking Pigeon People Peeper" experiment.

The procedure began with the professor sending out a dozen students with cameras and enough print film—this occurred nearly 50 years ago—among them for about 1,500 pictures in total. The students were told to walk around campus, in and out of buildings, snapping pictures left and right, up and down, close by and far away; just use up the film as this was not an art project.

When all the pictures were developed and printed, the professor and the students had a long meeting in which they first classified every picture as either a *people* picture or a *no–people* picture, and tossed each picture into the appropriate pile. All the *people* pictures somehow had people in them (e.g., tall, short, slim, fat, alone, in groups, clothed, near naked, adults, children, close up, far away, whole or part, and any combination of these in any of numerous positions or activities). None of this applied to the *no–people* pictures, of course. At the end of the meeting, they had two piles of pictures, a people–picture pile and a no–people picture pile. What was the value of two piles?

Stimulus classes. The answer requires something new. Recall our discussion of response classes. Well, welcome to *stimulus classes*. A class of stimuli is any bunch of stimuli that shares something in common. In the case of these two piles of pictures, every picture differed from every other picture in many, *many* ways. But for each pile, all the pictures in the pile shared something in common. In one pile *all* the pictures had people in them while in the other pile *none* of the pictures had people in them. This made each pile a separate stimulus class.

Rounding the numbers to keep the example simple, each pile had about 750 pictures. The students pulled out the 250 pictures on the literal top of each pile. They would use these as training stimuli, leaving some 500 pictures in each pile that they reserved for the crucial last step of the study. The students then proceeded to implement training contingencies with their six pigeon subjects (each of which is a separate experiment under the study's single subject design; for details, see Ledoux, 2014, Chapters 6, 8, & 9). They would present two photos at the same time to the bird, one *people* picture and one *no–people* picture, alternating position in a randomized fashion to prevent picture–type position from becoming part of the contingencies.

The two different picture types each served a different stimulus function. The *people* pictures served as S^{Ev}s that evoked a peck–the–people–picture response (which produced the reinforcer of a grain of food). The *no–people* pictures served as S^{A}s (so no reinforcer followed responses of pecking them). As a result of this evocation training, the pigeons became very accurate (around 95%) at pecking at *people* pictures, and not pecking at *no–people* pictures. (Some of their errors could have been speeding errors, as pigeons are quite fast. Under conditions of some food deprivation, the pecking response started as soon as the gate began to rise to expose the stimulus pictures and, if some part of the S^{A} picture bore similarities to persons, then that would evoke the response, the momentum of which could lead a pecking response to connect with the incorrect picture.)

Now, so far, this is only a description of a sophisticated evocation training procedure, sophisticated in that the training was based not on two *stimuli* but on two *stimulus classes*. Then things got very interesting.

The next step was to dig out all those reserved pictures, the ones which no pigeon had ever seen before, and present each of them as stimuli, in randomly mixed pairs, to each pigeon subject. Would the birds continue to peck accurately only at *people* pictures, and not at *no–people* pictures, when all of the pictures were novel pictures that had never appeared before? Indeed, that is what happened. The birds' accuracy dropped only a couple of percentage points. We would have to say that the birds had "got" the concept of people. Their responses correctly applied the concept to the pictures.

Also, I understand that one particular picture caused a lot of errors for the birds. So the experimenters inserted it as a challenge picture several extra times, and the birds kept pecking it even though the human group of experimenters had classified it as a *no–people* picture. Out of curiosity, someone examined this picture with a magnifying glass and found, in an upper story window in a background building, a person visible. Think about that. The contingency training that these pigeons had received "got" them the concept of people better than the human experimenters had "got" it.

More importantly, *can we say what happened* such that we can say "the birds got the concept of people?" Yes, we can. The differences between the stimulus classes were differentially evoking the birds' responses. At the same time, the similarity that each stimulus shared with all the other stimuli within each stimulus class induced generalization within each class. Remember, each stimulus class had either all *people* pictures, or all *no–people* pictures. So we observed stimulus *generalization occurring **within** each stimulus class* (i.e., all class members of each class affected responding the same way) at the same time as we observed differential *evocation occurring **between** stimulus classes* (i.e., members of one class evoked pecking while members of the other class did not). And such contingencies affect people more often than pigeons. (For more information, see Whaley & Malott, 1971.)

Conclusion: Seeing Ghosts

Stimulus–control concepts also help us with another complex response phenomenon, "seeing ghosts." Ghosts are spooky enough, and seeing them can be a quite real phenomenon, but probably not in the traditional manner that superstitious cultural contingencies condition. As clearly an exercise in interpretation, and taking into account what we know about contingency–produced behavior, we can speculate about some realistic circumstances that could typically produce the behaviors that we label as "seeing ghosts." (Trying to get research dollars, to help fund studies with laboratory analogs of these interpretations, could be the real ghost story.)

Where to begin? Reports of ghosts were more common prior to science winning increasingly more battles with superstition in the 1900s. Indeed, in that century, science and technology took off literally like a rocket. The changes in computer (and other gadget) technology that people currently witness in their lifetime, while amazing, may never equal the changes that people born in the 1890s witnessed during their lifetimes compared to those who came before them. These changes included going from foot or horse or horse–drawn buggy as the most common means of transportation, for several past millennia, to people walking in space and driving a horseless buggy on the moon. Such changes included going from hand–carried letters as the most common means of communication, for several past millennia, to cell phones, the internet, and email. (For a fascinating tale elaborating on these and other categories of extensive change, *as well as the relative lack of change in dealing with behavior scientifically in that same period,* see Joe Wyatt's short novel, *The Millennium Man,* 1997.) So let's consider seeing ghosts in the late 1800s.

Think about the circumstances for people living in the late 1800s. Many people lived within multi–generation extended families, often in the same big house, often on a family farm or business, and often for their whole lives. Thus, very likely, complex but mostly unvarying and repeating patterns of stimuli would evoke similar patterns of responding, all melding into a substantial familiarity and routine. When family members died, however, the stimuli that their presence had provided were removed from the equation. Not so, though, with all the stimuli that, for years to decades—depending on your age—had inevitably occurred along with the presence of those deceased family members. Being paired with this presence, these stimuli not only inevitably participated in the stimulus complexes evoking everyone's behavior while they were alive, but also these stimuli *continued* to participate in evoking everyone's behavior. Since some of the behaviors that these stimuli *shared* in evoking involved often previously seeing the now deceased person, no one should be surprised that these stimuli might on occasion again evoke the behavior of seeing the deceased. Now though, with that person dead, traditional superstitious contingencies induce describing the phenomena as "seeing a ghost," whereas scientific contingencies induce describing the phenomena as current stimuli evoking seeing behaviors similar to what previous stimuli evoked.

Here is a possible example. You live in those times, and in circumstances similar to the ones we described. More precisely, you are now in your late fifties, your mother is in her late seventies and your grandmother—your mother's mother—just died at nearly one–hundred. Your grandmother had always run the household in your experience, although she had "allowed" your mother to handle an increasing number of tasks over the last three decades, just as your mother has been "allowing" you to handle some routine tasks in the last couple of decades. The family house is, of course, filled with many other folks of all ages and various, often overlapping relationships (brothers, sisters, cousins, aunts, uncles, cousins, spouses, and I must mention cousins). All played various roles; all had a share of indoor or outdoor chores, and all affected everyone else's behavior.

Now, ever since you can remember, you knew bedtime had arrived for everyone when you saw a flickering light on the stairway. This was your grandmother making her way, with candle in hand, up the stairs to her room, and checking on everyone else as she went. Nearly every night, for as long as you can remember, you saw the flickering light on the stairway and you would then see her as she passed your door across from the top of the stairs.

But now she is gone. Or is she? Your mother never adopted that nightly check–everyone routine. Yet occasionally at night you see a light flickering on the stairway, and about once each month since her passing, *you see your grandmother* where she used to pass your door across from the top of the stairs. And then, just as in so many times before, she is gone from view. Remember, however, that seeing is a behavior under the control of the same laws of nature that control all behavior.

One need not posit a "haunted house" as the only explanation of your apparitions, your seeing behaviors. Other explanations are more plausible, more parsimonious, even if still interpretive or speculative. What we know about the why and how of behavior supplies at least a reasonable and scientifically grounded, interpretive alternative to the haunted house notion, and may dovetail well with related accounts from other natural sciences.

The stimuli that your grandmother's physical presence and interactions had provided, as part of everyone's contingencies, are now gone. However, *all the other stimuli* that had been present, when your grandmother had been present and interacting and going about the routines that her contingencies compelled, *are still present.* Past contingencies had made these stimuli part of ongoing contingencies, as function–altering stimuli and evocative stimuli and consequential stimuli. After all, life goes on.

Many stimuli had been involved in the past when your *grandmother's presence* evoked seeing her. And now some of these stimuli continue to evoke the behavior of seeing your grandmother even in her absence. Even the stimuli of the presence of *only a vague light* on the stairway can evoke your grandmother–seeing response. That vague light may only be someone else going by candlelight back to bed after visiting the newly installed, indoor flush toilet downstairs, at about the time at night when everyone has gone to bed, which is when you had so often in the past seen your grandmother disappearing at the top of the stairs. Unsurprisingly, that vague light at that hour again evokes the response of seeing your grandmother.

You are not seeing a ghost. Without implying any ghostly inner agents in you, you are simply again behaving the seeing, under different evocative stimuli, that you behaved in the past under the evocative stimulus of your grandmother's presence.

You will need to think more about those possibilities, because space limitations here allowed only a speculative, bare–bones interpretation. (Pun intended, I guess. As I write this, Halloween is only a couple of weeks away. That shows the subtle impact that such contextual, and multiple–control variables can have on our behavior.)♣

References (with some annotations)

Fraley, L. E. (2008). *General Behaviorology: The Natural Science of Human Behavior.* Canton, NY: ABCs.

Fraley, L. E. (2013). *Behaviorological Rehabilitation and the Criminal Justice System.* Canton, NY: ABCs.

Ledoux, S. F. (2014). *Running Out of Time—Introducing Behaviorology to Help Solve Global Problem.* Ottawa, CANADA: BehaveTech Publishing.

Whaley, D. L. & Malott, R. W. (1971). *Elementary Principles of Behavior.* Englewood Cliffs, NJ: Prentice–Hall. This is a classic primer on behaviorological science and applications that students found fun to read. Some of them could not put it down!

Wyatt, W. J. (1997). *The Millennium Man.* Hurricane, WV: Third Millennium Press. The bibliography may contain additional descriptive comments on this fascinating story.ᑉ

Chapter 9
Operant Behavior—II:

How can variables that follow a response affect that kind of behavior? Some operant causal variables act *after* behavior (i.e., *postcedent* variables). ...

*W*hat makes so many people around the world rise early, work or study or perhaps create (e.g., art) all day, recreate a bit in the evening, then sleep much of the night, only to repeat these patterns the next day, generally five days each week—with different but still relatively fixed patterns on the weekends—for most of the weeks each year, year after year? This question describes lots of behavior, though without much detail, and human behavior certainly contains so many more possibilities. At the end of this chapter, you will know how much it has expanded your possible answers.

To really get a grip, however, on what may well be the most extensive set of contingencies responsible for human behavior, one needs to become conversationally familiar with the most common types of contingency variables and their effects. The last chapter helped you grasp the antecedent variables, the ones coming before behavior. Now we turn to the postcedent variables, the ones that come after behavior. We call the two main categories *reinforcers* and *punishers.* The point of this chapter revolves around arranging conversational familiarity about the various types of reinforcers and punishers, how they work, and what effects they cause. Several subsequent chapters then elaborate a range of processes that provide procedural interventions using many of these antecedent and postcedent contingency variables. Applying these procedures can result in more of the kinds of behavior society and individuals demand as beneficial now as well as for human survival later (e.g., in child rearing, education, work, entertainment, science, and even global problem solving).

Becoming conversationally familiar presents little difficulty. Indeed, if you want difficulty, then study meteorology. The huge number of interacting variables needed merely to predict tomorrow's *weather*—which even weather scientists cannot currently *change*—vastly exceeds the number of accessible variables needed to predict a substantial number of everyday human behaviors. Many of these contingency variables remain relatively easy to change. And *changing these variables* is behavior that happens with everyone everyday as an ordinary part of life. By knowing more about behaviorology, the behavior that changes these variables becomes more based on science instead of coincidence or superstition, which leads to more beneficial improvements in behavior.

Basic Contingencies in which Behavior Occurs

To begin building your conversational familiarity, here is a little preview of reinforcers and punishers as postcedent contingency variables. Recognize that reinforcers remain our focus, because the side effects of punishers cause problems—covered in a later chapter—that make punishment generally ill–advised. In any case postcedent variables divide into various types according to four criteria that determine these four characteristics: (a) the status of these variables as reinforcing or punishing, (b) their status as consequences or coincidental "selectors," (c) their status as added or subtracted, and (d) their status as unconditioned (i.e., primary) or conditioned (i.e., secondary).

While we will mention all four of the criteria that determine these four characteristics, which can form various combinations, the first characteristic holds the most important information for understanding behavior. Examples will follow later in the chapter. Here, however, we will be brief.

Our first criterion concerns this question: Did the occurrence of the postcedent stimulus lead to more of the kind of behavior it followed, or less, in the future? If more, then we call the stimulus a reinforcing stimulus (i.e., a reinforcer). If less, then we call the stimulus a punishing stimulus (i.e., a punisher). This effect on behavior is our most important piece of information.

Our second criterion concerns this question: Did a response *produce* the postcedent stimulus—reinforcing or punishing—or did the stimulus happen coincidently? If a response produced the stimulus, then we call it a consequence. If the stimulus happened coincidently, we call it a coincidental selector.

Our third criterion concerns this question: Did the postcedent stimulus occur as an addition to the situation, or as a subtraction from the situation? If it occurred as an addition, then we call it an added reinforcer or punisher. If it occurred as a subtraction, then we call it a subtracted reinforcer or punisher.

Our fourth criterion concerns this question: Did the postcedent stimulus require previous conditioning to be effective? If it needed no previous conditioning, then we call it an unconditioned—or primary—reinforcer or punisher. If it required some previous conditioning (i.e., pairing with another stimulus) then we call it a conditioned—or secondary—reinforcer or punisher.

Those four criteria–driven characteristics summarize the most common kinds of contingencies causing behaviors. Before moving to combinations and examples, here are some other considerations.

"Added" and "Subtracted"

Instead of added or subtracted to describe both reinforcing and punishing postcedent variables, some natural scientists of behavior use the older terms *positive* or *negative*. Both "added" and "positive" describe variables that the contingency *presents* after a response occurs, while both "subtracted" and "negative" describe variables that the contingency *withdraws* after a response occurs. In either case, presentation or withdrawal, if that kind of response subsequently occurs *more* often or maintains, we speak of reinforcers, and if that kind of response subsequently occurs *less* often, we speak of punishers.

People would have little difficulty being clear about positive and negative reinforcers, and about positive and negative punishers, if "positive" and "negative" only had the meanings of present and withdraw. This, however, is not the case, because these words also have connotations in everyday language usage that compete with the technical "present" and "withdraw" usage. In everyday usage *positive* connotes good or pleasant while *negative* connotes bad or unpleasant. And the everyday usage occurs far more often in everyday experience. As a result people get confused. They have some difficulty with the concept of a *negative (Bad?)* reinforcer strengthening behavior. They have even greater difficulty with the concept of a *positive (Good?)* punisher weakening behavior. They have trouble imagining much that is positive about punishment. The resulting confusion can take years to overcome, and even some professional authors still occasionally get the terms wrong due to the confusion. As a solution to such problems, the related contingencies compelled behaviorologists to adopt the terms *added* and *subtracted* in place of positive and negative, because these words only reflect what has happened to the stimuli (added and subtracted) when they produce their effects.

Hang on. What is wrong with just using "presented" and "withdrawn?" Actually, these words would work well. However, the contingencies that drive the use of the symbolic notation for contingencies have induced using "+" and "−" and this gives using "added" and "subtracted" some slight advantage.

"Nature" and "Natural"

We are examining some natural variables (i.e., variables in nature) that experiments have demonstrated as adequate natural–science explanations regarding behavior. Also, throughout this book, as natural scientists, we deal with variables in nature that are functionally related to behavior. So we should explicitly revisit the words *nature* and *natural,* and clarify our usage of them from more common usages. One use of *natural* pertains to a general notion of *nature* as just the "great outdoors," what you experience on vacation "away from it all," mountains and forests and rivers and skies and plants and wild animals. Another use extends this initial notion to include *everything*—asphalt and cities and airplanes and cell phones, all matter and energy—*except behavior.* As natural scientists, however, we mean much more than any such usages. For starters *we include all things behavioral* as well, because no characteristic about behavior warrants excluding it from the rest of everything else in the universe. Furthermore, for us the word *natural* (e.g., as in natural variable, natural science, natural scientists) refers to the fundamental approach, to *all* real aspects of the universe, that involves searching for and applying objective and measurable principles, concepts, and laws. All of these—with respect to *everything* including behavior—are *of nature* as in affording no abiding place for mysticism or superstition as independent variables. This also means that mysticism and superstition are also researchable topics of natural behavior science (i.e., behaviorology). As well, saying, "all behavior is natural," is to say that all behavior is amenable, at its level of analysis, to the same laws of nature (i.e., natural laws)—of energy and matter and cumulative complexity—that affect everything else at all their levels of analysis.

So now, in this chapter, we examine some natural variables that affect behavior. We begin by considering the concept of contingencies among environmental and behavioral variables in general, both antecedent and postcedent, all of which involve energy exchanges traceable between physical and physiological structures and processes. Then we focus on facets of the most common variables on the postcedent side of our behavior formula, the variables and processes that we call *reinforcers* and *punishers* as well as *extinction*.

Environment–Behavior Contingencies

We introduced the concept of contingencies with what we called the "A–B–Cs" (antecedents–behaviors–consequences) of behaviorology. These contingencies pertain mostly to operant behavior and operant conditioning. After appreciating the convenience of the A–B–C initials, we introduced "postcedents" in place of consequences, because, like antecedent, postcedent is merely a term of placement in time. Thus we have A–B–P. The "A–B–P" formula improves to explicit contingencies such as "S^{Ev}–>R–>S^R" (with R as a response, and S^R as a reinforcing stimulus). We call this a "three–term contingency." It includes two "two–term" contingencies ("S^{Ev}–>R" and "R–>S^R") that we have seen before. When we put real variables into the places held by these symbols, the contingency formulas help us see the functional connections among the variables. This provides a basis for understanding, predicting, controlling, and interpreting behavior with natural–science contingencies rather than with mystical or superstitious agents (e.g., stars or selves).

Contingency Characteristics

Subject to the usual experimental verification, the most common antecedents in operant contingencies appear to be the stimuli that evoke responses, and the most common postcedents appear to be reinforcing stimuli. While we focus on postcedent stimuli in this chapter, a quick review of their connection with antecedents cannot hurt. Let's review these connections in terms of evocative stimuli and reinforcers.

With a relevant conditioning history, evocative stimuli work by transferring, to neural receptors, energy that triggers changes in nervous–system structure (on a moment by moment basis involving the more substantial energy resources of the body). These changes include the neural firings that constitute the mediation of neural and motor behavior. Meanwhile reinforcers work also by transferring, to receptors, energy that triggers changes in the nervous system. These changes, however, are altered structures *that are more enduring* and thereby leave the body different, such that the reoccurrence of relevant evocative stimuli more readily—which we observe as more frequently—evoke the behavior again, all this also involving the substantial energy resources of the body. Thus, the effects of the energy traces that we call reinforcers *provide the conditioning history* that makes S^{Ev}s evoke behavior. That is, the reinforcing stimuli functionally feed energy back into the organism's nervous system, changing it so that the now different nervous system mediates behavior differently. When the evocative stimulus that had evoked the behavior before confronts the organism again, the nervous system mediates the same appropriate behavior more easily or more quickly. Stimuli called punishers work like those called reinforcers except that the enduring nervous system structural changes leave the S^{Ev}s *less* effective as evocative stimuli so that the behavior occurs less frequently.

Put another way, the reinforcers functionally feed energy back into the organism's nervous system, *changing* it physically enough to make a difference, so that the now different organism mediates behavior differently. When the type of antecedent stimulus that had evoked the response before confronts the *now changed* organism again, the energy exchanges compel the nervous system to mediate the same kind of response more readily or quickly. Stimuli earn the title *reinforcer* when we observe their effect as enhancing the functional evocative control that the antecedent stimulus exerts on the behavior. We cannot call a stimulus a reinforcer until *after* we have observed this effect. And we call the process *reinforcement*. We see the behavior occurring more often, and we describe that by saying the consequence reinforced the behavior. Similarly, stimuli earn the title *punisher* when we observe their effect as diminishing the functional evocative control that the antecedent stimulus exerts on the behavior. We cannot call a stimulus a punisher until *after* we have observed this effect. And we call the process *punishment*. We see the behavior occurring less often and we describe that by saying the consequence punished the behavior.

Note that we find observing the energy trace connecting postcedent stimuli to the nervous system generally more obvious and easy to follow—more accessible—than the energy trace connecting antecedent stimuli to the nervous system. For example, from the back of a classroom, we can easily see a student at the front facing the class and reciting various numbers each of which is followed by one or another student raising a hand in a reinforcing "thumbs up" gesture that produces similar gestures from the whole class. Tracing the energy from these postcedent gestures to the reciting student's eyes is relatively easy; everyone in the room sees the gestures. But what evokes the number reciting? Finding out takes investigation, which may show that some class members are holding various simple multiplication problems, printed on cards (with answers on the back) toward the reciting student. These problems are providing the stimuli evoking the recited numbers. But finding this out takes considerable effort (i.e., is less accessible) particularly since the light energy reflecting off the cards into the reciting student's eyes is not as directly observable (e.g., is not line–of–sight) as are the thumbs–up gestures. The reciting student sees the problems on the cards, but no one else can see them. Similarly, in applying a therapeutic intervention, successfully arranging a postcedent, reinforcing stimulus occurrence after a target behavior is generally easier than successfully arranging an antecedent evocative stimulus for that behavior. For the evocative stimulus to be as successful in affecting the behavior as the reinforcer, separate behaviors such as head orienting toward, or eye contact with, the evocative stimulus may require conditioning first. Adding such steps is more complicated than presenting a food reinforcer after the behavior.

Fact and theory differences. To talk about some theories, we should first review what we mean by reinforcer and punisher. "Reinforcer" is the term we use when we not only observe a stimulus occur right after a behavior but we also observe that behavior subsequently occurring *more* often. "Punisher" is the term we use when we not only observe a stimulus occur right after a behavior but we also observe that behavior subsequently occurring *less* often. Using these terms derives from experimental demonstrations. We cannot technically call any stimulus a reinforcer or punisher until after the required experimental demonstration of the definition–related outcome (more behavior, for "reinforcer," and less behavior, for "punisher"). This leaves the status of stimuli as reinforcers or punishers determined according to observed facts.

Restricting reinforcers and punishers to observed fact differs from *theories* about how reinforcers or punishers work, which rely on the self–correcting nature of science for improvement. For scientific legitimacy, any such theory needs to include the physiological mediation of energy exchanges between the environment (internal or external) and behavior. And experimentation must be capable of proving any such theory false. This is good, because experimental findings help us deal more effectively with the world around us, improving the quality of our lives and the chance that we (i.e., our species) will survive.

Science, Economics, and Contingencies

In science, however, no finding is ever the last possible word, because we always have additional details to examine, and we can always investigate back another link in the functional chains leading to any event, at any relevant level of analysis. Thus we might at any time also discover an entirely new phenomenon, with implications for other phenomena that we might have thought we already understood. In the experience of scientists, past contingencies always provide plenty of reasons to suspect that some completely new and beneficial discoveries are likely always awaiting scientific attention. And all this is part of what we mean by the self–correcting nature of science.

Some effects of economics on discovering functional relations. Nevertheless, due to economics, scientific discovery and self–correction can be expensive processes. Since the scientific research that leads to discovery also involves at least testing, replication, and development—all requiring sources of monetary support—discovery can carry quite an economic price tag. Yet past reality shows that funding always has limits. We always reach some point at which uncovering and accounting for the remaining independent variables, responsible for a behavior, becomes too costly, at least for that moment (or decade or century).

Generally we have to settle for knowing just enough about behavior phenomena to deal with them at an acceptable level of effectiveness. For example, we may not yet have teased out every variable partly responsible for every percentage point of the observed variations in human–subject responding on certain schedules of reinforcement. But we should know enough to enable managing those schedules effectively when an intervention requires it.

Similarly, throughout this book we describe many known independent variables of behavior that we understand functionally at that kind of acceptable level. Skinner and others reported the original discovery of these variables in books, and in journals like *Journal of the Experimental Analysis of Behavior* (which started in 1958). All that research continues. However, when the next step needs to go beyond the available funding, we may face being unable to account for the remaining variables, and variation, due to our economics–induced ignorance, rather than to any capriciousness about behavior, or to any of those inadequate explanations we surveyed in previous chapters. As research funding continues, so too the search for further functional details continues.

The generic meaning of "contingencies of reinforcement." We describe each of the wide range of functional relations between environments and behaviors as a contingency, but we lack a separate name for every possible contingency type. Many postcedent contingencies involve reinforcers while others involve punishers. Some antecedent contingencies involve evocative stimuli while others also involve function–altering stimuli. And the list goes on. Rather than name each type, we use a convenient verbal shortcut for all these contingencies. We generically refer to them as *contingencies of reinforcement.* We manage the small but inherent confusion simply by pointing out, repeatedly if necessary, that this term covers *all* behavioral contingencies including the majority far beyond those that directly involve only reinforcers. For example, the term "contingencies of reinforcement" even covers extinction contingencies.

One reason we can use such a generic term is that all these contingencies share in common the fundamental participation in the exchange of energy traces between internal and external environments and behaviors. This occurs via the natural nerve and muscle processes that we call *mediating* behavior. This physical reality underlies each and every variable and behavior in each and every contingency that we consider.

Postcedent Contingencies

Early in this chapter, we covered four criteria questions whose answers led to particular characteristics of various postcedent stimuli. We did not, however, provide examples then. The time has come for a quick review of those postcedent stimuli, now with examples.

Jargon Problems and Solutions

Stimuli, events, and processes. Certain problems with some terms demand our attention. For one, we tend to speak of stimuli as if they were *things,* and in some ways this makes sense. But the reality is more complex. To respect the interconnection between behaviorology and physiology, the term *stimulus* actually refers to an event, an energy change that affects a functioning receptor cell (or bundle of such cells) at a nervous system entry point. We can call the thing, related to that energy transfer, a stimulus. But this is a convenience, a verbal–shortcut.

The difference between things and events arises in another context as well. For example, we usually speak of the thing, money, as a reinforcer. However, money is not always a reinforcer; sometimes it is a punisher. To get right whether some instance of money is a reinforcer or punisher, technical accuracy—which we sometimes need—demands that we clarify what is *happening* to the money, and that means addressing money the *event,* rather than money the *thing.* For instance, if we *present* extra money to the high school senior who mows our lawn because she did a particularly good job (and find that she takes extra care on future occasions) then that *presentation* of money constitutes a reinforcer. On the other hand, if a police officer *withdraws* some money from your wallet because you were speeding—this is what happens in some jurisdictions (although later you can argue your case in court)—and we find that you speed less on future occasions, then that *withdrawal* of money constitutes a punisher. The example of electric shock works the opposite way. Normally we think of shock as a punisher, but only the *presentation* of shock is a punisher. Thus we can also honestly say that shock is a reinforcer, but only in that the *termination* of shock is a reinforcer. As a result of these examples, we can see that any particular *thing* can end up as either a reinforcer or a punisher, depending on what happens to it, depending on the added or subtracted status of it as an event in contingencies.

We should also consider how we linguistically denote things and events and processes and procedures. In the case of postcedents, we use the word ending "–er" to denote *things,* hence reinforc*er* and punish*er.* Sometimes we treat events this same way, but the form "–ing event" is common also, as with reinforc*ing event* or punish*ing event.* On the other hand, for *processes* or *procedures,* we use the ending "–ment," hence reinforce*ment* and punish*ment* as processes or procedures. By the way, *processes* occur with or without human behavior involvement. The term, *procedures,* however, distinctly implies, even requires, human behavior involvement.

Postcedent Stimuli as Types of Reinforcers and Punishers

With those jargon problems solved (and almost put behind us) let's review some types of postcedent stimuli, now with examples. Note that a stimulus might earn several partially overlapping designations. In this way you greatly increase your conversational familiarity with some of the contingencies that cause behavior. (If you like full–blown, picky details, see the Glossary.) Our little review repeats our four criteria questions in slightly different forms.

(1) Does subsequent behavior increase or decrease? The answer determines whether you call the stimulus that follows the evoked behavior a *reinforcer* or a *punisher.*

When a stimulus follows an evoked behavior, and that kind of behavior subsequently occurs at the same level (which we call maintenance) or more often, then we call the stimulus a reinforcer. For example, as a result of a teacher placing a star on a young pupil's just completed, and correct, sheet of simple math problems, the pupil asks for and completes several more sheets right away. Thus, the stars earn the title of reinforcer.

On the other hand, when a stimulus follows an evoked behavior, and that kind of behavior subsequently occurs less often, then we call the stimulus a punisher. An example would be a dog on a chain barking loudly and trying to attack a letter carrier who crosses a side lawn instead of staying on the sidewalk. As a result the letter carrier stays off the lawn by using the sidewalk in the future. Thus, the dog's barking and attack earn the title of punishers.

(2) Does the evoked behavior produce the reinforcing or punishing stimulus, or does the stimulus occur coincidently? The answer determines whether you classify the reinforcing or punishing stimulus that follows the evoked behavior as a *consequence* (i.e., of the reinforcer or punisher variety) or as a coincidental selector (i.e., selecting behavior to occur more, or less, often).

When the evoked behavior *produces* the stimulus, then we call the stimulus a consequence. For example, when a dark room evokes flipping a light switch that produces a well lit room, then we call the well lit room a consequence. (If flipping a light switch occurs regularly when rooms are dark, then we call the well lit room a consequence of the reinforcer variety.)

However, if the stimulus merely occurs by coincidence after the evoked behavior, and the frequency of that kind of behavior changes, then we call the stimulus a *coincidental* selector (i.e., a *coincidental reinforcer* or a *coincidental punisher* that the evoked behavior *did not produce*). For example, consider that a poor exam score easily evokes the behavior of sad, downcast eyes on the walk home. On that walk you happen to spot a $20 bill that you cheerfully grab from under a pile of snow where it was sticking out. You would not have seen the $20 bill while looking straight ahead. Then, as a result of finding the $20 bill, you cheerfully walk with downcast eyes the rest of the way home (ready, of course, to spot any other $20 bills). The exam–score–evoked behavior of downcast eyes *did not put* the $20 bill under the pile of snow. Its presence was a function of some other set of events and so was merely coincidental with the downcast eyes. However, as a result of collecting it, your downcast–eyes behavior increased, even though this behavior did not produce (i.e., did not put in place) the $20 bill. Thus we call the $20 bill a coincidental reinforcer. By the way, we call the increased downcast–eyes behavior *superstitious* behavior, which a later chapter will cover in detail.

   (3) Did the postcedent stimulus occur as a presentation or as a withdrawal after the evoked behavior? The answer determines whether you classify the reinforcing or punishing stimulus that follows the evoked behavior as an *added* stimulus or as a *subtracted* stimulus.

   When the reinforcing or punishing stimulus occurs, after the evoked behavior, as a presentation (with or without human behavior involvement) then we describe the stimulus as *added.* For example when Jack rather loudly called out Jill's name from across the fruit and vegetable aisles, she *smiled and waved* to him, and subsequently he began calling her name rather loudly even when she was right next to him. This *smile and wave* classify as *added* reinforcers.

   On the other hand, when the reinforcing or punishing stimulus occurs, after the evoked behavior, as a withdrawal (with or without human behavior involvement) then we describe the stimulus as *subtracted.* An example would be a *reduction of penalty points* (perhaps that you earned from disciplinary infractions in class, such as cell–phone texting during a lecture). These penalty points were to be reduced by completing extra pages of practice word problems. However, you helped a student after class so your teacher reduced them. Subsequently you helped students after class regularly. *Reducing the penalty points* classifies as a *subtracted* reinforcer. Notice that this subtracted–reinforcer involves an event rather than just a thing. This expands the range of what can constitute postcedent—or even antecedent—stimuli in contingencies.

   What about *added punishers?* For example, on another occasion when Jack quite loudly called out Jill's name from across the several fruit and vegetable aisles, she *scowled at him and shushed him.* Subsequently he no longer called her name so loudly. *Scowling* and *shushing* classify as *added* punishers.

   What about *subtracted punishers?* An example would be a child's friends simply *leaving the room* when the child starts to cheat at the card game they are playing, and subsequently the child no longer tries to cheat. *Leaving the room* classifies as a *subtracted* punisher.

   (4) To have its affect of making the evoked behavior happen more often or less often, does the postcedent stimulus have to undergo some conditioning first? The answer determines whether you classify the reinforcing or punishing stimulus that follows the evoked behavior as *unconditioned* (i.e., as *primary)* or as *conditioned* (i.e., as *secondary).*

   What if the answer to question four is no…? If the answer is "no, the postcedent stimulus need not undergo any conditioning…," then we classify the reinforcers and punishers as *unconditioned.* Here are some examples.

   As an example of an **unconditioned reinforcer**, let's say you fail to unplug a lamp with a burned out bulb before changing the bulb. If, when changing the bulb, your hand contacts both the new light bulb base and the socket at the same time, then you receive an electric shock. (*This* classifies as an added punisher. Why?) However, the *shock stops* as soon as the quick movement of your hand breaks contact with either the bulb base or the socket. Subsequently your hand moves quickly away from metal lamp parts whenever it even brushes against them. The *shock stopping* needed no prior conditioning to increase your hand–removal behavior. So it classifies as an unconditioned reinforcer.

That example actually features the same "stimulus" functioning as both an added unconditioned punisher and as a subtracted unconditioned reinforcer. The same stimulus, electric shock, works as an added unconditioned punisher *when it starts*. However, it works as a subtracted unconditioned reinforcer *when it stops*. This shows us that *what happens* with a "stimulus" determines how it affects behavior and gets classified. Most stimuli operate similarly. Some work as added punishers when presented, and then work as subtracted reinforcers when they stop. Others work as added reinforcers when presented, and then work as subtracted punishers when they stop.

Also, note that for people, types of added unconditioned reinforcers are often roughly summarized by counting them on the fingers of one hand as food, water, sex, salt, and attention. We count salt separately from food, because people who are otherwise food–satiated often continue through half a bag of salty snacks after a meal, suggesting that the reinforcing capacity of salt endures beyond the satiation of food intake. Similarly, we can observe other animals, also otherwise food–satiated, still walking miles to a salt–lick location.

As another example of an **unconditioned punisher**, while you cut some roses for indoor appreciation, *the thorns prick your hands*. Subsequently you only admire the roses on the bush, rather than cutting them for indoors. The thorn pricks needed no prior conditioning to reduce your behavior. So they classify as unconditioned punishers. (Is this example an added, or a subtracted, unconditioned punisher?)

What if the answer to question four is yes...? If the answer is "yes, the postcedent stimulus needs to undergo some conditioning...," then we classify the reinforcers and punishers as *conditioned*. We get conditioned reinforcers and punishers when we apply the conditioning process by pairing neutral stimuli (i.e., those having no effect on the behavior) with, respectively, other reinforcers and punishers. Again, this characteristic is sometimes needed for analysis, and at other times it is not so needed. Consider some examples.

Here is an example of a **conditioned reinforcer.** A costume party invitation evokes wearing to the party a rather ugly, wildly colored scarf, one that had been hanging around the back of your closet for some years. At the party many wonderful verbal compliments occur about the wildly colored scarf. The scarf subsequently appears around your neck every other day at work for weeks. For years the scarf had languished in your closet as a neutral stimulus. Then at the party it got paired with all those compliments. Without that conditioning the scarf would likely still be languishing in your closet. Due to that conditioning, however, the scarf has become a conditioned reinforcer.

An example of a **conditioned punisher** would be a variation of our previously discussed *occurrence* of penalty points following disciplinary infractions such as cell–phone texting during class, with fewer such infractions occurring subsequently. Originally, points were neutral stimuli having no effects on behavior. However, after lots of extra written work got paired with the points, they became conditioned punishers. (Other procedures produce better outcomes than this problematic one, as a later chapter clarifies.)

Now, you have greatly increased your conversational familiarity with some basic contingencies causing behavior. So here is an outline–style summary.

A postcedent outline with symbols. Let's outline postcedents so that you can see both their hierarchical relationships and their interrelationships. This outline includes the common symbolic notation for each of the four basic reinforcers and the four basic punishers. In this notation, "S" indicates "stimulus." Upper case "R" indicates "unconditioned reinforcing" while lower case "r" indicates "conditioned reinforcing." Upper case "P" indicates "unconditioned punishing" while lower case "p" indicates "conditioned punishing." Also "+" and "–" respectively indicate added and subtracted. For example, "S^{r-}" reads as "subtracted conditioned reinforcing stimulus."

Note that, as in mathematics, we seldom write the "+" sign in the symbolic notation. It only appears here (and only in parentheses) for easier understanding. Furthermore, since the eight symbolized terms in this outline apply as much to *coincidental selectors* as to *consequences,* one can conclude that these eight actually represent a total of 16 types of postcedent stimuli:

I.   Antecedents (the other category of variables).
II. Postcedents.
   A. Selectors (a word that here means "affecting subsequent behavior").
    BOTH 1. Consequences. ***AND*** 2. Coincidental selectors.
    (a). Reinforcers.
      (1). Added reinforcers.
        [a]. Added unconditioned reinforcing stimulus [$S^{R(+)}$].
        [b]. Added conditioned reinforcing stimulus [$S^{r(+)}$].
      (2). Subtracted reinforcers.
        [a]. Subtracted unconditioned reinforcing stimulus [S^{R-}].
        [b]. Subtracted conditioned reinforcing stimulus [S^{r-}].
    (b). Punishers.
      (1). Added punishers.
        [a]. Added unconditioned punishing stimulus [$S^{P(+)}$].
        [b]. Added conditioned punishing stimulus [$S^{p(+)}$].
      (2). Subtracted punishers.
        [a]. Subtracted unconditioned punishing stimulus [S^{P-}].
        [b]. Subtracted conditioned punishing stimulus [S^{p-}].

(In Ledoux, 2014, you can find a *Postcedent Matrix* on p. 158. Also, you can find a branching *Postcedent Tree Diagram* on p. 159. For more details see Ledoux, 2015, pp. 199–204.)

Postcedent Contingencies that Change Behavior

Due to your growing conversational familiarity with basic contingencies, we can now look at some further details about three fundamental postcedent processes that change behavior. Here are more details regarding reinforcement, extinction, and punishment. Recall, however, that we sometimes simplify phrasing by using verbal shortcuts implying that stimuli reinforce or punish people. Actually, though, reinforcers and punishers only affect behaviors.

Other Types of Reinforcers
Various basic and applied researchers have at times used a number of other terms to describe some aspect of reinforcers, often as a counterpoint to some other reinforcer characteristic. Let's consider a couple of these terms here, noting that some folks use these terms to describe punishers as well.

Physical/tangible versus social/verbal reinforcers. We sometimes use the terms *physical* or *tangible* to refer to reinforcers that have more than a fleeting presence. For instance a bag of cookies, a bottle of wine, a beautiful painting, even a 50–dollar bill all have some continuing presence (unless your 50–dollar bill is paying for petrol) and they often leave some sort of record of their occurrence. For example the painting might hang on the wall for years, and cookies leave crumbs. In contrast a mother's attention, a teacher's praise, a peer group's interactions, or a colleague's compliment all have a fleeting presence. They occur quickly and just as quickly are history, often without leaving any record of their occurrence. And yet these fleeting postcedents can have as potent effects on behavior as their more concrete cousins have. In any case we sometimes use the terms *social* or *verbal* when describing these fleeting reinforcers. For both the enduring and fleeting reinforcers, the usual terms apply as well (i.e., added or subtracted and unconditioned or conditioned).

Coincidental reinforcers. As a quick review, we label a reinforcer as *coincidental* when the preceding, reinforced response did not produce the reinforcer. Some other natural process coincidentally led to the occurrence of the reinforcer. It happened coincidentally, hence the label *coincidental reinforcer.* And we use the term *superstitious behavior* for the behavior that such coincidental reinforcers condition.

Generalized (conditioned) reinforcers. Recall that an otherwise neutral stimulus becomes a conditioned reinforcer when it is regularly present when another reinforcer—usually an unconditioned reinforcer—occurs. This pairing, this conditioning, makes the previously neutral stimulus effective as a *conditioned* reinforcer.

That effectiveness, however, relies on the effectiveness of the unconditioned reinforcer. As long as the unconditioned reinforcer functions effectively, then the conditioned reinforcer also functions effectively. If some process, such as deprivation of the unconditioned reinforcer, increases unconditioned–reinforcer effectiveness, then conditioned–reinforcer effectiveness also increases for any stimulus paired with the unconditioned reinforcer. Similarly, if satiation of the unconditioned reinforcer reduces unconditioned–reinforcer effectiveness, then conditioned–reinforcer effectiveness suffers also. Conditioned–reinforcer effectiveness relies on the deprivation *level* of the unconditioned reinforcer with which it shares some pairing.

For example, if you usually pair small (i.e., pea–sized) pieces of dog biscuit, which are unconditioned reinforcers, with the *clicks* of your "clicker"—what as children we used to call a hand cricket—then the clicks become conditioned reinforcers. Using the clicker then makes your dog–tricks training sessions long and productive, because the "click" reinforcer can occur *immediately* after the selected behavior, with the food morsel coming next. Such "Clicker Training" currently involves the most effective set of practices for training other animals. For example see O'Heare, 2015, and the clicker training books by Karen Pryor and her colleagues (e.g., see Pryor, 1999, 2001). However, what if you run out of dog–biscuit morsels? If you then instead pair cubic–inch chunks of steak with the *clicks* of your hand cricket in a training session, then that session will not be very long or very productive, because your dog will rather quickly become food satiated. As a result the clicks will lose their reinforcing function, and your dog will go bed down and sleep off that great steak meal.

One way to get around the problem of conditioned reinforcers losing effectiveness (e.g., due to the occurrence of unconditioned–reinforcer satiation) is to pair the conditioned reinforcer with several, even many, other reinforcers. In this way one or another of those other reinforcers is likely always to be effective due to some functional level of deprivation. Hence the conditioned reinforcer also remains effective. In this circumstance we call the conditioned reinforcer a *generalized* reinforcer. (Actually, it is a *generalized conditioned reinforcer*, but by convention we leave out the middle word.)

As an example of a generalized reinforcer, think of a stimulus, in this case a tangible stimulus, that our culture essentially pairs with just about everything, at one time or another, legal and illegal, so much so that this stimulus has earned a designation as "the root of all evil." What could that be? If you saw dollar signs ($$$) you are right on the money.

Also, we treat the stimuli that serve as the "tokens" in token economies the same way (e.g., at home and in classrooms, and even in some workplaces). We pair the tokens with many other reinforcers, making the tokens function as generalized reinforcers. These could be buttons, poker chips, stars, check marks, tickets, or any number of other items.

More about Extinction

While we mentioned extinction in earlier chapters, we cover several additional aspects of extinction here. These include comparisons of the extinction process with the process of forgetting and the procedure of preclusion. Since both extinction and punishment decrease behavior, we discuss their connections more thoroughly when we move on to discussing punishment.

One type of extinction is *respondent extinction,* and we use this full two–word term, "respondent extinction," when we talk about this kind of extinction. While respondent extinction is not our focus at this point, it is the kind of extinction in which a respondent conditioned stimulus becomes unable to elicit the conditioned response, because it occurs too often *without* the unconditioned stimulus ever again occurring with it. As a result it returns to the status of a neutral stimulus with respect to the response.

However, the type of extinction we focus on here is operant *extinction,* a process for which we simply use the single word "extinction." This kind of extinction is one of the operant environmental–change contingencies that brings about the reduction of—and ultimately, if the process continues, the cessation of—the behavior of concern. In this kind of extinction, the reinforcers cease. That is, the reinforcers that have at least occasionally been accompanying or following occurrences of the behavior of concern *stop occurring.* They no longer accompany or follow occurrences of the behavior. As a result, a decrease occurs in the rate of the behavior. This decrease eventually leads to the cessation of the behavior as extinction continues.

Let's connect antecedent and postcedent processes with extinction. Without any reinforcers occurring, the effectiveness of the evocative stimulus diminishes, and ultimately it no longer evokes the behavior of concern, leaving the behavior of concern extinguished.

Here is an example of an experience with the extinction process that a student, now colleague, of mine had. Meghan Curry, who had already completed several behaviorology courses, reported this experience in class in 2012. She was at a park one day helping a mother with a rather precocious toddler around two years old. They had been there for a couple of hours and the child was thirsty. That is, a couple of hours of water deprivation had produced two related effects. The water deprivation (a) had momentarily raised the effectiveness of water as a reinforcer, and (b) had momentarily raised the effectiveness of any stimulus that evokes any drink–procuring behavior, such as a water fountain. In essence, the water deprivation momentarily raised the likelihood of any behavior that in the past had produced the water reinforcer.

In any case, Meghan was not surprised when, as they passed a water fountain, the thirst compelled the child to say, "I want some water." However, besides providing water, this water fountain also functioned in a non–standard way. Any sideways pressure on the spout would swivel it, sending the stream of water in another direction.

Now, the child could not both reach the flow–control button and drink (or swivel) at the same time, so Meghan helped him by pushing the flow button for him. When the water started, though, he would start to swivel the spout toward her; a quick swivel could get her wet. We can hardly resist the implication that getting her wet would be a more potent reinforcer of his behavior than drinking the water, at least for the moment. Perhaps, at some point in the past, getting water on a helper got paired with other reinforcers thus making it a conditioned reinforcer. We seldom have the kind of detailed access to past events that would enable us to specify when and how any such conditioning had occurred. Meghan, however, always released pressure on the button when he started to swivel the spout, which stopped the water flow before she got wet. For lack of any further reinforcements, however, after just three such unsuccessful swivelling attempts, the swivelling extinguished and the child simply drank the water coming from the spout.

All this gets even better. Having observed this sequence, the child's mother asked Meghan how she so quickly got him to just drink and stop swivelling the spout at her, and all without scolding him. Meghan's immediate reply was "What?..." Her button–off responses, from previous conditioning in misbehavior situations, had been automatic, that is, under *direct stimulus control* of the overt components of the situation. The mother's later inquiry evoked some covert observing responses repeating her timely button releases, and these responses chained to some overt responses reporting those timely button releases to the mother (e.g., saying that she simply released the flow–controlling button each time he began to turn the spout toward her, which kept any reinforcers from following swivelling). We will cover such processes in more detail in the chapter on consciousness.

Extinction Versus Forgetting

Both extinction and forgetting are *processes* that differ in an important way. In the extinction process, stimuli initially continue to evoke the previously reinforced behavior, which thus continues to occur. But no further reinforcers follow those occurrences, thereby weakening the evocative function of the antecedent stimuli. As a result they fail to evoke the behavior, which thus finally ceases to occur (i.e., extinguishes). We saw this happen in our swivelling–water–spout example. However, in the *forgetting* process, *no stimuli evoke the previously reinforced behavior* and so no further reinforcers occur *because the behavior fails to occur* (although the reinforcers would occur, if the behavior did occur). In the forgetting process, the neural structures that otherwise mediate the behavior gradually undergo normal and cumulative physiological changes that degrade the evocable mediation of that behavior until finally any still–occurring, previously evocative stimuli cannot evoke the behavior, which then cannot occur. For example, if educational contingencies conditioned you to recite a particular poem by rote in high school, and since then essentially no stimuli have ever evoked another recitation of that poem, then the stimulus of this sentence now is unlikely to succeed in evoking the complete rote recitation. It may even prove insufficient to evoke the name of such a poem. The responses have become unavailable due to the normal and cumulative physiological changes that degrade the evocable mediation of that behavior.

Extinction Versus Preclusion

Preclusion differs from extinction in a way similar to forgetting. While forgetting is a *process,* and extinction can be either a process or a *procedure,* preclusion is an intervention procedure that we can apply to help solve some types of behavior problems. Basically, preclusion takes advantage of the forgetting process. In *preclusion* the practitioner makes environmental changes that preclude any stimuli from evoking the previously reinforced behavior of concern. As a result the neural structures that otherwise mediate the behavior gradually undergo normal and cumulative physiological changes that degrade the evocable mediation of that behavior. Finally the behavior cannot occur, although the use of the preclusion procedure seldom continues for this long.

Why use this procedure? Because the practitioner may lack access to the usual reinforcers of the behavior of concern, and so may not be able to change the environment in ways that prevent those reinforcers from occurring. Hence the practitioner could not employ the extinction process as a procedure. However, with the preclusion procedure, the practitioner need not stop the reinforcers that usually follow the behavior of concern, because the preclusion procedure keeps the behavior from occurring. If the behavior cannot occur, then it cannot produce the reinforcers. Remember, *in extinction stimuli evoke the behavior* (and it occurs) at least initially, but it produces no reinforcers. However, *in preclusion,* as in forgetting, the behavior *does not occur,* because no stimuli evoke it (so of course no reinforcers follow it).

For example, let's say that the behavior of concern is a high school sophomore's incessant tossing of crumpled paper sheets in high arcs toward the corner waste basket while the teacher is writing lesson points on the board. Of course, many coercive techniques are available for attempting to manage this kind of problem. The results seldom solve the problem. Indeed, coercive techniques usually multiply the problems. Still, some teachers will readily practice coercive techniques even with less obnoxious problems than this one. But better options are available that avoid the even more obnoxious side effects of coercive practices. (We address all these topics in a later chapter).

One of those other options is a version of the *preclusion* procedure. A teacher has precious little access to changing the reinforcers for the tossing behavior, reinforcers often of the *bootleg* kind that come from all over the classroom in the form of peer attention. (Bootleg reinforcers are reinforcers that usually are not a part of, and often interfere with, an intervention.)

A preclusion procedure sidesteps all that. The teacher could, nonchalantly and without missing a beat in her presentation, simply move that waste basket to an out–of–sight location, perhaps under her desk. This step precludes the waste basket inviting, I mean, evoking, crumpled paper tossing. Given the short time frame of this example, we can only speculate whether or not precluding *this* evocative stimulus would lead to forgetting the tossing response. I rather doubt that *that* would happen. But this procedure would still solve this teacher's problem (at least until other evocative stimuli occur that lead to another sample of sophomore silliness surfacing).

More about Punishment

Both the extinction and punishment processes bring about reductions in behavior. However, the different ways in which they operate can cause some confusion. Let's consider some more aspects of punishment before comparing it with extinction to reduce the confusion.

In many ways the reinforcement and punishment contingency processes demonstrate flip sides of the same behavioral coin. They produce opposite changes in behavior and involve opposing respondent effects. And the effect of a stimulus switches from one side of the coin to the other (e.g., from reinforcer to punisher) simply by "occurring" the opposite way. For instance, a stimulus that reinforces when it starts becomes a punisher when it stops. As an example, *receiving* dollars immediately after mowing the neighbor's lawn increases (reinforces) the mowing behavior, whereas *losing* dollars through a fine immediately after jaywalking decreases (punishes) the jaywalking behavior. In this example the dollars *occurred* in opposite ways, but the dollars are still dollars and themselves do not change (with the real changes occurring physiologically, as we have described).

Actually, a more accurate description of the effect of receiving those dollars (after mowing the lawn) is the strengthening of the function of the neighbor's tall grass as an evocative stimulus for the mowing behavior. Similarly, the effect of losing some dollars (through a fine after jaywalking) is the weakening of the function of a street clear of traffic as an evocative stimulus for the jaywalking behavior. Remember, the stimuli do not change. The functions change due to energy traces changing the body physiologically.

Other Types of Punishers

Having already looked at basic punishment contingencies (i.e., added and subtracted punishers, and unconditioned and conditioned punishers) now let's look at other punisher types. The other terms that some basic and applied researchers use to describe some aspects of reinforcers also pertain to punishers. Thus you will sometimes hear of *physical* or *tangible* punishers. At other times you will hear of *social* or *verbal* punishers. Of course, the use of these does not replace the names for the usual punisher types. Furthermore, while the pairing of some conditioned punishers can occur with many other punishers, no consensus has yet emerged about a "generalized punisher" term. Similarly, while punishers can occur that the preceding response did not produce, no consensus has yet emerged about a "coincidental punisher" term. Additional basic laboratory research will address these considerations in due time.

Punishment and Extinction

Another area that needs more research is a certain confounding between punishment and extinction. But this problem presents particular methodological difficulties. Here is basically how this problem presents those difficulties.

Behavior that comes under a punishment contingency already occurs, which means that it is already under a reinforcement contingency that may have generated it and in any case now maintains it. (Of course, behaviors are usually under more complex sets of contingencies, but we will ignore that for now, to keep coverage of this problem relatively simple.) Thus, with the reinforcement contingency still operating, any punishment contingency decreases the behavior by working against the reinforcement contingency. We describe this outcome as resulting from a sort of *algebraic summation* of the effects of the competing contingencies (see Skinner, 1953, Chapter 14, pp. 218*ff*). However, this means that if the punishment contingency ends, the behavior will rebound due to the continued operation of the reinforcement contingency, which in some cases starts by producing a response rate higher than the rate prior to the start of punishment, a phenomenon that some call "over recovery."

On the other hand, if the punishment contingency begins and at the same time the reinforcement contingency ends, then the punishment need not compete with the reinforcement. Generally in this situation, the punished behavior not only reduces a little more quickly than when the contingencies are in competition, but it also is less likely to recur should the punishment contingency stop.

The confusion stems from that lack of reoccurrence of the behavior, when the reinforcement contingency ended early and then the punishment contingency ends. Why does the behavior fail to recur? Does it fail to recur due to the effectiveness of the punishment? Or does it fail to recur due to the lack of reinforcement (i.e., is the failure to recur due to the extinction contingency)? This question ("With no reinforcement contingency, is failure to recur due to a punishment contingency or due to an extinction contingency?") remains for ongoing research to answer. Natural sciences *always* have questions to answer.

Conclusion

This chapter began with a question about how to account for the day by day behavior patterns of much of humanity. I think that even now you are in a good position to begin providing some reasonable answers to that question. Please try it for your own pleasure. This indicates an increasing understanding of the contingency accounts of why human behavior happens.

Since the question mentioned art as an important component of general human behavior, let's elaborate on art just a little. Here is part of how I described a scientific take on art in a previous book (Ledoux, 2016): Any dictionary provides a range of definitions of "art." These are the definitions that ground most discussions and books on art. The realities of my "career" life as a professor of behaviorology, however, induce a different kind of definition. As the discipline that studies why behavior happens, behaviorology can provide definitions of art that, while not actually inconsistent with the standard definitions, are also consistent with the findings of natural sciences.

One such definition of art would state that, scientifically, art is the novel products of, and the conditioned production of, responding induced by a wide range of environment–behavior contingencies, in an equally wide range of media. These contingency causes may or may not produce reinforcing effects from the uses of the art (i.e., its functions) but they indeed produce emotionally reinforcing effects, for others as well as the artist, that typically evoke the human verbal response of "beautiful."

We can now add that the term "reinforcing effects" in this context refers to effects that make the art–production responses and art–appreciation responses occur again later under similar conditions. Behaviorology can say so much more, not only about art and life but also about scientific answers that begin to address many ancient human questions, and that can contribute to solving global problems as well. The more everyone knows about behaviorology, the less likely it can be misused. Later chapters begin to provide such answers for some of these ancient questions and problem solutions.♣

References (with some annotations)

Ledoux, S. F. (2014). *Running Out of Time—Introducing Behaviorology to Help Solve Global Problem.* Ottawa, CANADA: BehaveTech Publishing.

Ledoux, S. F. (2015). *Origins and Components of Behaviorology—Third Edition.* Ottawa, CANADA: BehaveTech Publishing.

Ledoux, S. F. (2016). *Beautiful Sights and Sensations—Small Collections of Native American and Other Arts.* Canton, NY: ABCs.

O'Heare, J. (2015). *The Science and Technology of Animal Training.* Ottawa, CANADA: BehaveTech Publishing.

Pryor, K. (1999). *Clicker Training for Dogs.* Waltham, MA: Sunshine Books.

Pryor, K. (2001). *Clicker Training for Cats.* Waltham, MA: Sunshine Books.

Skinner, B. F. (1953). *Science and Human Behavior.* New York: Macmillan. The Free Press, New York, published a paperback edition in 1965.℘

Chapter 10
Operant Behavior—III:

Can a precise behavior develop from a bunch of different responses? A postcedent variable consideration, "response class," and the related process/application, "differential reinforcement" …

*Y*ou play basketball well. Returning rebounds earns you many points. Your free throw results, however, remain dismal. You practice and get better, although you think that outcome is magical. Would you like to know what really happens during practice to produce improved performance? Even the simplest application of contingencies, the process that we call differential reinforcement, involving only reinforcement and extinction contingencies, plays a role in the success of practice in producing improved performances.

Our understanding of why human behavior happens begins now to go in the additional direction of engineering applications and interventions. We want to know more than merely about why behavior happens, more than merely about what causes human behavior, more than merely about contingencies. We want to know about *changing* the contingencies, the causal variables, in ways that can solve behavior problems in both local and global arenas. And we want to make sure that this scientific knowledge gets used by anyone or everyone only for the good of each and all of us. The best assurance that this happens arises from everyone knowing this science at least to the same level that they know other natural sciences from their high school education.

For this additional direction of engineering applications and interventions, we begin with the simple process of *differential reinforcement* and explore how this becomes a procedure. Later chapters explore some related contingency applications of slightly greater complexity.

Supporting Concepts

Our explorations would benefit, however, from a quick review of two supporting concepts. One concerns the difference between behavior and responses, which leads to "response classes." The other involves the status of attention, which is one of the most common kinds of reinforcing consequences.

Behavior and response differences. In practice, we must differentiate between behavior and responses. While *behavior* is the general term, we use the term *response* for particular instances of a behavior. Responses occur under the control of currently operating variables. This makes every response a new, and thus different, response. This could be a problem, because science usually deals with repeatable phenomena. We find, however, that our subject matter also involves repeatable phenomena through *response–class* members.

A class of anything means a group of different things that each share something in common with all the rest. Groups of responses naturally fall into response classes on the basis of sharing the same effect on the environment, the same stimulus consequence, the same reinforcer. That is, we call a bunch of responses a response class when all the responses differ in topography but produce the same postcedent effects. Thus members of response classes show repeatability, so *our experimental analysis operates on the level of response classes.* For example, each time you strum your guitar on a G^7 chord, the strum differs from other instances in topography. Although only ever so slightly compared to other instances, nevertheless the hand position differs, the hand pressure on the strings differs, the angle of the hand in relation to the strings differs, and so on. Yet in each case, the postcedent sound that you and others hear is essentially indistinguishable from the other strumming instances (unless other contingencies are producing greater differences in responses *and their consequences* that actually constitute different response classes). Sharing those same consequential sounds thus makes all those instances of strumming into repeating members of a single response class that we might simply call a "strumming response class," a response class that we could study.

As another example, consider the many different responses which could produce an open door. These could include turning the handle with a hand or any one of your four fingers, which already constitutes five different members of the door–opening response class. One can also open a door with a thumb or an elbow or a knee or even a mouth—no, not that way. (Well, I suppose you *could* open a door that way.) But I am referring to opening the door with the *words,* "Please open the door" when an appropriately conditioned audience is present. Regardless of the differences in topography in all of these responses, all of them are members of a door–opening response class, because they all produce an open door.

Some contention over attention. Some contention exists over attention as an *unconditioned reinforcer.* Unconditioned reinforcers are stimuli that function as reinforcers without prior conditioning. On the other hand, *conditioned reinforcers* are stimuli that do not function as reinforcers until they have occurred along with other reinforcers. Such pairing (i.e., conditioning) makes them function as reinforcers. I find that distinguishing between *attention* as an unconditioned reinforcer, and *praise* as a (verbal) conditioned reinforcer, helps people respond correctly to the distinction between unconditioned and conditioned reinforcers.

As a verbal behavior, praise can affect a child as a reinforcer only after a year or so of accumulating the effects of verbal–behavior (i.e., language) conditioning. Attention, however, affects a child as a reinforcer much earlier, probably as an unconditioned reinforcer. That is, attention might affect a child as a reinforcer without any prior pairing with other reinforcers (which is the kind of conditioning that would otherwise make attention a conditioned reinforcer). Confusingly, verbal comments, as in our example of praise, also have attention components.

Even though praise and verbal compliments and comments may be inseparable from some element of attention, they can easily be shown to be conditioned reinforcers (as they depend on prior verbal–behavior conditioning). However, while attention *can* be conditioned (i.e., attention often does occur with other reinforcers) trying to show that attention *must* be conditioned is much more difficult. Parents who attend in various ways to young infants when they cry are often left distraught when crying increases "for no reason" (as they describe it, although the actual reason involves the attention coincidently reinforcing the crying). Was such attention already paired with other reinforcers, making it a conditioned reinforcer? Perhaps. Perhaps not.

However, in my experience, describing "attention" as an unconditioned reinforcer and "praise" as a conditioned reinforcer definitely helps people distinguish between attention and praise. These descriptions also help people distinguish between unconditioned reinforcers and conditioned reinforcers. This educational consideration is presently valuable even if the self–correction of science might later show that "attention is a conditioned reinforcer."

The Place of Procedures and Interventions

The concept of response class participates in a contingency process that can also serve as a procedure. In earlier chapters we considered many basic properties and characteristics of contingencies causing behavior. We also described some details of many functional antecedent and postcedent variables along with brief forays into the interrelations of these variables with each other and behavior.

Those details included the processes of stimulus control including response generalization and stimulus generalization along with evocation, evocation training, evocative stimuli, function–altering stimuli, and some of the ways that all of these interrelate (e.g., concept formation). Elsewhere we stressed the basic nature of behavior as natural neuro–muscular behavior or just neural behavior, whether walking or talking or emoting or thinking or remembering or any other behaviors, operant or respondent. In all cases behavior lacks any explanatory need for the mysticism or superstition of any inner agent. Instead we focused on behavior both (a) as the always–new product of current stimuli, and (b) as real events that serve stimulus functions for other behaviors. We even had some fun with ghosts, or at least with one scientific, interpretative accounting for them.

Now we introduce some of the combinations of those basic contingencies, combinations that provide us with processes that operate in more complex ways and so produce more complex behaviors and more complex changes in behavior. We also consider how to harness these processes as intervention procedures to begin the contingency engineering of our environments to interact with physiology (i.e., bodies) to produce needed behaviors.

Of course contingencies have always compelled humans to engage in such engineering practices with respect to behavior. But mystical foundations limited both our understanding and the results. Perhaps now, upon the natural–science foundation of behaviorology, we can all increasingly partake of interventions with less damage from misunderstandings and misuse. And more benefits accrue from the improved effectiveness that typically occurs when our efforts deal with behavior naturalistically, scientifically, without mystical input.

Differential Reinforcement

We begin our contingency–engineering foray into procedures and interventions with *differential reinforcement.* This application "merely" involves a rather simple combination of reinforcement and extinction processes. We use this name, differential reinforcement, because the procedure involves reinforcing some responses differently from other responses.

Actually, to describe how differential reinforcement works, we must invoke another previously introduced concept, *response class.* A response class refers to a group of responses that have in common the same effect on the environment. The classic notion is that the members of a response class all produce the same kind of reinforcement. So, how does the differential–reinforcement procedure work? How does it involve the concept of response class?

Differential reinforcement works by providing one member of a response class with a different consequence from the consequence that follows other response–class members. *Differential reinforcement involves reinforcing the response–class member that constitutes the behavior of concern while extinguishing (i.e., not reinforcing) all the other response–class members.* As a result this procedure makes the particular behavior of concern occur more often, *so long as it already happens at least occasionally.* If this behavior of concern never occurs (i.e., if it is not at all occurring currently) then we cannot reinforce it. So we could not apply differential reinforcement. We would then need some other procedure to make it occur more often. Consider some examples.

Differential Reinforcement Examples

Very likely we are all interested in helping others. Sometimes we do this just by talking. If we understand differential reinforcement, we are likely to be of greater help, naturally and automatically, even in these ordinary, everyday encounters. Here is what can happen.

Helping a friend. Say you are talking with a friend who makes a range of verbal responses related to the personal effects of some hard times. These responses include comments that indicate some comprehension of the problems he is experiencing and some reasonable steps toward solutions, plus comments that reasonably clarify his feelings about himself and any involved others. On the other hand, his responses also include comments that indicate some pessimism, some confusion, some apprehension, and some mis–comprehension regarding his problem–related behavior (including feelings).

Now, any of your friend's verbal responses could evoke verbal responses on your part that, at least as attention, could function as reinforcers for your friend's responses (i.e., your friend's responses could all be members of the same response class, all producing attention from you). However, I severely doubt that you would want to reinforce *all* of those responses. Indeed, I am sure you would not want to reinforce certain ones, as that could be harmful. What would instead be helpful? Since these responses are already occurring, we bring differential reinforcement to the rescue.

You would be helping your friend even if your main reactions only involved following his problem–comprehension responses, and his emotion–clarifying responses, with some warm and empathic verbal and postural responses of understanding and agreement on your part. Meanwhile, *you would also be helping your friend* if you followed his pessimistic, confusing, apprehensive and mis–comprehending responses with verbal and postural responses of relative indifference or silence, which essentially means ignoring (i.e., extinguishing) these problematic, negative responses. Reinforcing these negative responses would only further encourage them (i.e., condition them) which would be harmful for your friend.

With these steps you are actually applying differential reinforcement. You might even witness some relatively immediate verbal indication of improvement on your friend's part. The general lines of this example also apply at the professional level (see Truax, 1966). If your friend's problems were severe, he would likely need more help, perhaps even from an applied behaviorologist. Many applied behaviorologists prefer to practice under a label different from "applied behaviorologist," as this label puts the stress on the *basic science* rather than on engineering applications. So they prefer to practice under the label ABA (i.e., Applied Behavior Analysis) and many have achieved Board Certification as BCBAS (i.e., Board Certified Behavior Analyst).

With some folks, mystical or superstitious verbal behavior occurs due to psychology related contingencies. These folks have claimed that behavior–focused procedures, like differential reinforcement, only affect "symptoms" and ignore the more deeply rooted problems that they assume and insist must exist. These "deep" problems—that the machinations of some miscreant inner agent putatively causes—supposedly cause the symptoms. And these "real," deep–rooted problems will surely manifest again through new symptoms in a sort of "symptom substitution" scheme. However, such claims are grounded in mystical assumptions, and credible substantiation of these claims has not occurred. (Again, see Truax, 1966. This 50–year–old article remains a worthy read, although readers must recognize that it comes from near the end of the time period during which the natural science of behavior and psychology were sharing their history. For more on that period, see Fraley and Ledoux, 2015.)

Note that you may not have needed the educational contingencies of behaviorology study to tell you how best to help your friend. The everyday contingencies of experience can induce some correct responding in such situations. You absolutely needed behaviorology, however, to tell you not only why those were good steps to help your friend, but also to tell you how to figure out whether or not some other possible steps for your friend would be harmful, and why, and what the alternatives might be.

Nevertheless, the reader should be aware that trying to help oneself or others with procedures grounded in natural–science data can violate the draconian laws that are on the books in some nation states regarding who can legally help others. This particularly applies in states that maintain no–longer–appropriate legal sanctions favoring guild interests like those of psychologists whose discipline cannot qualify as natural science. Check your state's statutes and be careful to obey the laws, even while we all work to change them to better reflect reality. Meanwhile, let's consider a more complicated example of differential reinforcement.

Helping a professor. This example would benefit from some visual imagination responses on your part, responses that this statement is designed to begin evoking for your benefit. The example starts with a human resource problem (i.e., not enough faculty). This problem once left a colleague teaching an overbooked *Introduction to Behaviorology* course (BEHG 101). The course roster showed 160 students in a large lecture hall that had a broken sound system. The professor could project her voice adequately. However, the acoustics of the room made it difficult for the 40 students in the left rear quadrant of the room (from the professor's perspective) to hear well unless the speaker was standing well to the left of the podium, near the edge of the blackboard; from this location everyone could hear her well.

Due to the large class size, the professor had switched from her usual interactive classroom activities, which involved applications of the very science she was teaching for conditioning new behaviorological repertoires. She switched to the lecture method, even though the lecture method had become somewhat obsolete long ago with the invention of the printing press, and even more obsolete more recently with the invention of programmed instruction.

(See Holland & Skinner, 1961, for an excellent example of a programmed–instruction textbook that introduces the natural science of behavior. Also, see Shuler & Ledoux, 2017, for a behaviorology–terminology update for the Holland & Skinner program. Combining these resources may make the best programmed–instruction introduction to basic behaviorological science.)

Returning to the example, the professor lectured for a two–hour period each week, and her lecturing style had her roaming all over the place around the podium. Note that this included occasionally spending a little time lecturing from a position on the far left side of the podium, the location from which all the students could hear her well. After each lecture the students met in groups of 40 for the third class hour for review, questions, and quizzes. Each quadrant met with one of the four graduate teaching assistants (TAS) who were working with her on the course.

The students in that left–rear quadrant, however, were an unhappy lot, because they could not hear the professor very well. When they complained to their TA, she suggested that they talk to the professor about possible solutions to this problem. Since the professor was a bit famous, these young freshmen could not really see themselves following this advice. But another possible solution occurred to the TA when she began to review the last lecture with them. In that lecture the professor had described differential reinforcement, and… Wait. What's *that?* Since the students had been unable to hear the professor clearly, they really had little clue about what differential reinforcement was. So the TA first had to review differential reinforcement for them very thoroughly, and that was when they hatched a solution plot. Here is what happened.

Together, the TA and the students defined the professor's lecturing response class. This step is needed to enable changing the contingencies on the lecturing–response–class member of concern. In this case they defined it *positionally.* That is, they defined each of seven locations, from which lecturing occurred, as members of the response class of lecturing. These seven positions were at the podium, just right and just left of the podium, middle right and middle left of the podium, and far right and far left of the podium. Then they arranged to reinforce only the one member of this response class that enabled everyone in the room to hear the professor. This member involved lecturing from the far–left location (from the professor's perspective; i.e., on the professor's far left). They would ignore lecturing from all the other locations. They would extinguish all the other response–class members.

All the TAs then coordinated their groups in a cooperative demonstration of differential reinforcement. Of course they needed an effective reinforcer. Since it was a state university that could only pay its professors poorly, they considered that throwing paper airplanes made from dollar bills might work, but as students they too lacked dollar bills. Then they thought of attention, and they laid their plans. At the next lecture, when the professor arrived, everyone had cast their eyes downward toward their desk, and moved their pens slowly as if disinterested and doodling. They repeated this performance at any time the professor was anywhere except near her far left side of the podium, near the edge of the blackboard. However, whenever the professor was at that sweet spot at her far left side of the podium, a position from which everyone could hear her, all heads would snap up, every eye wide open with interest, and every pen taking copious notes on everything the professor said.

Would that kind of attention reinforce the professor's behavior of lecturing from that location? They thought it could work, because that response at least occasionally occurred already. The professor's routine involved starting at the right edge of the blackboard (i.e., the left edge from the student's perspective) and putting a topic on the board, and turning around to comment on that topic at length before putting up the next topic. Gradually she worked her way through the topics and across to the other side of the room. What happened?

The first time she got to the sweet spot, on her third or fourth topic, she had put the topic of superstitious behavior on the board and then turned around and saw 160 bright-eyed, smiling and most-attentive faces. This was certainly an improvement. She had earlier noted the apparent gloomy mood of the class, and had wondered what caused it and, more importantly, how to fix it. Apparently a great deal of interest in superstitious behavior existed. So instead of the couple of minutes that she usually spent at that location, and on that topic, before returning to the other side of the room and beginning on another topic, she spent just over ten minutes on that left side talking about superstitious behavior, with extra examples. And the bright, beaming faces remained, and the pencils kept scratching notes.

By the way, the ten minutes on that topic provided an unplanned example of what kind of behavior? If you said, "superstitious behavior," then you are correct. The attention indeed reinforced lecturing from that location as planned. Witness the increase from the usual three minutes to over ten minutes there. But the attention also *coincidentally* reinforced talking about a particular topic as well, in this case the topic of superstitious behavior. As I am sure you recall, we use the term *superstitious behavior* to describe behavior that coincidental reinforcement conditions.

Well, this sequence played out a couple more times, with the students' heads returning to the disinterested position as soon as the professor moved from that sweet spot. But each time that her routine took her back there, the heads popped up again and she spent a longer time there, finally lecturing for the last 35 or 40 minutes of the period from that spot while she erased the old topic from the board behind her and wrote the new topic in the same place. The students had successfully differentially reinforced the member of her lecturing response class that involved her lecturing from that particular spot, a position from which everyone could hear her. The term we use for such outcomes is *response differentiation.* The procedure differentiated this one response, this one response-class member, from all the others.

At the end of the class, with a TA's help, the students thanked the professor for speaking from the acoustically helpful spot, and for this real demonstration of differential reinforcement. Say…What? After getting over her initial surprise—and at least partly humorous threat to flunk them all—she agreed. It was a real, and realistic, application of the differential reinforcement procedure. The students also found that, like so many of the better professors, she was indeed very human and approachable. They need not have gone to such lengths to solve the classroom sound problem. She made the necessary phone call (as this was in the days long before the invention of email) and the problem got fixed right away.

While I have changed some details for pedagogical reasons, a professor of my acquaintance reported in a textbook the kind of experience that this example describes (see Whaley & Malott, 1971, p. 65–66). However, what if a different professor had taught this class, a new professor who stayed near the podium in front of the students, a professor who never ventured over to the side of the room where all the students could hear the professor well? Then the students *could not use* this simple differential–reinforcement procedure, because the target behavior of lecturing from that side of the room never occurs. Are any other procedures available to manage environmental variables that could produce the behavior of lecturing from that side? Yes, at least one is available, and the short name we use for it is *shaping*. You use the shaping procedure when the target behavior never happens. We cover the shaping procedure in the next chapter.

Conclusion

Before going on to discuss the shaping procedure, however, here is an answer to the question of what role the differential reinforcement process plays in the success that practice shows in producing improved performances. In particular, how does differential reinforcement relate to improving your basketball free–throw results through practice?

The answer contains several components similar to those that we saw in our other differential reinforcement examples. Every attempt in free–throw practice constitutes a member of the response class of "basketball free throwing," because every attempt produces the same consequence of moving the basketball toward the hoop. As in many situations involving skill practice, however, differential reinforcement already operates. Only certain response–class members produce reinforcement. In this case these members have the particular characteristics responsible for making the ball go *through* the hoop rather than just toward it. This reinforcement nearly instantaneously follows the appropriate response–class members. As a result the number of these select members increases while the number of lower quality members gradually reduces, because they lack reinforcing consequences (i.e., they extinguish). This is not magical; this is science. Before long, your free throw results become superb rather than dismal. Good practicing! ♣

References (with some annotations)

Fraley, L. E. & Ledoux, S. F. (2015). Origins, status, and mission of behaviorology. In S. F. Ledoux. *Origins and Components of Behaviorology—Third Edition* (pp. 33–169). Ottawa, CANADA: BehaveTech Publishing. The authors, as participant observers of the reported events, first privately published this piece in 1992. They had written it over the previous five years after the meeting, in May 1987, at which a group of nine leading natural scientists of behavior moved for complete separation and independence of their discipline from psychology, under the behaviorology label. The version cited here is essentially unchanged from the original private publication. As a member of the group, I recorded the meeting and prepared a transcript, which you can find (at www.behaviorology.org) in (2013) *Journal of Behaviorology, 16* (1), 3–13.

Holland, J. G. & Skinner, B. F. (1961). *The Analysis of Behavior.* New York: McGraw–Hill.

Shuler, M. & Ledoux, S. F. (2017). Behaviorology terminology adjustments for *The Analysis of Behavior* by Holland and Skinner. *Journal of Behaviorology, 20* (1), 3–15.

Truax, C. B. (1966). Reinforcement and non–reinforcement in Rogerian psychotherapy. *Journal of Abnormal and Social Psychology, 71,* 1–9.

Whaley, D. L. & Malott, R. W. (1971). *Elementary Principles of Behavior.* Englewood Cliffs, NJ: Prentice–Hall. Students found this classic primer on behaviorological science and applications fun to read. Some of them could not stop reading it!✌

Chapter 11
Operant Behavior—IV:

Can a very different behavior develop from an apparently unrelated response? The postcedent "shaping" process/application ...

*W*elcome to your local, shooting–sports rifle range. The operation of various contingencies in your own history and current setting have led you here. Among the many contingency possibilities, perhaps getting good scores on paper targets has become a more potent reinforcer than getting tin cans to move occasionally by plinking at them before recycling them. Or maybe a friend or sibling or parent or son or daughter has gotten really excited that their target scores are improving, and asked about your scores. Or perhaps *their* progress in safe rifle marksmanship skills would reinforce your behavior, or your participation would reinforce their behavior, and so on. The problem, however, is that you have only ever shot at tin cans. You have never before shot at paper targets with your little .22 caliber rifle. At the standard 50–foot target distance, the black circle in the target center, at only 1.5 inches across, looks awfully small. Nevertheless, the standard for an expert marksmanship rating is ten targets each with a score of at least 40 out of a possible 50, while standing and taking five shots offhand. The behavior of putting five holes into that black dot that add up to a score of 40 or more, while standing and shooting offhand, is definitely not currently happening. Indeed, when you first tried these targets, your scores averaged 8 out of 50. What contingencies could possibly lead to the kind of improvement that moves you from 8 out of 50 to 40 out of 50? Before considering a specific answer for this question, let's consider the general procedure that applies in situations like this in which *the behavior of concern is not currently happening.* "Shaping" is the short name for this procedure.

Shaping

When we previously covered the differential–reinforcement procedure, we noted that it only worked when the behavior of concern *already* occurred with some minimal frequency. This left us wondering what could help if the behavior of concern was not occurring at all. We even promised to extend our classroom sound–problem example to this circumstance. Before changing, however, from an example of simple differential reinforcement to an example of *shaping,* let's consider what shaping involves.

As with differential reinforcement, the shaping procedure involves only combinations of the reinforcement and extinction processes. But differential reinforcement, which we sometimes call *simple* differential reinforcement, and shaping involve these two processes in slightly different ways. With simple differential reinforcement, extinction and reinforcement occur simply according to response–class membership. Reinforcement follows the member that comprises the target behavior while extinction follows the other members. In shaping, however, *the response class changes.* This happens each time reinforcement works on the current, occurring, response–class member that is most like the non–occurring desired behavior. In essence the contingencies that operate in shaping involve *the repeated use of differential reinforcement, on changing response classes, for conditioning a behavior that initially is not occurring.* Let's see how that plays out.

If the behavior of concern, which we can also call the target behavior, *is* occurring, then we need not start a shaping procedure. We apply simple differential reinforcement to strengthen the target behavior that initially occurs at least occasionally, because we *can* reinforce it since it is already occurring. But when the target behavior never occurs, we cannot follow it with a reinforcer. Then, we need to apply the shaping procedure, which does not require an already occurring target behavior.

The full name for the shaping procedure is *the method of successive approximation,* because it is the method of reinforcing responses that ever more closely approximate the desired, or target, behavior. You must, of course, start with a behavior that is currently occurring, because you can only reinforce a behavior that is occurring. But which one? At a minimum, after you have identified your target behavior, the initial occurring response that you start reinforcing should share some characteristic or dimension with your target behavior. Then, along this dimension, you reinforce those particular response versions that ever more closely approximate the target behavior.

However, to really describe shaping and how it works, we need to clarify that notion of "response versions." It comes from the concept of *variation in behavior.* When particular instances of a behavior happen, instances that we call responses, these instances are never absolutely identical. Generally, trained observers, and often untrained observers as well, can readily note the difference, the variation, between responses (i.e., the difference is usually of sufficient magnitude to evoke recognition and other relevant responses from observers—no inner agents are ever necessary or involved).

If we select a current, initial behavior that shares some dimension with the target behavior, then, along this dimension, some of the small variations in the response instances will be slightly less like the target behavior while others are slightly more like the target behavior. We can shape the current behavior a little toward the target behavior by reinforcing those slightly more similar variations while extinguishing (e.g., by ignoring) the slightly less similar variations. This is a use, our first use, of differential reinforcement in shaping our target behavior. How do we move on to additional uses of differential reinforcement?

As a result of that first use of differential reinforcement, the center of the variation range of the current—but now somewhat different—behavior shifts toward the target behavior. We then change the criteria for reinforcement to the side of the range of the newly current behavior that is still, and now even more, similar to the target behavior. This is not the behavior we reinforced under our first use of differential reinforcement. We consider this a new behavior, and we then use the differential–reinforcement procedure *again* to reinforce the slightly more similar variations that are on this side of the range. Meanwhile we extinguish the slightly less similar variations that are on the other end of the range. This is *another* use of differential reinforcement.

We then repeat that process of both changing the criteria for reinforcement, and applying differential reinforcement, again and again, until the resulting behavior *is* our target behavior. We have then shaped this target behavior (which, remember, had not been occurring) using the method of successive approximation. This involved the *repeated* use of the differential–reinforcement procedure on the newly shaped variations of each newly current behavior, each of which more closely approximates the target behavior along some dimension.

All that helps us see why, taking behavior variation into account, one of the easiest ways to define shaping is as *the **repeated use** of the differential–reinforcement procedure to condition a series of new response variations toward an initially non–occurring target behavior.* Also, note that the change in the criteria, regarding which responses get reinforced, can happen with every reinforcement, as is often the case for laboratory studies. Or, it can happen only after responding has stabilized in each differential reinforcement use, as is often the case for applied interventions. That is, the change from the use of differential reinforcement with a member of one particular response class, to the next use with a member of a slightly different response class, depends on criteria that reinforcement providers change. They can change criteria after *each* reinforcement occurs following a response. Or they can change criteria after *several* reinforced responses occur that establish some stability for each intermediate behavior between the initial and the target behaviors. To see shaping at work, let's start by changing our classroom sound–problem from an example of the simple differential–reinforcement procedure to an example of shaping in which the criterion regarding which responses get reinforced changes after several reinforcers stabilize responding.

Shaping Examples

Reconsider the question that ended our original classroom sound–problem example: What if a different, less experienced professor was teaching this class? This new professor stays near the podium in front of the students. He never ventures over to the side of the room where all the students can hear the professor well. With this professor the students could not use the simple differential–reinforcement procedure, because the target behavior of standing on that side never occurs. In this case we bring *shaping* to the rescue.

Students shaping professor behavior. We pick up the example when the whole class is in the classroom for the professor's next lecture. The whole class is then about to implement the intervention that the TA and students from the acoustically challenged left–rear quadrant have designed. To improve the consistency of applying the criteria changes that move from one differential reinforcement contingency to another, the sound–deprived students' TA is sitting down in the first row right in front of the podium, and everyone is keeping an eye on her for the cues regarding when to provide the reinforcer and when not to. These cues include when to wait, because a criterion change requires a closer approximation before reinforcement occurs. When her head goes up all wide–eyed and smiling, then the heads of all the students go up all wide–eyed and smiling.

Also, the students and TA have previously seen this new professor lecture from the center of the podium as well as while *leaning* to the left of, and to the right of, the podium (and only from these positions). So the students and TA have defined the professor's initial lecturing response class as consisting of these three members (i.e., lecturing from the center, left, and right of the podium). Since only the left–of–podium response–class members more closely approximate the target behavior of lecturing from the edge of the blackboard at the professor's far left, these are the initial responses that they reinforce when these responses occur.

Here is how the two–hour lecture class with shaping might go. We base this on the reasonable assumption that the successful reinforcer of the beaming–and–smiling–faces in our previous and *real* example would continue to function as a reinforcer for the behavior of this new professor. At first, each time the professor leaned to the left, the attention from all the eager faces beaming and smiling would occur. Some (perhaps much) back and forth movement of the professor might occur, along with a similar amount of heads–up and heads–down movement of the students. Still, we would see the reinforcing effect appear in the increasing time that the professor spends leaning or, more likely, standing to the left of the podium while lecturing.

A change in the criterion for reinforcement would become appropriate when standing to the left of the podium had become somewhat stable. At that point, the leading TA would wait for some *additional* leftward movement before going heads up with the reinforcer. Why could this additional leftward movement happen? Because when the left–of–podium standing responses stabilized, variation is still present. *Some variation is always present.* The professor now occasionally lectured a little farther to the left of this new position, as well as a little back to the right of this new position. Then, using differential reinforcement *again,* a new lecturing position would stabilize, and another reinforcement–criterion change would become appropriate.

The students would use differential reinforcement repeatedly, each time on slightly farther left positions than each new position that had just stabilized after several reinforcements. The reinforcements that the class delivers contingent on closer approximations gradually shape the positional lecturing responses across the room. They change the responses from the room center over to that left–edge position where everyone can hear the professor speak. This was a behavior that initially was not occurring. Yet now it occurs, thanks to the particular balance of reinforcement and extinction in the shaping procedure.

Note that the class had to work with whatever initial response was available. If the professor had not leaned to the right or left near the podium, but had only *looked* toward the right or left, then the class *would have had to start by reinforcing looking to the left,* because that would have been the initial response that most closely approximated the target behavior. I am sure you can imagine some of the intermediate steps that shaping would require to go from that point to the point where our example started.

People and pigeons shape each other's behavior. Now let's consider another example of shaping. This time the criteria, regarding which responses get reinforced, change after each reinforcement *as the "shapee's" behavior also shapes the "shaper's" behavior.* That is, after each reinforcement, you see additional response changes that more closely approximate the target behavior before another reinforcement occurs. This example also shows an increase in complexity in that shaping occurs with respect to the behavior of a literal flock of subjects, pigeon subjects, rather than a single individual. This happened, in this case, through the effects of an intermittent reinforcement schedule because, while any member of the flock could get the reinforcer, only one member actually gets any particular reinforcer. Here is how it worked.

This shaping example occurred in France. My wife and our son, Miles, both play English Handbells. Some years back she signed up herself and our son, who was 18–years–old at the time, to participate in the first ever *massed* English Handbell concerts in France. Over 100 ringers (who all paid their own way) and some 500 instruments were on the week–long tour with several concerts. I went along both for the music and to accompany our daughter, Susannah, who was nine–years–old at the time.

This shaping example occurred on the day of the last concert, which was at the *Eglise de la Madeleine* in Paris. During the afternoon, while the ringers were inside the church for a pre–concert practice, Susannah and I passed the time outside in the church yard. Unfortunately, very little was happening that might engage her there, although about a dozen pigeons flocked around elsewhere in the yard. They were initially rather unapproachable, always about ten feet away, and always keeping that distance if we moved toward them or away from them.

Then, recalling the small bread roll that I had put in my coat pocket from lunch, in case hunger struck one of us during the afternoon, I asked Susannah if she would like the pigeons to be more friendly. When she responded positively, I suggested that we use the roll to shape some pigeon approach behavior. Before reading how we shaped that behavior, ask yourself what steps you would take. We have already covered everything you need.

After turning the bread roll into lots of half–inch pieces, we sat on a bench and observed the pigeons, noting the ten–foot distance that they kept away from us. Then I suggested to Susannah that she toss a bread piece half way between us and the pigeons. As expected, the birds' excitement level increased when she complied, but they would not come close enough to get the piece. Briefly, however, they came closer by one or two feet. After retrieving that piece, she tossed it just within the closest edge of the milling flock's perimeter. Of course, they went for it, but the result was also that the flock's perimeter had moved closer to us. And this movement reinforced (i.e., shaped) our bread tossing. Remember, all of us—you and I and the pigeons—are all behaving organisms with behavior under control of the same natural laws. Our responses of tossing bread toward the flock of birds, when their movements were slightly closer to us, reinforced those movements, compelling even closer–to–us movements. Meanwhile these same closer–to–us movements reinforced our bread tossing, just as the professor's left drift had reinforced heads up responses.

For several minutes we repeated that kind of shaping interaction, with the criteria for the next reinforcer changing each time to require movements that made the perimeter closer to us. By the end of only those several minutes, the pigeons covered Susannah from head to knee (as she was still sitting on the bench). Luckily they were rather polite pigeons, as they covered her only with themselves; they dropped nothing unpleasant on her. When we had distributed all the pieces of bread, the pigeons lost interest in us. However, the perimeter of their continued movements nearby had shrunk to about five feet, and they were still maintaining that closer distance when the time came for us to go inside about a half hour later.

So we see that shaping is a fun as well as a potent process. Yet all that is happening is that reinforcers are following some behaviors while other behaviors are extinguishing. Now, reinforcement and extinction (each alone, or together as in differential reinforcement) are natural processes that can occur as helpful procedures. And shaping also is a natural process, one that again we can use as a procedure to help improve problem behaviors.

Due to potential dangers, however, please resist trying to shape wild pigeons. These were Paris church pigeons, quite acclimated to the proximity of people. While pigeons lack teeth (which makes them far safer as shaping subjects than squirrels) they can peck. Instead, I recommend shaping the usual range of fun performances, and healthful routines like exercise, in your domesticated companion animals (i.e., pets). They make great initial shaping subjects. The best pet training protocols, which are applications of behaviorology to generating and maintaining behaviors of other animals, involve the previously mentioned "clicker training." A number of good references are available (see O'Heare, 2014, 2015, and Pryor, 1991, 1999a, 1999b, 2001; also see Johnson, 2004, and Kurland, 2000).

Professionals shaping client behavior. Before leaving the differential–reinforcement and shaping topics, let's revisit, and go beyond, the reference I gave you in the differential–reinforcement chapter as an example of the application of these procedures at the professional level. That reference was Truax, 1966, an article addressing the inherent *directiveness* of Rogerian "non–directive" psychotherapy. Truax showed that Rogerian therapy was indeed directive, which accounts for its helpfulness, because the experienced therapist's feedback reinforces, even shapes, appropriate client responses in ways similar to our earlier helping–a–friend example. A truly random (i.e., non–directive) feedback pattern would likely reinforce more of a client's more common negative comments, which could easily make the client's problems worse, a rather unethical alternative. This topic was one of the facets of the "Rogers/Skinner debates." The Truax article and these debates are part of the history that behaviorology and psychology shared. This history happened before psychologists gave the attempts of proto–behaviorologists, to change psychology into a natural science, an adamant refusal (i.e., in what psychologists called the cognitive revolution). Ultimately this led to the organizing of behaviorology as a discipline separate from and independent of psychology.

The debates between Carl Rogers (the developer of Rogerian psychotherapy) and B. F. Skinner (the original behaviorological scientist) are legendary with respect to the incommensurability between natural sciences like behaviorology and non–natural–science disciplines like psychology. Rogers represented so–called humanistic psychology, a psychology that presented itself as proud of seeing science (i.e., natural science, philosophically and methodologically) as largely irrelevant to human nature and human behavior. Such a position makes this psychology school perhaps the clearest version of psychology's general anti–science allegiance to the mysticism of inner agents that conveys secular–religion status on the psychology discipline. Other humanists, however, openly acknowledge the relevance of natural science to human nature and human behavior. We see this in the (American) Humanist Society's naming of B. F. Skinner as the *Humanist of the Year* for 1972 (Bjork, 1993, p. 220) after the publication of his book, *Beyond Freedom and Dignity* (1971).

From Consequences to Evocatives

We have now tackled procedures that involved arranging *consequences* in particular ways to produce particular outcomes. These consequence–related procedures included the processes of differential reinforcement and shaping.

Procedures that involve arranging *evocative stimuli,* in particular ways to produce particular outcomes, are also possible. In upcoming chapters, we tackle two of these, the backward–chaining procedure and the fading procedure.

Conclusion

When we left you at the rifle range at the start of this chapter, we were asking what contingencies could possibly lead to improving your scores from 8 out of 50 to over 40 out of 50. Our new–found familiarity with the method of successive approximation leads us to investigate how this method applies to shaping your rifle marksmanship skills enough to qualify you for the Expert rating. How could shaping apply to this task?

Actually that question was addressed many decades ago, although the designers of the shaping steps then were probably unaware that they were invoking the shaping procedure. Nevertheless, they laid out a carefully sequenced set of steps that succeeded admirably in shaping the rifle marksmanship skills of large numbers of adults and children around the country and across many years. And, of course, it still works.

Those steps took into account both the increasing difficulty in shooting positions and the increasing difficulty of gradually raising required minimum scores across sets of ten targets. The shooting positions are prone (i.e., lying flat on the ground), sitting, kneeling, and standing. Standing is actually the most difficult position from which to shoot accurately due to its inherently greater instability. So the shaping steps begin with the prone position and gradually ramp up to sitting, kneeling and, lastly, standing. Meanwhile, the minimum qualifying score, for targets at each position, also gradually increases.

To move from one level to the next, the score on each of ten targets must reach or exceed the minimum score for that level. Here are the shaping requirements for each level, using the name of the award achieved for completing the requirements for each level:

❧ *Pro–Marksman:*	Prone, with minimum scores of 20.
❧ *Marksman:*	Prone, with minimum scores of 25.
❧ *Marksman First Class:*	Prone, with minimum scores of 30.
❧ *Sharpshooter:*	Prone, with minimum scores of 35.
❧ *Sharpshooter Bar 1:*	Prone, with minimum scores of **40**.
❧ *Sharpshooter Bar 2:*	Sitting, with minimum scores of 30.
❧ *Sharpshooter Bar 3:*	Sitting, with minimum scores of 35.
❧ *Sharpshooter Bar 4:*	Sitting, with minimum scores of **40**.
❧ *Sharpshooter Bar 5:*	Kneeling, with minimum scores of 30.
❧ *Sharpshooter Bar 6:*	Kneeling, with minimum scores of 35.
❧ *Sharpshooter Bar 7:*	Kneeling, with minimum scores of **40**.
❧ *Sharpshooter Bar 8:*	Standing, with minimum scores of 30.
❧ *Sharpshooter Bar 9:*	Standing, with minimum scores of 35.
❧ *Expert:*	Standing, with minimum scores of **40**.

The gradual shaping of marksmanship skills shows clearly in that sequence of steps. No wonder this shaping program succeeds in producing such refined skill levels. How long might this whole process take? A minimum of 140 qualifying targets would complete the process. Most people, however, require more than this minimum to qualify for the expert rating. They may need to complete 150 to 200 targets, with five shots per target. One can manage ten targets in each one to two hour range session without undue fatigue. So 20 range sessions could be typical. Even at a hobby level of one range session each week, the full program may take less than six months to complete. This leaves you with a life–long skill and the prospect of many more safe range sessions, along with far more accurate tin–can plinking.

The programs of martial arts schools provide another relatively common example of intuitively applied shaping. The gradually increasing skill requirements for the belt ranks (e.g., yellow belt, green belt, brown belt, and black belt) and the various steps between ranks, constitute a sophisticated set of carefully arranged shaping steps.

Before moving on to other contingency topics, let's ask an important informational question, because we respect full disclosure. Where did that shooting–behavior shaping program originate? Experts at the National Rifle Association designed the program as part of their concern to improve safe firearm handling along with marksmanship skills, especially among young people or those new to the shooting sports. Their program for pistol skills shares shaping steps similar to those for rifle skills. I experienced the fun and success of these programs personally in the late 1960s as part of my scheduled high school extra–curricular activities. These activities involved contingencies that continue to shape my responses regarding civic responsibilities and how these interact in support of helping to solve local and global problems. (Flying is fun too, but I found riflery to be less expensive.)

Let's now shift our exploration of contingences to other processes applicable to problem solving. Over the next couple of chapters, let's turn to *chaining*, and then to *fading*.✲

References

Bjork, D. W. (1993). *B. F. Skinner: A Life*. New York: Basic Books.

Johnson, M. (2004). *Clicker Training for Birds*. Waltham, MA: Sunshine Books.

Kurland, A. (2000). *Clicker Training for Horses*. Waltham, MA: Sunshine Books.

O'Heare, J. (2014). *Science and Technology of Dog Training*. Ottawa, CANADA: BehaveTech Publishing.

O'Heare, J. (2015). *Science and Technology of Animal Training*. Ottawa, CANADA: BehaveTech Publishing.

Pryor, K. (1991). *Lads Before the Wind: Diary of a Dolphin Trainer*. Waltham, MA: Sunshine Books.

Pryor, K. (1999a). *Don't Shoot the Dog! The New Art of Teaching and Training (Revised Edition)*. New York: Bantam Books.

Pryor, K. (1999b). *Clicker Training for Dogs*. Waltham, MA: Sunshine Books.

Pryor, K. (2001). *Clicker Training for Cats*. Waltham, MA: Sunshine Books.

Skinner, B. F. (1971). *Beyond Freedom and Dignity*. New York: Knopf.

Truax, C. B. (1966). Reinforcement and non–reinforcement in Rogerian psychotherapy. *Journal of Abnormal and Social Psychology, 71,* 1–9.∾

Chapter 12
Operant Behavior—V:

Can several responses connect together to make a set or chain of response? The postcedent "chaining" process/application …

$\mathcal{M}$eet Clarice, a very traditionally trained pianist. Many past and current contingencies conspire to compel her continued participation in competitions for musicians. After all, she is on her way to a professional career as a concert pianist. Her idol is the famous pianist Clara Schumann, and her favorite composer is Fanny Mendelssohn.

For some months now Clarice and her accompanist, a violist, have been practicing strenuously on a complicated piano and viola piece for a big competition that she would like to win. Several weeks before the competition, however, her violist friend slipped and broke his arm. Sadly, this ended his participation in this competition. Insufficient time remains, though, to find and practice with another violist. Clarice could drop out of the competition, which would bring relief to her competitors. Instead, she selects an equally appropriate solo piece for the competition. But how can she become excellent with the new piece in the little remaining time?

Let's leave Clarice waiting for the moment while we turn to some contingency arrangements that might help solve problems like hers that need an efficient repertoire expansion procedure. Such contingency arrangements exist, going by the name *backward chaining*. But they seem somewhat counter intuitive, which is worth exploring along with how backward chaining works.

As background, we continue to consider, in some detail, various combinations of basic behaviorological concepts, methods, and principles. These combinations compile into more sophisticated processes and procedures. Recall that "procedures" is simply our term for processes that happen through the natural occurrence of ordinary events, including the responses of other behaving organisms such as other people. These procedures happen through processes involving contingencies with reinforcing outcomes that benefit various individuals and sometimes society. At the professional level, we use the term "interventions" for these procedures.

We already covered two postcedent contingency–arranging procedures, the *differential reinforcement* procedure, and *the method of successive approximation* procedure that we more commonly call *shaping*. Both of these combine the two simple postcedent processes of reinforcement and extinction in ways that bring about beneficial behavior changes. In elaborating these two procedures, we also revisited the two concepts of response class and variation in behavior.

Now, we continue our coverage of combinations of behaviorological basics. This time, however, we focus on two potent *antecedent* environment–change procedures. These procedures arrange *evocative* stimuli in particular ways that produce particular behavior–change outcomes that are also of considerable benefit. Here we focus on what we call the *backward–chaining* procedure. In the next chapter, we focus on what we call the *fading* procedure.

Backward Chaining

The term "chaining" describes the process of operant responses linking together into a sequence. As mentioned in previous chapters, behaviors connect, or *chain,* to each other. Sometimes responses chain together merely by the *occurrence* of one response being the stimulus that evokes the next response, as with some covert response chains. At other times responses chain together through connections with the stimuli that separate them, as with some overt response chains. This happens both to neural behaviors (e.g., thinking, daydreaming, and other consciousness behaviors) as well as to neuro–muscular behaviors (e.g., putting on pants or playing a solo piece for cello or any other musical instrument). The chaining process leads to chains of stimuli and responses that we call stimulus–response chains. What makes these connections and chains?

Some stimulus–response chains are rather fleeting and might occur only once in any particular or exact sequence. Other chains, however, are more fixed and repeatable. *Current* contingencies often establish the components of fleeting chains. For example the contingencies of an approaching holiday, and its traditional complex recipes, compel the chain of thinking responses, even during driving, covering in turn each of the necessary ingredients to go on a shopping list. Meanwhile your driving responses remain under direct stimulus control. On the other hand, *prior but repeating* contingencies often establish the components of more fixed and repeatable chains. Examples would include the contingencies that result in the chains producing musical compositions, and the contingencies that control the building of the chains occurring when playing the composition.

Some fleeting chains, particularly chains of neural responses (e.g., thinking about shopping for those recipe ingredients) happen as each response, as a real event, functions as the stimulus that evokes the next response. However, in many chains of neuro–muscular responses, each response produces a distinct stimulus that both reinforces that response and evokes the next response. We call this phenomenon, of a stimulus both reinforcing one response and evoking another response, the *dual function of a stimulus.* One interpretation of musical improvisation is that at least some parts or aspects of improvisation involve this kind of fleeting chain. Each playing response (of a musical note or phrase) produces a stimulus sound that both reinforces the response that produced it and also shares in evoking the next response, which produces a stimulus sound that both reinforces the response that produced it and also shares in evoking the next response, and so on. Other contingencies control the length of the improvisation chain. These include both the contingencies surrounding a music pattern, and—at least in traditional jazz—the contingencies determining when the occasion for the next player's improvisation begins.

Under contingencies that drive applying helpful interventions, our concern here stresses chains in which the stimuli and responses explicitly link through the dual functions of stimuli. Let's examine this phenomenon more closely before turning to chaining procedures involved in problem solving.

Dual Function of a Stimulus

The dual–stimulus–function phenomenon arises from the process that produces many evocative stimuli. Recall that in this process, these stimuli are regularly present when another, reinforcing stimulus follows a response. As a result the presence of these stimuli comes to evoke the response, and we then call these stimuli evocative stimuli. *In addition,* and as an inherent part of this process, many of these stimuli–becoming–evocative–stimuli are also pairing with (i.e., occurring at the same time as) the reinforcers. This pairing makes these stimuli also function as conditioned reinforcers. Each such stimulus then serves two functions, the functions of both evocative and reinforcing stimuli, which we call the dual function of a stimulus. These stimuli now function both as a conditioned reinforcing stimulus (for a previous behavior) and as an evocative stimulus (for the next behavior). Indeed, we define *dual function of a stimulus* as a stimulus serving two functions as a result of that stimulus being regularly present when a reinforcer follows a response. Being regularly present when a reinforcer follows a response makes the stimulus function as an evocative stimulus and as a conditioned reinforcing stimulus.

We can diagram a stimulus–response chain. The chain shows how each stimulus, in serving two functions, links responses together into a chain. Here is one way to make such a diagram (indicating functions in temporal order):

$$\ldots \rightarrow S^{r/Ev} \rightarrow R \rightarrow S^{r/Ev} \rightarrow R \rightarrow S^{r/Ev} \rightarrow R \rightarrow S^{r/Ev} \rightarrow R \rightarrow S^{r/Ev} \rightarrow \ldots$$

Both functions appear in the single term "$S^{r/Ev}$." As an S^r each stimulus functions to reinforce the previous response. At essentially the same time, as an S^{Ev} each stimulus also functions to evoke the next response.

In building a chain, stimuli and responses can link together in either direction. They can link *forward,* from the start of the chain to the end of the chain. Or they can link backward, from the end of the chain to the start of the chain. Each direction can be helpful in some cases, although any evoked link only leads to later links. Also, backward chaining has demonstrated greater practicality for interventions.

Indeed, the most valuable applied possibilities seem to surround the procedure of building stimulus–response chains "backwards." Before covering this backward–chaining procedure, let's first consider "forward" chaining.

Forward Chaining

As a procedure, forward chaining links stimuli and responses into a chain by starting at the beginning of the chain and building links toward the end of the chain. Such forward chaining can leave backward chaining seeming counter intuitive, because forward chaining builds by following, and intuitively generalizes from, the commonly experienced sequence and flow of events in time. Forward chaining works well for simple or short chains with a small number of quick or discrete responses. For example, educational contingencies usually build the short chains that we call word spelling through forward chaining. We usually reinforce each letter response starting with the *first* letter of a word that someone is spelling, and each letter shares in evoking the next letter, and then the next letter, until all the appropriate letters are present in the correct, reinforced sequence.

Longer or more complicated chains, however, benefit from starting with the end of the chain and building links backward, toward the beginning of the chain. We first put the current response under control of an evocative stimulus. This S^{Ev} then becomes a new reinforcing stimulus (i.e., S^r). We use this new S^r to add a new response *before* the current response by conditioning (i.e., reinforcing) the new response with the new reinforcing stimulus. Let's elaborate how this works to build chains backwards.

The Backward–Chaining Procedure

As a procedure backward chaining has provided many benefits in a range of application and intervention areas. This kind of chaining works best when both the stimuli and the responses in the chain are explicit and directly accessible. Again, as a procedure, backward chaining links stimuli and responses into a chain by starting at the end of the chain and building links toward the beginning of the chain. Therein also resides the source of the benefits of backward chaining. These involve the stable, repeatable occurrence of the response chain due to the inherent *repetition* (i.e., practice) of all the remaining chain links during chain building. How are chains built?

You start with the final response, which is usually one that the main, often unconditioned, reinforcer follows. Then, using standard evocation training, you reinforce this response only in the presence of a particular stimulus. As a result this stimulus becomes an evocative stimulus for the response, *and also as a result it becomes a conditioned reinforcing stimulus that can condition another response by following the other response.* When it follows the other response, which it reinforces, *it also then evokes the original response.* Together, these two responses are now linked by that single stimulus occurring between them.

Repeating that whole procedure builds another response onto the chain, which then has three linked responses. Repeating the procedure again adds another link. For an open chain, the more often you repeat the procedure, the more links you accumulate.

Let's use our technical symbols. (You can skip this paragraph if you like.) You have a response (R_A) that produces a reinforcer (that here is an unconditioned reinforcer, S^R) in the presence of a stimulus (S_1). This makes S_1 (as S_1^{Ev}) evoke R_A. The reinforcer's occurrence in the presence of S_1^{Ev} also makes S_1^{Ev} become a conditioned reinforcer, S_1^r. You then arrange a prior response (R_B) to precede and produce S_1^r, which reinforces R_B *and* evokes R_A due to the dual functions that such stimuli serve. Next you arrange for that S_1 to follow the prior R_B *only* in the presence of another stimulus, S_2. Soon S_2 (as S_2^{Ev}) evokes R_B which produces the reinforcing S_1^r that also evokes (as S_1^{Ev}) R_A, which produces the unconditioned reinforcer. This is a two–response chain.

Here is what that sequence looks like. The dual function appears as a "dual" stimulus in two forms—one above the other—having the same number:

$$S_1^{Ev} \rightarrow R_A \rightarrow S^R$$
$$S_2^{Ev} \rightarrow R_B \rightarrow S_1^r$$

Observe how this diagram (and the next one) reads left to right as usual in terms of time flow, and *bottom to top* in terms of response progression, but *right to left* in terms of chain–construction procedural steps. Of course, you can add any number of additional stimulus–response links to the chain by repeating the procedure. If a chain inherently has a limited number of responses (i.e., a *closed chain* such as a speech with a start and an end) then that number automatically limits how often you repeat the procedure. An *open chain* means that other variables control the limit. Here is a generic diagram of an open chain (as long a chain as I can fit across the page) in which the responses are lettered, with the *last made* (but first constructed) response being letter **A**, and each dual stimulus is numbered, with the *last made* (but first constructed) $S^{r/Ev}$ being number **1**:

$$S_1^{Ev} \rightarrow R_A \rightarrow S^R$$
$$S_2^{Ev} \rightarrow R_B \rightarrow S_1^r$$
$$S_3^{Ev} \rightarrow R_C \rightarrow S_2^r$$
$$S_4^{Ev} \rightarrow R_D \rightarrow S_3^r$$
$$S_5^{Ev} \rightarrow R_E \rightarrow S_4^r$$
$$\ldots S_5^r$$

What would the next steps be, with respect to $S_5^{r/Ev}$? Also, note that the end of this chain, from the $S_2^{r/Ev}$ to the S^R, is identical to the two–response chain that we diagramed earlier. Both diagrams exemplify the definition of backward chaining. In the definition, *backward chaining* is the procedure of taking advantage of the dual function of a stimulus to build chains of responses by linking them with stimuli. You strengthen the function of the evocative stimulus for a response by reinforcing the response in its presence, and then using this stimulus, which has also become a conditioned reinforcer, to strengthen another response before it, and repeating this process from end to beginning until the chain is complete. Each stimulus functions both as an evocative stimulus for the next behavior and as a conditioned reinforcing stimulus for each added "previous" behavior. As we describe the examples, recognize that, especially of the examples that involve "practice," the procedure stresses that every new link or practice run includes *always proceeding to the end of the chain* regardless of where in the chain a practice run begins. This step supports success so much that we will repeat it.

Let's now consider some examples of backward chaining. You might find drawing your own diagrams of the parts of each example beneficial.

Some Backward–Chaining Examples

These examples include open and closed chains, fleeting and fixed chains, with both children and adults, both humans and (in one example) other animals. Applying a couple of them can even improve your own performances.

Barnabus the rat. Our first example describes the 14–response chain that made Barnabus the rat famous half a century ago. People thought his performance was unbelievable, and they thought the training for it must have been nearly impossible. Perhaps some praise, or at least appreciation for the effort, should go to the people who were a natural part of Barnabus's contingencies. We call them his conditioners. Then again, perhaps such praise is inappropriate, not only because they, like everyone else, lack inner agents who earn the praise by telling their host bodies the right things to do, but also because the task is actually not quite as hard as the public presumes. It was not easy, but neither was it nearly impossible, as any good scientifically informed and clicker–expert animal trainer will tell you and show you.

To grasp Barnabus's act, imagine a four–foot tall, four–story house that would make any doll envious. It was open across the front so people could observe the performance. The inside contained not only floors and rooms but also some connecting structures and other items that served as evocative and—through dual–stimulus functions—reinforcing stimuli for the Barnabus act. These included a light, spiral stairs, drawbridge, ladder, chain, pedal car, tunnel, straight stairs, tube, elevator, string (attached to a flag and the elevator), buzzer, response lever, and food pellets. Note that some of the appropriate responses likely required some skillful *shaping* on the part of Barnabus's conditioners (e.g., raising the flag, which lowered the elevator, and pedaling the car).

The act begins with Barnabus in the house waiting near the spiral staircase. The stimulus that evokes the first response is a light. When it comes on, Barnabus is off, up the staircase. He continues through the whole stimulus–response chain, ending up eating a small food pellet near the bottom of the staircase, where the light is now off. When it goes on, he will again be off.

Before considering the components that comprise the stimulus–response chain that constitutes Barnabus's act, let's first use simple agential language to run through the sequence that his audience observes. Of course, as you know, no inner agent inhabits any body, including Barnabus's, telling it what to do. The contingencies have established the response sequence.

Barnabus begins on the bottom level. When a light comes on, he runs up a spiral staircase to the second level. On this level a small platform connects to an upright drawbridge. He pushes down the drawbridge and crosses it to another platform, which connects to a ladder. He climbs the ladder to the third level where he pulls a chain that produces a small pedal–car. He climbs into the car and pedals it through a tunnel that ends at the bottom of a stairway. He then climbs the stairway to the fourth level and runs through a tube ending at an elevator. He enters the elevator and pulls a string that simultaneously raises a flag and lowers the elevator taking him back to ground level. Just after exiting the elevator, he hears a buzzer and goes to a lever on the wall and presses it. Then he moves to a nearby food tray and eats the pellet of food that it contains. Then he waits, probably grooming himself and catching his breath, until the light comes on again, and off he goes again.

Now let's list the components of the stimulus–response chain that constitutes Barnabus's act. Note that the first response on the list is the last response that his conditioners conditioned. The last response on the list is the first one that they conditioned. And remember that the procedure involves Barnabus always running through *all* responses, from the current new response to the last response at the end of the chain, each time they are conditioning or practicing a new stimulus–response link at the then start of the chain. The list shows most stimuli as $S^{r/Ev}$s; their appearance/occurrence reinforces the previous response and evokes the next response. Here is the chain:

S^{Ev}:   The light goes on (for a few seconds, perhaps due to a timer set to give Barnabus a rest between runs) immediately evoking…

R: Climbing the spiral staircase which leads to a platform on the next level where…

$S^{r/Ev}$:   *The presence of* an upright drawbridge evokes…

R: Pushing down the bridge and crossing it (after which it springs back up) which leads to another platform where…

$S^{r/Ev}$:   (The presence of) a ladder evokes…

R: Climbing the ladder, which leads to where…

$S^{r/Ev}$:   A chain evokes…

R: Pulling the chain, which makes…

$S^{r/Ev}$:   A car appear that evokes…

R: Climbing into the car in which…

$S^{r/Ev}$:   Pedals evoke…

R: Pedaling, which takes the car through a tunnel, which ends at…

$S^{r/Ev}$:   The bottom of some stairs, which evokes…

R: Climbing the stairs, which leads to…

$S^{r/Ev}$: A tube, which evokes…

R: Running through the tube, which leads to…

$S^{r/Ev}$: An elevator, which evokes…

R: Entering the elevator, where…

$S^{r/Ev}$: A string (attached to a flag and the elevator) evokes…

R: Pulling the string, which raises the flag while lowering the elevator down through all floors, which leads to…

$S^{r/Ev}$: An open access on the ground floor that evokes…

R: Exiting the elevator, during which his steps trip a pressure switch in the floor setting off…

$S^{r/Ev}$: A buzzer, which evokes…

R: Pressing a nearby response lever, which releases a pellet of food that, when it hits the food tray, produces…

$S^{r/Ev}$: A sound, which evokes…

R: Stepping to the food tray in which the now visible…

$S^{r/Ev}$: Food pellet evokes…

R: Eating the food pellet, which produces…

S^{R}: The unconditioned reinforcing effects from food consumption, with the pellet tray conveniently close to the light and the spiral staircase where, when the light comes on again, the chain begins again.

Barnabus may not need a standing ovation, but observing his thrilling performance, over and over again, could easily evoke one. Note that we have broken down the end of his performance into his last three stimulus–response components (from the buzzer to the food eating) although his conditioners would likely have conditioned these three as a unit. Indeed, neither the buzzer nor the lever need to be present. Exiting the elevator could trip a switch that drops a pellet of food into the tray. Seeing the tray as he exits the elevator reinforces leaving the elevator and evokes approaching the tray and eating the food. Near the end of training, his conditioners could have faded out the buzzer and tray (and we cover fading in the next chapter).

On the other end of the chain, which is certainly long enough, the complexity of the doll house could have enabled an even longer chain, giving this chain the character of an open chain. Regardless of the length (i.e., the number of stimulus–response links) Barnabus's conditioners followed the procedure we previously described to condition each response in relation to the relevant evocative and reinforcing stimuli. Your describing each of the procedural steps that they took for each link can be tedious but beneficial.

For the record my first exposure to this example was, as a student, from its appearance in Whaley and Malott (1971, pp. 298–300). These authors regularly referred to the original sources for their examples, in this case, Pierrel and Sherman (1963) at Brown University where the conditioning of Barnabus took place. As with many examples in this book, I have for decades shared the examples with my students. Let's consider another example that the Whaley and Malott book inspired. This is a human example, which can generalize to help anyone working with or around young children.

Putting on pants. Have you ever tried to teach a toddler to put on his pants by himself? With your lifetime of conditioning experience, putting on pants is now quite routine. For the toddler, however, it can be a major undertaking. Depending on the method, the outcome can lead to very strained or very pleasant relations. To support pleasant relations, use backward chaining. Here is how backward chaining applies to this closed chain (i.e., a chain with definite beginning and ending points).

You start, of course, with the *last* component of the chain. So you put the pants on the child and pull them *almost* all the way up, say, to just below the child's rump. Then you provide a verbal evocative stimulus for the child's response, perhaps by saying, with plenty of enthusiasm, "Pull up your pants." You might need to put the child's hands on the top of the pants when you say that. At first you might even need to mold the response by taking the child's hands while they are grasping the pants and helping pull the pants up. Supplementing the usual reinforcer (i.e., the pants being on) with more added attention and verbal reinforcers can be helpful here. If the interaction seems fun (for both of you) then you can remove the pants and restart and run through the process a couple more times. Or you can just let progress accumulate with each normal occasion for putting on pants.

When saying "Pull up your pants" effectively evokes the child's pull–up response from this initial position, you are both ready for the next chain component. Next time, you pull the pants up only to the above–knee level and then say, "Pull up your pants." That will evoke the pulling up response which will put the pants at the below–rump level, a position that now *already functions* as the evocative stimulus for pulling the pants the rest of the way up.

The remaining chain components accrue similarly. When each new position effectively evokes the child's response of pulling his pants up such that previous positions then also evoke further pulling until the pants are on, then you move on to the next chain component, then the next, until you can hand the child his pants and simply say, "Put on your pants," and the pants get put on. A typical set of chain components includes the pants at below–rump level, at above–knee level, at above–ankle level, on only one ankle, and simply handing the pants to the child, or even letting the child remove the pants from a drawer. Rather than being a bother for both of you, backward chaining makes the task both successful and fun.

Through several examples that they cast in terms of teaching developmentally delayed children various skills, Whaley and Malott (1971, p. 300) inspired that putting–on–pants example. In every case the examples followed the backward–chaining procedure that we previously described to condition each response in relation to the relevant evocative and reinforcing stimuli. Let's consider another human example, one which can help any player of a musical instrument.

Performing a piece of music without the sheet music. Have the contingencies operating in your past ever produced musical–instrument playing beyond the beginner's level? If so, are you ever under contingencies to perform a solo piece in a setting where the presence of sheet music is inappropriate (e.g., a solo concert setting or a competition)? In preparing to perform this kind of closed chain, would traditional forward chaining help most, or would backward chaining help most? In both cases you first analyze your piece for the boundaries of its inherent sections and sub–sections, not only for the usual musical reasons but also because these become your chain components. Let's explore both forward and backward chaining in these circumstances, especially in terms of the likely outcomes that might occur during and after your performance.

In the simpler forward–chaining procedure, you begin to practice your piece by playing through its first chain component several times, repeating a couple of times any little parts that give you any trouble. Then you play through the second chain component several times the same way, followed by playing to this point from the beginning. You then repeat this whole procedure for all remaining chain components.

Seemingly all prepared, and a little nervous, you start to play your piece in the concert hall. The beginning is easy; after all, you have practiced it more often than any other part, an inherent characteristic of the forward–chaining procedure. As you get farther into the piece, however, little errors creep in, but as a professional you push on. Still, the piece is getting more and more difficult. Why? Because the farther you go, the more you get into less and less practiced material. These contingencies force far more focus on the technical aspects of playing just to get the notes right, while the artistic aspects gradually all but disappear. You are far more nervous near the end of your piece than when you started, and the audience seems more tempted to throw tomatoes than roses. Such are possible outcomes of forward chaining.

Even though it seems more complicated, let's try backward chaining instead. In this procedure, you begin to practice your piece by playing through its *last* chain component several times, (always repeating a couple of times any little parts that give you any trouble). Next you play through the second–to–last chain component several times, *each time playing through the last component as well.* You always play through *to the end of the piece,* every time you play any chain component. Then you play through the third–to–last chain component several times, as usual playing to the end of the whole piece. And you repeat this whole procedure for each and every remaining chain component.

This time, seemingly all prepared, and a little nervous, you start to play your piece in the concert hall. The beginning is easy. After all, you were practicing the beginning off stage just before your stage call. This time, however, as you get farther into the piece, your backward–chaining practice pays off as your performance gets ever better. It gets better, because the farther you get, the better you play, as you have practiced all later parts more than earlier parts, an inherent characteristic of the backward–chaining procedure. Now *these* contingencies induce far more focus on the artistic aspects of playing, because your practice procedure already polished the technical aspects. You are far less nervous near the end than when you started, and you are playing better than at the beginning. Possibly more importantly the audience seems more appreciative of your technical and artistic prowess, and more prepared to throw roses, and perhaps even money.

Again, the better outcome in this example followed the backward–chaining procedure that we previously described to condition each response in relation to the relevant evocative and reinforcing stimuli. I first discovered that many musicians already followed this procedure when I asked my spouse, Nelly Case about it. (She is now an emeritus professor of music at the Crane School of Music at SUNY–Potsdam.) She pointed out that many voice and other instrumental performance faculty long ago came under practical contingencies to practice pieces through the backward–chaining procedure, and to pass this procedure on to their students, although they may not have known this scientific name for it, or *why* it works.

Is music playing the only activity to benefit from backward chaining? As a sample to show that this procedure can benefit many more activity areas, let's briefly consider one last human example, one which can help anyone with a public–speaking engagement.

Delivering a speech without a text (or even an ear piece). Have the contingencies operating in your past ever required delivering a speech that at least appears extemporaneous (i.e., no written text, and no ear piece/ teleprompter)? As was the case with the last example, in preparing to perform this kind of chain, backward chaining would be very helpful, certainly more helpful than forward chaining, and for similar outcome reasons. This chain could be either a closed chain or an open chain, depending only on the contingencies regarding the most appropriate length of your speech.

Of course, your speech will not be extemporaneous. It will be well rehearsed. You begin to practice your speech by reading the last sentence, as the last chain component, several times (although some use paragraphs as chain components, but that is more difficult). Ultimately you are trying to recite the sentence without reading it or, more precisely, until the first word or two of the sentence evokes its full recitation. Next you read the sentence before that several times, again trying to recite it without reading it, *but each time always also reciting—without reading—all remaining sentences through to the end of the speech.* That is, you practice until hearing each sentence without reading it reinforces reciting that sentence *and evokes reciting the next sentence* (i.e., in simple terms, each sentence leads you into the next sentence). You repeat these steps for each and every remaining sentence *back* through the speech until you can start the first sentence, which you have practiced the least, and then chain through all the sentences to the end, which you know so well, as you have practiced it the most by always reciting the speech through to the end. Before long the first words of the speech are sufficient to evoke the full recitation of the whole speech. Then, normal contingencies induce other characteristics of good public speaking to occur that can endear you to your audience. Of course, we have always been saying "you" as an agent–free verbal shortcut.

My first exposure to this example occurred through personal experiences, the contingencies of which definitively demonstrated the increased reinforcing value of backward–chaining practice. Your contingencies will predictably generate similar outcomes in many activity areas.

However, continually recognize that when applying backward chaining with practice repetitions, such practice must include always proceeding *to the end of the chain* on each practice run regardless of where in the chain a practice run begins. This key element of the procedure maximizes the benefits of backward chaining.

In the next chapter, we consider the procedure that we call "fading." Both backward chaining and fading emphasize the antecedent side of contingency–engineering applications.

Conclusion

Remember Clarice? We left her, as a budding concert pianist, waiting while we considered some contingency arrangements that might help solve her problem of needing an efficient repertoire expansion procedure. We covered such contingency arrangements under the name *backward chaining*. How would backward chaining apply in her case?

Rather than you or I just repeating the steps of backward chaining, here is what happened. Clarice called the professor who had served as her mentor and advisor while she was completing her doctorate. The professor had recently come across a thorough description, with research outcome data, of backward chaining, similar to what we covered earlier, and told Clarice all about it. Although skeptical, she was also desperate, given the little time remaining before the competition. So she worked out exactly what the steps needed to be, and followed them correctly in all her practice time before the competition. The result amazed her. While she "only" took second place, she remains appreciative of backward chaining, and teaches this practice procedure to all of her own students.♣

References

Whaley, D. L. & Malott, R. W. (1971). *Elementary Principles of Behavior.* Englewood Cliffs, NJ: Prentice–Hall.

Pierrel, R. & Sherman, J. G. (1963, Feb). Barnabus, the rat with college training. *Brown Alumni Monthly,* 8–12.ↄ

Chapter 13
Operant Behavior—VI:

Must a stimulus remain unchanged to be effective?

An antecedent process/application, "fading" …

*L*ate one summer a dozen children turned up among the kindergarten and first–grade cohort of a local school. These children showed not only little or no language (i.e., verbal) behavior but also a range of other behavior deficits and excesses, some severe. The school reported that typically only one or two such children show up each year in the entering cohort. Out of concern both for these children and to discover why the number was larger than usual, the school administrator called the behaviorology (not the psychology) department at your university for help, and the department chair directed the call to you. The administrator harbored a reasonable fear of a connection between these problems and one of the civil or industrial hazardous waste dumps near your city. While that would require an additional level of intervention, we are more interested in any appropriate behaviorological process that could help these children. Before addressing these specific concerns, let's first investigate the "fading" procedure, which emphasizes antecedent contingencies.

The Fading Procedure

Some procedures share the characteristic of *gradual change,* which evokes part of our descriptive responses regarding these procedures. We have already seen the kinds of gradual change that the *shaping* procedure brings about. Shaping starts with one response and gradually, as we apply the procedure, changes that response along some dimension (e.g., topography) to a different response.

Fading also involves gradual change and, like shaping, can occur either as a process or as a procedure (i.e., an intervention). While shaping gradually changes characteristics of responses, fading gradually changes characteristics of stimuli, leading to different effects of evocative stimuli. The fading procedure starts with at least two stimuli that share one or more characteristics. Each of these characteristics can be similar or dissimilar across the stimuli. We refer to such similarities and dissimilarities as *dimensions* of the stimuli. Fading involves gradual changes of, or in, a stimulus (or dimension of a stimulus) that result in the *transfer of stimulus control* from one stimulus (or dimension of a stimulus) to another stimulus (or dimension of a stimulus). In some cases, the procedure actually involves literally fading out a stimulus, as we will see.

To keep our description of fading simple, it may imply that the whole stimulus gradually changes. However, far more commonly, only one or another dimension of the stimulus gradually changes. Before fading begins, one stimulus exerts control over (i.e., evokes) the response while the other stimulus is merely present along with this evocative stimulus. As the fading procedure unfolds, we make small changes in the evocative stimulus while it repeatedly evokes the response and the other stimulus remains unchanged.

At that point two possible outcomes exist. In one outcome the changes make the evocative stimulus into a different—sometimes very different— *yet still effective* evocative stimulus, while the other stimulus simply remains unchanged. As we will see in an example, this is a valuable outcome, especially if the final evocative stimulus form could not originally evoke the response.

In the other outcome, the changes make the evocative stimulus into a different—sometimes very different—stimulus, *but one that no longer serves an evocative function,* while the other, unchanged stimulus actually becomes the evocative stimulus. As we will see in a different example, this too is a valuable outcome, especially if the unchanged stimulus could originally not clearly evoke the response.

In both of those cases, whether changing the evocative stimulus form or making the unchanged stimulus evocative, the control that a stimulus exerted at the start of fading gradually changed into control that a different stimulus exerts at the end of fading. The point of the fading procedure is to get evocative stimulus control to transfer from one stimulus, or dimension of a stimulus, to another stimulus, or dimension of a stimulus.

Notice also that something "extra" often happens in fading. In the usual evocation training process that makes a stimulus function as an evocative stimulus, errors are common. We call each unreinforced response that occurs in the presence of only the non–functional stimulus (i.e., in the presence of the S^Δ, as we previously described) an error. In fading, however, the stimulus that evokes the response at the end of fading usually comes to evoke the response errorlessly, that is, without the errors that occur during regular evocation training. Fading can produce *errorless evocation.*

In the past we called this errorless "discrimination," but language contingencies led too many people to misuse the term "discrimination" with inner–agent implications (e.g., wondering *"Who* was doing the discriminating?") such that we rarely use it anymore; see Ledoux, 2014, Chapter 12, for details. Let's see if such errorless evocation, in our two examples of fading, evokes neural (i.e., non–agential) recognition responses on your part.

Two Fading Examples

Here are two examples of fading. Both examples involve two stimuli. Gradually changing one stimulus results in a stimulus–control transfer. In the first example, *the transfer is to the changed form of the stimulus that underwent gradual change.* In the second example, *the transfer is to the other, unchanged stimulus.* As a matter of record, Whaley and Malott also inspired both of these examples (1971, pp. 195–201).

Changing the evocative–stimulus form. Let's say you needed to teach a young child to respond to her name by pointing it out from among at least two possibilities. Through the fading procedure, you succeeded. Here is an example of how that could have happened.

You began by putting the child's name, Laurie, in white letters on a black card, and you put another name, Debbie, in white letters on another black card. Then you presented the cards to Laurie along with the stimulus, "Point to your name." Unfortunately, she merely took the cards and waved them around, or ran around the room with them, only rarely pointing to one or the other. However, you had earlier interactions with Laurie, and these showed that she likes small fish–cracker snacks. So you made giving her a fish cracker contingent upon her pointing to the card with her name on it. Actually the full contingency would involve four terms. The stimulus of "Point to your name" was to make the card with her name on it evoke the response of pointing (to this card) which would produce a fish cracker.

At first, under that contingency, responses like running with the cards or waving the cards or pointing to a card *all* occurred, in a mixed up fashion, under the general evocative range of all the various stimuli present in the situation. Before long, though, this contingency maintained pointing while gradually extinguishing the non–pointing responses. However, the stimuli continued to evoke pointing to *both* cards. Sometimes she pointed to the card without her name, and so no fish cracker occurred. At other times she pointed to the card with her name, and so a fish cracker occurred. But little else happened. She still pointed to both cards. The fish crackers were insufficient to make the card with her name on it exert enough stimulus control to generate pointing only to it while extinguishing pointing to the card that lacked her name.

Still, Laurie's responses now at least only involved pointing to the cards. However, even though you often reversed the position of the cards, roughly half of Laurie's responses were still errors with respect to her name. She seemed to be pointing to whichever of the cards was closer to her. Her name was not exerting stimulus control over her pointing responses.

At this point you invoked the fading procedure. You took the white letters of her name off the black card, and moved them to a white card of the same color as the letters; this card duplicated every aspect of the other name's black card except color. In addition, so that Laurie would not satiate on fish crackers, you began providing a fish–cracker reinforcer for only some of her correct responses. These now involved pointing to the white card with her white–lettered name on it. Within just a few trials, she was consistently pointing to the card with her name on it, although *the overall whiteness was the stimulus controlling her responses,* not her name.

Next, you instituted a series of gradual changes. Every time she pointed to the correct card on 40 consecutive trials, you moved the letters of her name from the current card to a new card of a *slightly* darker shade. And you found that she continued to make correct responses after every such shade change.

After 11 or 12 gradually darkening shade changes, the card with her name, *Laurie,* on it—in white letters, as always—was *the same shade of black* as the card with the other name, *Debbie,* on it, which was the card–name combination that you had never changed. More importantly she still continued to point to her name card. Fading succeeded.

The stimulus control that the white–name–on–white–card stimulus originally exerted, on Laurie's pointing behavior, gradually transferred to the white–name–on–black–card stimulus that ended the sequence of gradual changes. The only difference between the two cards at the end of this sequence was the difference between the names, *Laurie* and *Debbie.* Now, instead of whiteness controlling her pointing responses, her name *Laurie* (in white letters on a black card) controlled her pointing responses.

Also, throughout all the changes to ever darker card shades, Laurie's responding was errorless. When her name card and letters were both white, the whiteness was the stimulus controlling her responses. As the color of this card gradually changed to darker shades, no errors happened even as the stimulus control was shifting from the overall card color to the pattern of the letters of her name. The final *Laurie* stimulus (as white letters on a black card) came to evoke her pointing responses errorlessly, an example of errorless evocation.

In this name–card fading example, at least one early version of which really happened (see Whaley & Welt, 1967) stimulus control transferred to the *changed* form of the stimulus that underwent gradual change. Now let's look at a different fading example. In this next example, stimulus control transfers to the *unchanged* stimulus.

Making the unchanged stimulus evocative. Let's say you needed to test the hearing of children who lack language behavior. Through the fading procedure, you succeeded. Here is an example of how that could have happened.

Like so many of our human behavior examples, this one actually happened (see Meyerson & Michael, 1964). Here, however, we recast it as the story of your behaviorological activities in solving the problem. As you recall from our chapter introduction, late one summer a dozen children turned up in the kindergarten and first–grade cohort of your local school, showing not only little or no language (i.e., verbal) behavior but also a range of other behavior deficits and excesses, some severe. The school reported that typically only one or two such children show up each year in the entering cohort. Out of concern both for these children and to discover why the number was larger than usual, the school called the behaviorology department at your university for help, and the department chair directed the call to you.

Your behavior started under control of the "little or no language behavior" part of the report. Considering that the natural–science analysis of verbal behavior (AKA language) clarifies the vital role of hearing in verbal–behavior conditioning, one of your first questions would have been "What do the children's hearing tests indicate?" However, since hearing tests at the time required at least minimal language skills to make the test steps work, you already knew the answer. "What tests? These kids lack language so we cannot give them hearing tests." Based on that answer, you knew that you would have to find another, and non–verbal, method to test these children's hearing.

Finding another method was vital, because the hearing test results were vital. If a child could not hear at all, or could not hear well, then medical analysis would be appropriate to discover the problem and solve it (e.g., with a prosthetic device or medication or surgery as required). If a child could hear, then a very different range of solutions would need investigation, usually involving far more extensive contingency engineering appropriate to the child's particular kind of difficult behavior excess or deficit.

You also knew that the longer you took to solve this problem, the sooner someone else might move, with little data, to put an as yet unjustified diagnostic label, such as "autistic," on these children. This would likely lead to inappropriate treatments (e.g., unnecessary drugs or psychological therapies or institutionalization; remember, this example happened in the early 1960s, before we had scientific treatments for autism).

Essentially, you wanted to find a procedure that gets the children to "tell" you, somehow, that "I hear a sound" and "I no longer hear a sound." These are the responses that various verbally instructed methods produce in standard hearing tests. Perhaps your recent work with the fading procedure induced your application of fading to help solve this problem.

With the cooperation of the school, you converted a small meeting room into a temporary experimental chamber. In it you set up, along one side, a false wall with some stimuli, a couple of manipulanda (i.e., things for the children to manipulate as responses) and two reinforcer sources. You would manage the stimuli, and maintain the supply of reinforcers, while sitting in the space behind the false wall. The manipulanda were two levers protruding from the wall. You set these several feet apart so that a child could not operate both at the same time. Below each lever you made a small hole in the false wall through which your reinforcer dispensers could drop various reinforcers (e.g., raisins, chocolate chips, peanuts, and various other consumables, or trinkets) into a small wall–mounted tray. Above each lever you installed a light. Turning a rheostat that you placed behind the false wall would dim both lights together in fixed increments. Three feet above the level of the lights, and between them, you also mounted a small speaker on the false wall. You wired the speaker so that it sounded a continuous tone whenever the left light was on, but was silent whenever the right light was on. Only one light was ever on at a time, and the tone was a mid–volume, mid–range tone. Your hearing–science colleagues said that if a child could hear at all, he or she would very likely hear this tone.

Let's consider the children's interactions with this apparatus. First, however, here is a diagram of how the stimulus set–up of this apparatus worked:

Initial S^{Ev} / S$^{\Delta}$ Diagram

Left lever S^{Ev}: Light on (& Tone **on**)   Right lever S^{Ev}: Light on (& Tone **off**)
(Left lever S$^{\Delta}$: Light off [& Tone off])   (Right lever S$^{\Delta}$: Light off [& Tone on])

	HIGH WALL SPEAKER	
LEFT LIGHT		RIGHT LIGHT
LEFT LEVER		RIGHT LEVER
S^{R} TRAY		S^{R} TRAY

To begin, you demonstrated this apparatus with each child, one at a time. You made a few presses on the left lever with the left light lit (while the light on the right was off) and at each press a reinforcer fell into the tray. After giving these to the child, you put the child's hand on the lever, and off the child went, eagerly working the lever, which produced reinforcers. After a moment, that left light went out as the right light came on. Following some unproductive left–lever–press errors, the child moved over to the now lit right light and began to operate the right lever, which now produced reinforcers. Each time the current evocative light went off, the other light came on at the same time. Coming on evoked switching levers and pressing some more. To avoid satiation, you changed the reinforcement schedule from a "CRF" (i.e., "continuous reinforcement") schedule, on which *every* lever–press response produced a reinforcer, to a VR–8 (i.e., a variable ratio–8) schedule, on which an average of eight responses produced a reinforcer (details in a later chapter).

Within a matter of minutes, the behavior of each child came under control of whichever light was on, switching quickly when the current light went off. But so far this was only a nice demonstration of stimulus evocation, with a lit light as the evocative stimulus. How did this help to get each child to "tell" you that "I hear the sound" and "I no longer hear the sound"? Here is how.

To get each child to "tell" you that, you implemented a fading procedure. Recall that whenever the left light is on, a sound plays over the speaker, and whenever the right light is on, no sound plays over the speaker, as our "Initial S^{Ev}/S$^{\Delta}$ Diagram" showed. So you started adjusting the rheostat to make the lights dimmer. Light intensity provided the dimension along which fading operated. And the children continued to respond correctly through every decrease in light intensity. Typically, you gradually faded the lights across half a dozen rheostat settings, spending some time at each setting, until the lights were not only completely off, but you physically removed them from the wall.

(This is not actually a necessary step in the fading procedure.) Yet each child still continued to respond appropriately, now pressing the levers according to whether the sound was playing over the wall speaker or not. After fading out the lights, here is how the remaining stimulus set–up worked:

Final S^{Ev} / S^{Δ} Diagram

Left lever S^{Ev}: Tone **on**   Right lever S^{Ev}: Tone **off**
(Left lever S^{Δ}: Tone off)   (Right lever S^{Δ}: Tone on)

HIGH WALL
SPEAKER

LEFT   RIGHT
LEVER   LEVER

S^R TRAY S^R TRAY

The children were finally telling you, "I hear the sound" and "I no longer hear the sound," according to which lever they were pressing. As the lights gradually dimmed, literally fading out to off, the evocative stimulus control that these lights had exerted on lever pressing gradually transferred to the wall speaker sound, or lack thereof. The lights gradually stopped evoking presses on a particular lever as the presence and absence of the speaker sounds came to evoke the children's responses. And this happened errorlessly.

Throughout the fading changes that made the lights dimmer, the children's responding was errorless. At the start, the lights were the stimuli evoking their responses. Once conditioning produced them, these responses continued virtually without error, even though by the end of fading only the presence and absence of the speaker sounds evoked these responses. The children made essentially no response errors as the stimulus control shifted from the lights to the wall speaker. The speaker–sounds stimuli came to evoke their responses errorlessly, another example of errorless evocation.

Did all your effort make any difference. You bet! As a last step, you shaped the wearing of standard earphones. Your hearing–professional colleagues then proceeded to run a standard hearing test with each child, who essentially told them, "I hear sound" or "I don't hear sound" by operating the appropriate lever. The results enabled your physician colleagues to make a proper diagnosis of each case where the test revealed hearing difficulties, and to move forward with appropriate corrective strategies. The school then placed all the children whose hearing improved into special language–delayed classes to "catch them up" as much and as quickly as possible. What about the couple of children who could hear but still showed severe behavior deficits or excessses? This is about the number that this school typically experienced turning up in each kindergarten or first–grade cohort. The school asked your behaviorology department team to make further evaluations of these children and to arrange for interventions appropriate for each child's particular difficulty. As for why a dozen children lacking language repertoires entered the school that year, we may never know. Unfortunately little evidence ever supported any particular possibility.

Recall that in the earlier name–card fading example, stimulus control transferred to the changed form of the stimulus that underwent gradual change. However, in this hearing–test fading example, stimulus control transferred to the unchanged stimulus. As with many of our examples that have actually happened, bigger–picture contributions followed. In this case the outcome of the Meyerson and Michael intervention (1964) contributed to hearing–science practitioners developing a standard method using fading as part of a procedure to test the hearing of anyone lacking facility in the particular language of the hearing–test provider.

Timeliness of Works Cited, and Conclusion

You probably noticed that nearly all of the works cited, in this chapter and some previous ones, occurred in the 1960s. Certainly other, more recent references are available. These, however, keep us in contact with some of the classic research that has occurred in this science over the last 100 years. For this reason we often provide older references along with our recent ones.✿

References

Ledoux, S. F. (2014). *Running Out of Time—Introducing Behaviorology to Help Solve Global Problem*. Ottawa, CANADA: BehaveTech Publishing.

Meyerson, L. & Michael, J. (1964). Hearing by operant conditioning procedures. *Proceedings of the International Congress on Education of the Deaf, 238*–242.

Whaley, D. L. & Malott, R. W. (1971). *Elementary Principles of Behavior*. Englewood Cliffs, NJ: Prentice–Hall.

Whaley, D. L. & Welt, K. (1967). Use of ancillary cues and fading technique in name discrimination training in retardates. *Michigan Mental Health Research, 1*, 29–30.✂

Chapter 14
Operant Behavior—VII:

Must consequences occur after every response?

Occasional consequences with four basic "schedules of

reinforcement" processes/applications ...

*T*his chapter covers several examples of reinforcement schedules. Examples of related contingencies, however, deserve some attention also. One concerns virtually everyone's typical reaction to a vending machine that fails to deliver the selected goods. You put in your money, push the button for your selection, and wait. Something is supposed to happen—the machine is supposed to deliver your goods—every time, on a continuous reinforcement schedule that we call a "CRF" schedule. If nothing happens, I doubt you think or say, "Oh, well; life is uncertain. I should have eaten dessert first. Since the universe operates magically, maybe next time the machine will deliver my cookies," and walk calmly away. No way! Instead, some other things typically happen. What happens, and why? As part of answering, let's delve more deeply into schedules of reinforcement and related phenomena.

Schedules of Reinforcement

Recently we have alternated between *postcedent* contingency processes (i.e., differential reinforcement and shaping) and *antecedent* contingency processes (i.e., backward chaining and fading). Now, we return to postcedent contingency processes by considering schedules of reinforcement. All of these processes also work as procedures (i.e., interventions) or parts thereof.

In our examples of reinforcement in past chapters, we generally stated or implied that reinforcers occurred after each response. This gives us the reinforcement schedule that we traditionally call CRF. It means that reinforcement occurs continuously. Occasionally we pointed out that reinforcers could occur after only *some* responses rather than after *all* responses. We resisted, however, adding this increased complexity until we could focus on it thoroughly. Now we can.

Now, in some systematic detail, we consider the basics of the schedule topic, the topic of intermittent reinforcement. Several all–natural (as no other kinds exist) contingency arrangements manage the occurrence of reinforcers after some, rather than all, responses. These comprise the intermittent–

reinforcement schedules. We begin by covering the four fundamental reinforcement schedules, followed by the building of schedule performance as part of a range of related schedule considerations.

Four Fundamental Schedules

The term *schedules of reinforcement* refers to the contingency patterns of reinforcer occurrence. Many reinforcers occur occasionally rather than after every response. These schedules provide a range of common contingency causes for the behavior patterns of humans and other animals, in both individual and social circumstances. These schedules also replace a host of putative inner-agent causes of behavior, which do not work anyway.

We consider four schedules as fundamental due to their combination of response and time factors. We differentiate these four into two categories. One category concerns the number of responses since the last reinforcer occurred. We use the term *ratio* schedules for the schedules in this category. The other category concerns the amount of time—plus a contingent response—since the last reinforcer occurred. We use the term *interval* schedules for the schedules in this category. The values of either ratio or interval schedule can be *fixed* or *variable.* The resulting combinations define the four fundamental intermittent schedules of reinforcement. They are fixed ratio (FR), variable ratio (VR), fixed interval (FI), and variable interval (VI).

As we describe each of those four basic schedules, notice that each produces its own characteristic response–rate pattern. For simplicity here, we will analyze each of the patterns from these four schedules in terms of just two characteristics of the response rate *under established (i.e., stable) responding.* One characteristic concerns whether the response–rate pattern appears generally high or low. The other characteristic concerns whether the response–rate pattern appears generally steady or unsteady. From observations of these response–rate characteristics of a behavior of concern, you can deduce the general schedule type likely controlling the behavior. Keep checking Figure 14–1 for these characteristics as we consider each of the four fundamental schedules in turn.

Note that Figure 14–1 employs data from stable (i.e., *not under construction*) schedules. We plot these data "cumulatively." Here is a brief partial description of "cumulative recorders" to enable reading Figure 14–1 effectively:

A cumulative recorder basically consists of a roll of paper (or its computer/ digital counterpart) unrolling at a constant rate under a pen. As the paper unrolls, this pen marks one tiny, equal step *up* for every response that occurs. This pen only goes up—accumulating responses, hence "cumulative record"— except when it has to reset to the bottom of the page so it can start up again. The width of the paper roll allows the response pen to step up a maximum of 400 responses before the pen resets to the bottom of the paper and starts up again. However, since the paper is always unrolling, this pen is always *also drawing a horizontal line.* These two motions result in an angled line that we call the *response slope.* That is, the two motions of always *horizontal,* as the paper unrolls under the pen as time passes, and sometimes *vertical,* up a tiny step for each and every response that occurs, draw various kinds of sloped lines recording responses in real time. Look at the various slopes in Figure 14–1. The response–slope line can directly provide the rate of occurrence for the behavior in responses per time unit (e.g., seconds). A steeper slope shows a higher rate, a shallower slope shows a lower rate, and a changing slope shows a changing rate.

When a response produces a reinforcer, the response pen also makes a tick, or hatch mark, off the response–slope line at the same step that records the response occurrence. Since the ink line is thicker than the space between response steps, a run of reinforcers, when every response is earning reinforcement, makes all the reinforcer hatch marks run together as a thick and solid black line. Figure 14–1 only shows some of these features. (For more details on cumulative recorders or cumulative records, analog or digital, see Ledoux, 2014, Chapter 8.) Also, note that discussing these schedules can seem tedious, *but they cause so much behavior that the tedium pays off in understanding.*

Ratio Schedules

When a reinforcer is contingent *only* on some number of responses occurring since the last reinforcer, we describe the schedule as a *ratio* schedule. This number of responses per reinforcer (i.e., the *ratio* of responses to reinforcers) can have a set value or a changing value.

Fixed–ratio schedules. If the value of this number remains constant, then we call the ratio schedule a *fixed–ratio* (FR) schedule. Let's set the responses–to–reinforcer ratio at 100 to 1 (although it could be any value). In this case, a reinforcer follows every hundredth response, 100 responses for each reinforcer. We denote this schedule by writing FR–100 (and, in a manner similar for all reinforcement schedules, we read FR–100 as "*ef–ar* one hundred").

The value, of course, could be a higher value or a lower value. It could even be one (i.e., a ratio of 1 to 1, one response per reinforcer). A ratio of 1 to 1 would constitute an FR–1, a schedule that we already know as the CRF schedule.

We note several characteristics of behavior on fixed–ratio schedules (see Figure 14–1). Most obviously, the cumulative record shows a step–like pattern. Also, the overall response rate, while not steady, is typically relatively high. We refer to such a behavior pattern as a "break and run" pattern. Once stimuli evoke the first response of a ratio, we observe the organism running through the remaining responses until a reinforcer follows the last response of the run.

After the reinforcer occurs, however, a pause in responding occurs before stimuli eventually evoke the start of another run. We call this pause the *post–reinforcement pause.*

What brings about those basic response pattern characteristics of a high and unsteady rate? The answer to this question can involve some complex variables as well as some simple ones that match the scope of this book.

As for the high rate, recall that reinforcer occurrence depends *solely* on responding. Thus, more reinforcers occur when more responding occurs. This means that the faster responses follow each other (i.e., the higher the rate) the faster reinforcers occur. This part of the schedule contingency could induce continually rising rates. Such rates, however, rapidly produce counter–controlling contingencies, like fatigue, which oppose continually rising rates and lead, by algebraic summation, to the observed, relatively high rate.

As for the unsteadiness of the rate, for starters recall that when responses produce no reinforcers, in the process of extinction, the response rate takes a hit, as the saying goes. Now, while we look at the fixed–ratio cumulative record—in Figure 14–1—and consider the fixed–ratio contingencies, would any responding right after reinforcer occurrence ever produce reinforcement? Since the answer is no, we should experience no surprise when no responding occurs right after reinforcement. However, since further reinforcement remains contingent on completing the next response run, we still experience no surprise when, after a pause, the next run commences. Other more complex variables actually have a greater hand in determining the post–reinforcement pause and its duration, and thus the unsteadiness of the fixed–ratio response rate. However, they go beyond the scope of this book.

The example of fixed–ratio schedules that most people easily recognize, possibly because they see it as making people work too hard, involves piece–rate pay (a pay method that FR *contingencies* produced long before this science discovered the schedule). With FR pay contingencies, regardless of what you are producing, your pay is neither by salary nor by the hour. Your pay, your reinforcer, accrues after completing a fixed amount of product, such as sewing five shirts or filling a bushel basket with apples.

Variable–ratio schedules. If the value of the number of responses per reinforcer changes, varying around some average, then we call the ratio schedule a *variable–ratio* (VR) schedule. Let's again set the responses–to–reinforcer ratio at 100 to 1. In this case, however, a reinforcer does *not* follow every hundredth response but instead follows every *average* of 100 responses. We denote this schedule as a VR–100. That is, on a "VR–100" schedule, a reinforcer would occur, not after *every* hundredth response—which would be an "FR–100" schedule—but after every set of responses, with each set *averaging* 100 responses. For instance, using one of several available methods to arrange this VR schedule, ten reinforcers would occur in 1,000 responses with each reinforcer following a set of responses comprised of between one response and 200 responses.

A few behavior characteristics stand out on variable–ratio schedules (see Figure 14–1). With the cumulative record showing only rare pauses, the overall response rate maintains a fairly constant steadiness at a relatively high rate. What brings about these response–rate characteristics? Regarding the high rate, again reinforcer occurrence depends solely on responding. Thus, the higher the rate, the faster reinforcers occur. Again, this happens until counter–controlling contingencies—fatigue is again a good example—oppose continually rising rates leading, by algebraic summation, to the observed, relatively high rate.

As for the steadiness of the rate, recall that occasionally the variation in ratio sizes includes the occurrence of two (and sometimes even three) reinforced responses *in a row,* or at least very close together. *And such occurrences have happened in the past* as the contingencies in the current VR schedule established stable responding. Contingency arrangements like this induce steady responding. That is, pauses or other delays under these contingencies reduce the amount of reinforcement and so become less likely to occur, which leaves a relatively steady response–rate pattern.

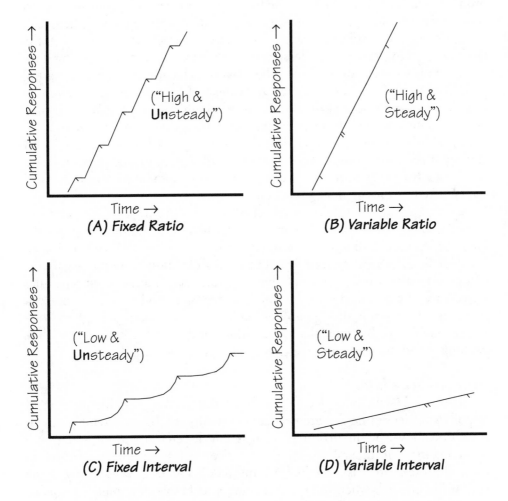

Figure 14–1. Stylized cumulative records showing a typical behavior pattern for each of the basic reinforcement schedules:
(A) Fixed Ratio,　　(B) Variable Ratio,
(C) Fixed Interval,　& (D) Variable Interval.

Sometimes people describe this pattern as a sort of intense *persistence* in responding, but recall our previous discussion about description versus explanation: No "things" called intensity or persistence reside agentially inside the body to cause the continuously steady, high–rate behavior pattern. Instead and at best, the intensity and persistence labels provide verbal shortcuts for longer descriptions of the observed variable–ratio response pattern. At worst, these labels comprise aspects of an inner agential description that is unneeded in accounting for the behavior.

Outside the laboratory vr schedules are common. The example of variable–ratio schedules that many people easily recognize involves games of chance such as gambling. The relatively rapid and steady response pattern that vr schedules produce readily evokes images of the behavior of players on traditional casino "one–armed bandit" slot machines. Hence we often refer to the vr schedule as the "gambler's" schedule. Considering the pattern of behavior observed in the laboratory on the vr schedule, let's examine the behavior of the gambling slot–machine player that you observe. The player's behavior involves a repeating pattern of *put coin in slot and pull handle down* then *put coin in slot and pull handle down,* and so on, again and again, at a high and steady rate, with only occasional interruptions from collecting the relatively rare monetary reinforcer. And this behavior pattern often continues, either on into the night (until closing, if the casino ever closes) or until the player runs quite short of funds (i.e., "loses his or her shirt" as the saying goes). Again, for such reasons, we call the vr schedule the "gambler's schedule."

Contingencies, like those in vr schedules, produced gambling centuries before science discovered and analyzed this schedule. Back then, as now, the laws of nature, including the laws of behavior, affected people, controlling all behavior. So even back then, as now, the vr schedule–induced response patterns compelled purveyors of games of chance intuitively to arrange vr schedules for control of the behavior of their players. However, today as back then, vr schedule effects—not the "gambling habits" of fictitious inner agents—are responsible for the behavior that often reduces individual citizen wealth while swelling government treasury coffers from lotteries and gambling taxes.

Interval Schedules

Those ratio schedules depend solely on the fixed or variable number of responses to determine the occurrence of a reinforcer. Interval reinforcement schedules, however, involve a reinforcer contingent *only on one* response occurring *after* an interval of time passes since the last reinforcer. During the interval no reinforcer is available. A reinforcer becomes *available* only at the end of the interval, when the interval times out. However, a merely available reinforcer remains undelivered. It only occurs after a response occurs. Thus, any responding during the interval remains irrelevant. That is, regardless of how many, or how few, responses happen during the time interval since the last reinforcer occurred, only the *first* response that occurs *after* the interval has ended produces the reinforcer. Even if this response is the only response to occur since the last reinforcer (i.e., even if no responses happened in the interval) or even if this response is but the last one of a series of responses all but this one of which occurred during the interval, *only this one response produces the reinforcer.* Although the reinforcer is contingent upon this one response, the reinforcer may coincidentally affect any other occurring responses. Meanwhile, the length of the interval since the last reinforcer, usually stated in minutes or seconds, can have a set value or a changing value.

 Fixed–interval schedules. If the value of the time interval since the last reinforcer remains constant, then we call the interval schedule a *fixed–interval* (FI) schedule. Let's set the interval at 100 seconds (although it could be any value, from a few seconds to many minutes or more). In this case, a reinforcer follows the first response that occurs after the 100–second interval times out (i.e., after each 100–second interval). We denote this schedule by writing FI–100 seconds. That is, on an "FI–100 seconds" schedule, a reinforcer would follow the first response to occur after each 100–second interval, with each interval starting with the delivery of the previous reinforcer.

 Here, with fixed–interval schedules, we again note several characteristics of behavior (see Figure 14–1). Most obviously, the cumulative record usually shows a scalloped pattern. The *overall* response rate, while not steady, is typically relatively low. What brings about those response–rate characteristics?

 We can view the generally low rate as resulting from a sort of competition between contingencies. This kind of competition applies as well to variable–interval schedules, so we will repeat some details then. For starters the reinforcement–maximizing aspect of the contingencies tends to exert a lot of control over responding. One way to maximize reinforcement (i.e., one way that collects every reinforcer as soon as it becomes available) involves a high rate of responding such that, as soon as an interval times out, a response occurs that collects the reinforcer. However, this wastes lots of energetic effort. The opposite alternative involves very low–rate responding in which *every* response collects a reinforcer with no wasted effort. However, the collecting response might occur long after the reinforcer became available, a wasted time period for the schedule, which means that some potential reinforcers go uncollected, thereby not maximizing reinforcement. The algebraic summation of the effects of these opposing contingencies results in a generally low overall response rate that neither wastes too much effort nor misses many available reinforcers.

 To see better how that works, consider that essentially in 600 seconds (ten minutes) six reinforcers could occur on an FI–100 second schedule. While FI responding actually looks different, responding that occurs once per second would collect every one of these six reinforcers, but at the cost–burden of over 590 unneeded responses, because only six of these responses actually produce reinforcers. Alternatively, responding that occurred rather minimally, say, at only once every 120 seconds (two minutes) would only collect five of the six possible reinforcers (83%), because each 100–second interval *only starts* after reinforcer occurrence. In this case each interval starts 20 seconds "late," and 600 seconds only has room for five such 120–second spans (100–seconds for the schedule interval plus a 20–second late response period). However, a relatively low but not minimal rate of, say, one response every ten seconds avoids about 90% of the wasted response efforts while collecting all six of the reinforcers when they become available. (Yes, you should and can check my math.)

As for the actual, unsteady, scalloped–looking FI rate pattern, our fixed–ratio concern about extinction crops up here also. Responding right after the reinforcer occurrence has never produced reinforcement, so we experience no surprise when no responding occurs right after reinforcement. However, while responding in the interval also produces nothing, the end of the interval still gradually approaches as time passes, and this induces a gradual increase in response rate such that by the time the interval ends, responding has reached a high rate for at least a short period. Since the interval times out during this high–rate series of responses, one of these responses produces the reinforcer immediately. However, this reinforcer, coincidentally occurring during high–rate responding, tends to make high–rate responding more likely, at least as the end of each interval approaches. This slow start, then a gradually increasing middle rate, and finally a fast–rate that leads up to reinforcement, all show up in the cumulative record as a scallop.

An example of a fixed–interval schedule scallop that many people can easily recognize involves behavior while waiting for a bus. You take the bus twice a day, to and from work. (Yes, I have been on this schedule also.) A bus arrives at your stop every 20 minutes but, as happens all too often, you arrive just in time to see one pulling away. That leaves you 20 minutes until the next bus arrives. So you take a seat, dig a good book (perhaps this book) out of your bag, and begin to read. Occasionally you look up, checking to see if the bus is visible. It would still be a couple of blocks away, which is good, as you need time both to put your book away and to dig out the required exact change. However, your looking–up–and–checking responses occur not on a regular basis but on a basis of gradually shrinking periods. For the first ten minutes or so, checking happens only every four, then two, minutes. For the next seven minutes or so, checking happens only every minute, then about every 40 seconds. In the last few minutes before the bus arrives, checking happens every 15 second, then every ten seconds. Finally the reinforcing bus appears. If we plotted your checking responses cumulatively, we would see that they formed a typical fixed–interval scallop, and this scallop would fairly closely repeat each time you arrived at the bus stop just in time to see a bus pulling away. Now, this example describes single intervals, each up to 20 minutes in duration, that typically occur about eight hours apart on each working day. This, of course, differs from the sequentially repeating intervals of a more typical FI schedule. Nevertheless, the example shares enough contingency similarities with typical FIs that they suit our pedagogical point about the kinds of contingencies that produce scalloped response–rate patterns. Also note that while for simplicity we cast this example in terms of *"you,"* the schedule contingencies account for your responding, not any inner agential "you."

Variable–interval schedules. If the value of the time interval since the last reinforcer changes, varying around some average, then we call the interval schedule a *variable–interval* (VI) schedule. Let's again set the interval at 100 seconds, but it is not fixed at 100 seconds. Rather the interval varies around an *average* of 100 seconds. We denote this schedule by writing VI–100 seconds. In this case a reinforcer does not follow the first response that occurs after every 100–second interval, because the intervals differ in length from one another. Instead, the reinforcer follows the first response that occurs after each different–length interval times out. These intervals vary in length from a couple of seconds to several minutes, but average 100 seconds. That is, on a "VI–100 seconds" schedule, reinforcers follow the first response to occur after each interval ends, with the intervals averaging 100–seconds. Thus, using one of several available methods to arrange this VI schedule, ten reinforcers could become available in 1,000 seconds, with the intervals ranging in duration from, say, two seconds to four minutes (i.e., 240 seconds). As usual each interval starts with the occurrence of the previous reinforcer.

Also as usual we note certain characteristics of behavior with variable–interval schedules (see Figure 14–1). Most obviously, the cumulative record typically shows a relatively steady pattern with a relatively low overall response rate. What brings about those response–pattern characteristics?

We can again view the generally low rate as resulting from a kind of competition between contingencies. Recall that reinforcer occurrence under interval schedules depends solely on a single response happening *after* each interval times out. Responding during the interval is irrelevant. However, collecting all possible reinforcers *is* relevant. Yet nothing obviously indicates the passage or end of the interval. If something did, that indication would evoke the one needed response that collects the reinforcer. Instead, one way to maximize reinforcement involves constant, essentially high–rate responding such that, as soon as the interval times out, whenever it times out, a response is occurring that collects the reinforcer. However, this involves a lot of wasted energy, something that nature generally abhors. The alternative involves very low–rate responding, such that every response collects a reinforcer. However, the collecting response might occur long after the reinforcer became available, which means that some potential reinforcers go uncollected. The combination of these opposing contingencies results in a generally low overall response rate that neither wastes too much energy nor misses many available reinforcers.

As for the steadiness of the rate, recall that occasionally the variation in interval sizes includes some short intervals that allow the occurrence of two (and sometimes even three) reinforced responses in a row, or at least very close together. Such occurrences will have happened in the past while the variable–interval schedule was establishing the stable behavior pattern of the current organism. Contingency arrangements like this induce steady responding. That is, scallops or other pauses or delays under these contingencies reduce the amount of reinforcement and so become less likely to occur, which leaves a relatively steady response–rate pattern.

An example of a variable–interval schedule that people can easily recognize involves hunting. Humans and other animals engage in various activities that provide food, such as hunting. Hunters tell me that one type of hunting involves waiting in a particular place and checking the different available paths for prey (and catching it when it comes along). The length of the interval, during which checking–the–paths–for–prey responses occur, may average several hours (e.g., a VI–5 hour schedule). But any particular prey–waiting time can range from a few minutes to longer than daylight. Multiple prey may appear on a path within minutes of each other, or no prey may appear all day long. Since prey may wander by at any moment, the checking–the–paths–for–prey responses occur in a steady pattern at a relatively low rate with respect to each available path, which is a typical variable–interval schedule–induced behavior pattern.

Response–Rate Patterns Predict Schedule Types

Recall that each of the four basic schedules produces its own characteristic response–rate pattern (see Figure 14–1). From these patterns you can predict, through deduction, which basic schedule is likely responsible for a pattern of behavior that you observe, at least as part of your starting point for analyzing a behavior of concern.

We analyzed each of the patterns from these four schedules in terms of two characteristics of the response rate under established responding. Remember, established responding refers to the stable responding that occurs when schedule performance is no longer under construction. Was the pattern response rate generally high or low? And was the pattern response rate generally steady or unsteady (e.g., stepped or scalloped)? Can you already see how to deduce the general schedule type likely controlling the behavior from observations of these response–rate pattern characteristics? What schedule characteristics correlate with which response–rate pattern characteristics?

When the answers to those questions begin to evoke correct responses, you have another reason to call the VR schedule the gambler's schedule. Why? Because we can cast the predicting of basic schedule types from response–rate patterns in terms of the gambler's schedule. You begin by gambling that you can recall the rate characteristics of the gambler's schedule (i.e., a high and steady rate). Then, you sort all the schedule contingency parts (i.e., ratio or interval, and fixed or variable) by all the behavior–rate pattern characteristics (steady/unsteady rate, and high/low rate). This gives you several conclusions: Ratios tend toward higher rates (while intervals tend toward lower rates) and fixed values tend toward unsteady rates (while variable values tend toward steady rates). As a result of this exercise, let's address just the gambler's VR schedule contingencies and rate characteristics. We see that the VR schedule characteristic of high rate correlates with the ratio contingency. And we see that the characteristic of steady rate correlates with the variable–value contingency. Variable ratio correlates with steady and high rate, respectively.

Observing the rate characteristics of a behavior of concern leads to deduction of the likely schedule controlling the responding. If you observe a relatively high and unsteady rate, then you deduce a fixed–ratio schedule (fixed, from the unsteadiness of the rate, and ratio, from the highness of the rate) as it is the schedule that most commonly shows the high but unsteady rate characteristics. Gambling that you can stay focused on the rate characteristics of the gambler's schedule (high and steady) go ahead and deduce several other examples that you can imagine observing.

VR Schedules and Comet Discoveries

Before moving on to other reinforcement–schedule considerations, let's touch on one of innumerable examples of reinforcement–schedule impacts on science. Over a period of more than 40 years, astronomer Alan Hale has sighted over 500 comets, one of which was a major comet that he co–discovered with Thomas Bopp in 1995. That comet, Comet Hale–Bopp, was his 199th sighting and, arriving in 1996, was also one of the really spectacular comets in the last 100 years. However, given my high northern latitude at the time, I must admit to personally enjoying Comet Hyakutake more, with its sky–spanning tail. It arrived a year later in 1997. Hyakutake's arrival also coincidentally correlated with my promotion to full–professor, which happened shortly after its arrival. The occasional occurrence of this kind of coincidental correlation in the past has, over the centuries, induced the superstition that comet arrivals definitively indicate the occurrence of notable events, good or bad.

Note again, though, that Comet Hale–Bopp was Alan Hale's 199th sighting of a comet. Was a comet visible every time he looked? Of course not! Hale looked often, every night that he was at his telescope, night after night (weather permitting). The contingencies on looking through his telescope involved no required time intervals. They only required the observing responses. Hence his behavior occurred under variable ratio schedule contingencies.

Other Schedule Considerations

Two additional areas of schedule considerations deserve our attention. One pertains to schedule characteristics, such as resistance to extinction. The other, which we tackle first, pertains to the building of schedule performances.

Building Schedule Performance

Three areas require comment regarding the building of schedule performances. These include (a) recognizing the difference between later, stable performance on a schedule, and the earlier, constant but gradual changes in contingencies that change behavior while building up a schedule, (b) the role of the extinction burst in schedule construction, and (c) the importance of stretching—while not straining—the schedule.

Schedule construction versus performance stability. We previously pointed out that our cumulative graphs of various schedule performances show the behavior pattern of performances that have stabilized on their respective schedules. In the laboratory, depending on the subject species, this stable responding may have taken an experimenter from days to months to establish. Behavior that occurs under the contingencies of *schedule building* in this period may differ widely from the stable behavior that occurs on the established schedule later. Also much schedule–induced behavior of concern outside the laboratory may exist only under changing, schedule–building contingencies.

The extinction burst. Data show (see, for example, Ferster & Skinner, 1957) that when a behavior ceases to produce reinforcement, the rate of that behavior momentarily *increases* before extinction sets in. We use the term *extinction burst* to describe this momentary increase in rate. Experimental researchers and contingency engineers often take advantage of the extinction–burst to build intermittent reinforcement schedules.

For example, say you have a laboratory research subject who has been responding steadily with presses on a telegraph key, with each press producing points that are worth money. Yet the study in which he is participating requires establishing stable key pressing on some intermittent reinforcement schedule, say FR–450. What would happen if you simply changed the contingencies from a reinforcer following every single response to a reinforcer following every 450th response? You are getting the hang of this if you said that you think your subject's behavior would extinguish long before ever getting to that 450th response the first time. Of course, this reminds us that just as we only reinforce behavior—not people—we also only extinguish behavior. We are not involved with extinguishing subjects, which is illegal, immoral, and so on.

Instead of going directly from CRF to FR–450, we gradually build the ratio up to FR–450. We describe this as gradually *stretching the ratio* by taking advantage of the extinction–burst. When you first stop reinforcing key presses at the beginning of stretching, your subject's pressing rate briefly increases. In short order the first 15 responses of an extinction burst may have occurred in the time he usually took to make five responses. At that point, before the burst rate decreases, you again provide reinforcement, and for a while you continue to reinforce every 15th response. As soon as this FR–15 schedule shows some stability, you can stretch the ratio some more, again taking advantage of an extinction–burst. This stretching goes on repeatedly until, after enough intermediate steps, you reach the FR–450 that the study requires.

Stretching versus straining the schedule. If, during the stretching steps, you go too many responses before again providing reinforcement, then you *strain the ratio* (i.e., start an extinction contingency) rather than merely *stretch the ratio*. In this case as your subject's key pressing extinguishes, he or she may become temporarily emotionally unstable and possibly damage the equipment under the usual exaggeration that occurs during emotional arousal. Either of these—the damage or the extinction—would unpleasantly delay your research.

For another example, consider some professional gamblers (e.g., poker players and pool hustlers). They come under a particular contingency that has been shaping their behavior since long before behavior science discovered the ratio–stretching process. This contingency involves letting the victim win while gradually stretching the victim's schedule ratio. This process keeps their "pigeon" victim playing—and losing—the game for longer periods. In the beginning, the professional gambler arranges for the victim's playing to produce small wins frequently on low–ratio VR schedules. Gradually, however, the victims lose more than they gain in their wins. Wins now occur only occasionally, and usually with little at stake, as the card sharks or hustlers gradually stretch the ratio, switching to higher–ratio VR schedules. Whether they stretch the ratio through skillful playing or cheating is beside the point (although skillful playing seems more common). In either case the high and steady response rate induced by the VR schedule countercontrols the effects of other contingencies that might otherwise induce a quicker exit by the victim from the games.

That kind of problem permeates today's casinos! The only way to reduce it long term involves everyone understanding this science, which provides the contingencies that indeed induce a quicker exit from such rigged games.

Extinction and Intermittent Reinforcement Schedules

Extinction relates to these intermittent reinforcement schedules in a couple of ways. One involves the *extinction burst,* which we already covered. The other involves a phenomenon that we call *resistance to extinction.*

Resistance to extinction. Numerous responses go without reinforcement on intermittent reinforcement schedules. Nevertheless, *the responding on these schedules is more difficult to extinguish* when compared to the relative ease with which responding extinguishes if reinforcement had been occurring after every response. Our measure is the number of unreinforced responses that occur during extinction. More occur after reinforcement had been intermittent than after it had been continuous. To describe the extended responding under extinction after intermittent schedules, we speak of *resistance to extinction.*

The general and counter–intuitive sounding rule describing resistance to extinction holds that *the more similar a schedule is to extinction, the more the responding on that schedule resists extinction.* The similarity to extinction derives from the number of *un*reinforced responses *occurring regularly on the schedule,* compared to extinction in which *all* responses go unreinforced. For example, FR–1,000 resembles extinction more closely than FR–10 on this basis. FR–1,000 has 999 *un*reinforced responses for each reinforced response (i.e., 99.9% *un*reinforced) versus only nine *un*reinforced responses for each reinforced response on FR–10 (i.e., 90% *un*reinforced). In this case we find that far more individual responses occur in extinction after FR–1,000 than after FR–10. The behavior that was on FR–1,000 *resists extinguishing* more than the behavior that was on FR–10. More accurately, FR–1,000 contingencies induce more stable mediation of the behavior than FR–10 contingencies induce.

Is FR–1,000 an extreme schedule? Not really. For example, to discover whether or not nature limits how large a ratio can get, some researchers continued to stretch the ratio for their pigeons' behaviors the way some business managers try to stretch the ratio of their employees' behaviors. Fortunately, for those employees with union representation, their union provides enough countercontrol to prevent the ratio from stretching beyond the point of hurting people. The researchers working with pigeons stopped before getting that far too. They did, however, get as far as FR–10,000. If you saw these pigeons working, you might swear that they have a compulsive personality, or that they are obsessed with pecking the key before them. Neither is the case, for pigeons or for humans whose behavior on similar schedules shows similar characteristics. Such inner–agent causes drive neither pigeons nor humans. The behaviors are driven by the contingencies. Another contingency on the experimenters that stopped the ratio stretching at FR–10,000 involved a particular discovery. Of course, the researchers free–fed the pigeons after each daily hour–long experimental session. And each time the pigeons pecked the key 10,000 times, they ate from a grain tray for several seconds as the reinforcer. Nevertheless, the researchers discovered that the number of calories that the pigeons burned while pecking the key 10,000 times for each reinforcer was more than the number of calories that they got from the grain they ate as a reinforcer. Without the free feeding, the birds would have starved.

What about extinction? The pecking of these birds (across daily sessions after which the free feeding continued) showed the most resistance to extinction. Under the extinction contingency, far more responses occurred with these birds than occurred with birds on FR–1,000, before responding stopped.

Regarding resistance to extinction, for more than 50 years, many researchers, called this the *partial–reinforcement effect* (PRE). They tried to discern what could make the occurrence of fewer reinforcers result in a greater number of responses. They have offered many suggestions. Most of these, however, arose during the period in which historical contingencies forced natural scientists of behavior to cohabit academic departments with psychologists (see Ledoux, 2015). Thus we are not surprised to find that many of these suggestions have included discredited inner–agent accounts. Of those that adhered to naturalism, the "response–unit" account may have the most empirical support. This account suggests re–conceptualizing a response unit as the complete behavior requirement that produces the reinforcer so that, for instance, one response is the response unit on a CRF schedule, and 100 responses comprise the response unit on an FR–100 schedule. That is, the "unit" that produces reinforcement is a multiple of the unit that usually we have called "a response." While this suggestion may go a long way to explaining the PRE, other suggestions remain in the research hopper. In any case these explanations need also to square with physiology, and its mediation of behavior.

Reinforcement Schedule Summary

Basically, reinforcers are postcedent stimuli whose occurrence produces increases in the frequency of the behaviors that they follow. *Schedules of reinforcement* are the patterns of intermittently occurring reinforcers. These schedules are defined either in terms of the *number* of responses since the last reinforcer occurred (i.e., *ratio* schedules) or in terms of the amount of *time*—plus a required response—since the last reinforcer occurred (i.e., *interval* schedules). The values of either type can be fixed or variable, thereby defining the four basic intermittent schedules of reinforcement: fixed ratio (FR), variable ratio (VR), fixed interval (FI), and variable interval (VI). For example, on an "FI– 30 second" schedule, a reinforcer would follow the first response to occur after each 30–second interval since the last reinforcer, while on a "VI–30 second" schedule, a reinforcer would follow the first response after each interval, with intervals *averaging* 30 seconds in duration. On a "VR–30" schedule, a reinforcer would occur, not after *every* thirtieth response—which would be an "FR–30" schedule—but after every set of responses, with each set *averaging* 30 responses.

Outside the laboratory VR schedules are common. They produce relatively rapid and steady response patterns. For centuries before science discovered and analyzed this schedule, these response patterns compelled purveyors of games of chance intuitively to arrange VR schedules for control of the behavior of their players. And still today VR schedule effects, rather than the "gambling habits" of fictitious inner agents, are responsible for much individual citizen wealth reduction as well as for swelling government treasury coffers from lotteries and gambling taxes.

Overall, schedule research has repeatedly led to several general conclusions. The most applicable include these three: (a) Many features of behavior emerge as the effects of the contingencies comprising particular reinforcement schedules. (b) Schedules with only subtle contingency differences often produce distinctly different response patterns. And, perhaps most importantly at the present time, (c) the direct effects of schedules of reinforcement reduce a wide range of putative inner–agent emotional and motivational causes of behavior to misleading redundancies that deserve avoiding. The tedium is worth it!

Conclusion

Now we are ready to revisit the vending–machine failure example from the start of the chapter. As you recall it concerns the typical reaction to a vending machine that fails to deliver the selected goods. You put in your money, push the button for your selection, and wait. Sometimes, however, nothing happens. As we already discovered, your thoughts and actions run far from "Oh well, life is uncertain… Maybe next time the machine will deliver my cookies." And few if any ever really walk calmly away. So, what happens, and why?

Actually, several other things happen. For many people, the earlier response of putting money into such offending vending machines extinguishes instantly. For some of them, this reaction generalizes to all vending machines. At least I have never witnessed someone continue to feed money into a malfunctioning machine, although one machine serviceman informed me that this does happen. Much more observable, however, is the extinction burst of the button pushing response. This response typically increases briefly in rate, and even in intensity. My machine–serviceman informant even reports that this extinction–burst button pushing sometimes breaks the buttons. Additional extinction–induced and emotionally exaggerated behavior also occurs, such as loud cursing and kicking the machine. Sometimes this hurts the machine. More often it hurts the foot. What behavior has the loss of money to offending vending machines caused in your case? Under the contingencies that this chapter's new information provides, the next time a vending machine fails, maybe just walking away makes a reasonable reaction, along with simply seeking a reimbursement from the machine's owner.

"Pre–Part II" Topic Extensions

An "Introduction to Behaviorology" textbook (i.e., Ledoux, 2014) would cover some further, technical–side, chapter–long topics before proceeding to all the additional general topics that we cover next, in Part II. Of course, interested readers can cover these topics later. The topics include laboratory methodology, practical methodology, experimental research, stimulus equivalence relations, elementary verbal operant analysis, and multi–level evolutions, among others.❖

References (with some annotations)

Ferster, C. B. & Skinner, B. F. (1957). *Schedules of Reinforcement.* Englewood Cliffs, NJ: Prentice–Hall. This comprehensive book describes the authors' multitude of laboratory studies researching a wide range of reinforcement schedule parameters. I understand that, for over a decade, it was deservedly one of the most widely cited references in the literature of the traditional natural sciences, and we still hold it in high regard.

Ledoux, S. F. (2015). An introduction to the origins, status, and mission of behaviorology: An established science with developed applications and a new name. In S. F. Ledoux. *Origins and Components of Behaviorology—Third Edition* (pp. 3–24). Ottawa, CANADA: BehaveTech Publishing. Originally written in 1990, this paper was delivered to several audiences before appearing (2004) in *Behaviorology Today,* 7 (1), 27–41.

Ledoux, S. F. (2014). *Running Out of Time—Introducing Behaviorology to Help Solve Global Problem.* Ottawa, CANADA: BehaveTech Publishing.☙

PART II

Complex Considerations and Contingencies, with Scientific Answers to Some Long–Standing Human Questions

Photo by Stephen F. Ledoux

Burrhus Frederic Skinner

(1904–1990)

Conversing at a convention in 1982

The products of the contingencies of his life established behaviorology.

Chapter 15
Aversive Controls Over Behavior:

Why is punishment easy, but unsuccessful in the long run? Inherent problems with, and alternatives to, aversive controls, especially in eight coercion traps ...

*T*his chapter flips the consequence coin to introduce some concerns about how *coercion* operates in contingencies controlling behavior. Coercion includes punishment, the threat of punishment, and even subtracted reinforcement. These three comprise the main components of aversive postcedent control. This control works in the opposite way—in both effects on behavior and emotional effects—of reinforcing postcedent control, particularly added reinforcement.

Control by coercion and control by added reinforcers has *always* been with us. For example over 3,000 years ago, the Hittites usually sold into slavery those who fought them and lost. In contrast to that aversive consequence, the Hittite Kings settled other captives on vacant land, gave them "supplies to start farms, and granted a three–year remission on taxation" for farming (Dise, 2009).

In spite of both kinds of control having been with us for so long, our pre–scientific human experience makes us see these two kinds differently. We overemphasize aversive control as if it were the *only* kind of control. Meanwhile we take little notice of control by added reinforcers. Our historical and cultural literature of freedom emphasizes ending aversive control but largely ignores, even denies, the vital effects on our existence of added–reinforcement control (see Skinner, 1955–56, 1971). This ignorance can lead to disastrous global outcomes (discussed later). Since we have emphasized added–reinforcement control in several chapters, we now cover coercion control.

Behavior control, which is ever present and always operating, comes in the two flavors that we call coercive control and positive control (i.e., *positive,* in its original meaning of *good)*. Both involve natural laws of behavior. To enhance successfully dealing with ourselves and the world around us, these laws of behavior require thorough investigation. Sharing the results with everyone supports some major contingencies that counter–control any contingencies inducing the misuse of these laws. As we saw through previous chapters, we have made some progress along these lines. Still, we have much to discover about the laws of behavior, including the troublesome laws involving coercion.

So, taking into account the effects and implications of what we have already discovered, here we cover the basics of aversive control, its problems, and its alternatives, in three respective sections. After covering escape and avoidance, we move on to eight coercion traps. Then we finish with some alternative ways of dealing with the kinds of behaviors that too easily otherwise evoke punishment and other coercive methods of control. First, however, we briefly review punishment, some of its problems, and why we so often and so easily respond to the behavior of others with punishment.

Punishment Review

Recall that punishment, as a postcedent process or procedure, occurs either as the addition, or the subtraction, of stimuli. The stimulus addition or subtraction functionally results in a reduction in the ongoing rate of the behavior that the stimuli followed. The rate reduces because the energy trace from these postcedent stimuli alters nervous–system structure. As a result the antecedent stimuli that usually evoke the behavior function less effectively. They no longer as readily evoke the punished behavior. Recall our example of getting shocked when replacing a burned out light bulb in a plugged in lamp. We call this added punishment. Future burned out bulbs will no longer as readily evoke the same behavior.

Unfortunately you see added reinforcement happening too rarely when you look around. Instead, nearly everywhere, you see the common kinds of coercion. You see subtracted reinforcement, in which some behavior produces the *end* of punishment (as ending hand contact with the lamp produces the end of the shock). Or you see the threat of punishment. Or you see actual punishment. When an adult offends, the offended takes a swing. When children act out, a parent threatens them or spanks them or both. When a pupil misbehaves, paddling or its legal substitutes occur. When countries disagree, war often follows. *We cannot recommend any of these as satisfactory solutions.* When contingencies drive secular–law violations, feelings of guilt—if the necessary emotional conditioning has occurred—and fines or jail follow. When contingencies drive religious–law violations, feelings of sin—if the necessary emotional conditioning has occurred—and penances, excommunication, or—some folks say—eternal damnation follow. The data on coercion and its problems compel my support for any contingencies that make you find such practices in need of improvement if not outright replacement.

Despite its wide use, punishment has many problems; here we mention only a few. We describe the use of punishment as *incredibly risky.* In the long run, it drives wedges between people (e.g., between parent and child, and between teacher and student). It induces disabling emotional anxieties, and exaggerations of some subsequent operant reactions. It compels those on the receiving end *to get away* from the punisher, *to stay away* from the punisher, *and to get even* with the punisher. In technical terms we refer to these last three major effects of punishment—getting away, staying away, and getting even—as *escape, avoidance,* and *countercoercion.*

We often consider countercoercion as counteraggression. Coercion is a form of aggression, and an attack induces those under attack to counterattack. The punisher usually bears the full force of the counterattack but, if that is not possible, then others or even objects bear the force of the counterattack. This often takes an exaggerated form due to emotional components. These effects can even flow through a community like a wave. A boss reprimands and threatens to fire an employee who goes home and yells at the spouse who yells at a child who yells at the cat that goes out and screeches at the neighbors who yell at the police officer over the phone about the screeching cat—and everyone lives happily… Oops; sorry, wrong ending.

Meanwhile, given such problems, we must wonder why we and other people so often and so easily respond to the undesirable behavior of others with punishment. Perhaps the biggest reason is that punishment *is* an effective process, when it occurs in certain ways, and especially in the short term. If a punishment of adequate (i.e., not excessive) *intensity* is *contingent* on a particular behavior, and *immediately* and *consistently* follows this behavior every time it occurs, then the punishment rapidly produces a sizeable decrease in the rate of the behavior that the punishment follows.

Due to that effectiveness, the immediate consequences of punishment affect its use far more than the long–range consequences affect its use. The immediate consequences of using punishment *reinforce its use,* but the long–range consequences are disabling. The long–range effects of punishment involve those problems of emotional effects and escape, avoidance, and countercoercion (among others). But the rate reduction of punished behavior *reinforces the behavior of applying punishment,* which we see later as increases in the rate of applying punishment. This accounts substantially for why we so often and so easily respond to the undesirable behavior of others with punishment. Other kinds of reinforcers can follow punishment use as well, such as when a manager praises a foreman for keeping staff noses to the grind stone through punishment (an added reinforcement for punishment use) or when a principal stops criticizing a teacher when the teacher "gets tough" with the students through punishment (a subtracted reinforcement for punishment use). These also increase the rate of punishment application, even though available alternatives would more effectively deal with the behaviors of concern.

If you are on the receiving end of punishment, however, the stimuli evoke those behaviors of escape (i.e., getting away), avoidance (i.e., staying away), and countercoercion (i.e., getting even). Let's consider escape and avoidance in more detail.

Escape and Avoidance

To appreciate escape and avoidance contingencies, understanding the disadvantages of some common terms helps. We should also understand some respondent outcomes of operant consequences.

People often use the term *aversive stimuli,* or simply *aversives,* for stimuli that qualify as punishers. However, these terms can cause confusion, because the terms can apply to either operant or respondent stimuli. Witness calling an added punisher, like a spanking, or a subtracted punisher, like a fine, an aversive, while also calling a stimulus that elicits negative emotional reactions, like the stink of a broken sewer pipe, an aversive. Sometimes the same aversive stimulus has both operant and respondent effects. For instance the stimuli from whacking your thumb with a hammer function as operant consequences (e.g., punishing sloppy swinging). They also function to elicit negative emotional reactions (e.g., anger) with subsequent exaggerated effects (e.g., you throw the hammer with force rather than merely drop it).

That is not surprising, is it? Operant consequences not only alter response rates, but they also elicit respondent effects. Added reinforcers not only increase response rates, but they also elicit positive emotional respondent reactions. On the other hand, punishers not only decrease response rates, but they also elicit negative emotional respondent reactions. What about escape and avoidance?

Escape

We call a response an *escape* response when the result of that response is the *termination* of *ongoing* aversive stimulation. Escape behavior works equally well for escaping from both *unconditioned* and *conditioned* aversive stimuli. However, we traditionally use the term *escape* for responses that terminate *unconditioned* aversive stimuli that are ongoing (i.e., that someone is currently experiencing). This usage leaves room for something we call "avoidance."

Avoidance

In contrast to escape, which has the single outcome of terminating ongoing aversive stimulation, we use the term *avoidance* to refer to a response that has two outcomes: (a) The response *avoids,* as in *prevents* or *delays,* the occurrence of an *unconditioned* aversive stimulus. (b) But the same response can also simultaneously occur in response to, and sometimes *terminate,* an ongoing *conditioned* aversive stimulus. Technically the second outcome makes it an escape response, except that the first outcome already makes us call it an avoidance response.

Escape and Avoidance Example

As usual, escape and avoidance actually involve additional, more complex realities. But you now have enough to appreciate an extensive example—the "Hefferline study"—that interrelates them with other important topics.

Whaley and Malott (1971) summarized a study that Hefferline, Keenan, and Hartford had published in *Science* in 1956. Even after over 50 years, this study bears re–description. The study provides an early experimental example of escape and avoidance with humans. In addition it provides an example of how, for humans, stimuli affect responses without either of these evoking human consciousness behaviors (e.g., awareness). Here is a verbal shortcut way to say this, although it requires ignoring the inherent but unsupported agential implications: We need not be aware of responses or stimuli for either of them to occur, or to affect (e.g., evoke or consequate) our further behavior.

Hefferline and colleagues arranged for 12 human subjects to participate in their study. Each subject took a turn in the comfortable seat in the test room and, for the 80 to 90 minutes of the experiment, listened to music. Noise occasionally sounded on top of the music. When the noise was happening, a tiny but measurable thumb twitch—too small to evoke awareness responses—turned off the noise for 15 seconds (i.e., escape conditioning). When the noise was not happening, the same kind of response *during the music* delayed the onset of noise for 15 seconds (i.e., avoidance conditioning).

The researchers arranged their study as a single–subject experimental design that followed an "ABCA" pattern of phases. The A phase began each subject's experimental session and lasted for five to ten minutes. During this phase music occurred without noise. This constituted a baseline phase for measuring and recording the thumb–twitch behavior of concern prior to any conditioning. The B phase came next, lasted for about an hour, and included the escape and avoidance conditioning. The C phase lasted for ten minutes after the hour–long B phase, and involved an extinction contingency. During this extinction contingency, the noise sounded continuously, but responding had no effect on it. The second A phase followed the extinction phase for a few minutes and, as a reversal phase, returned to the baseline conditions of music playing continuously without any interfering noise. (For details on single–subject designs, see Ledoux, 2014.)

To measure the behavior of concern, the experimenters attached three pairs of electrodes to the subjects, although *only* the wires attached to the thumb and edge of the left hand operated. The experimenters then observed and recorded the electrical voltage of thumb twitches through these wires. Thumb twitches showing between one and three microvolts constituted the behavior of concern. Neither the subjects nor the experimenters could directly observe this imperceptible level of thumb twitch.

The researchers set up four conditions, one for each group of three human subjects. The four conditions differed mostly in that each group received different instructions. The instructions to Group 1 were merely to listen through the earphones. The instructions for Group 2 added that a small, invisible response would turn off, or postpone, the noise, and that their task was to discover that response and use it. The instructions for Group 3 added even more, specifying that a tiny twitch of the left–thumb would affect the noise. The instructions for Group 4 were the same as for Group 3, but Group 4 got to watch a meter, for the first half hour, that indicated when a correct response had occurred.

Each subject also participated in an interview after the experimental session. The questions evoked each subject's verbal–reporting behavior about any motor behavior, and any neural behaviors of consciousness, that various aspects of the study had evoked.

Let's consider the outcomes for each of the groups. As we proceed, consider how fully lawful each outcome is. For Groups 1 and 2, the escape and avoidance conditioning proved effective. Tiny twitches occurred that escaped or avoided the noise. However, while Group 1 members thought that their behavior had no effect on the noise, two members in Group 2 reported giving up the search for the right response half way through the experimental hour. The third insisted that he had discovered and used the required noise–reducing response, which he described as involving hand–rowing movements, ankle wriggling, jaw displacement, exhaling, and waiting. We would describe such a pattern as superstitious behavior. The escape and avoidance conditioning failed for two members of Group 3; their "small" instruction–evoked thumb–twitch responses were not small enough. Indeed they were so large that they interfered with responses tiny enough to work. The conditioning succeeded for the third member of Group 3, but only because he misunderstood the instructions and tried to increase pressure gradually on an imaginary button, which allowed some reinforceable responses to occur. Unsurprisingly, given the presence of the feedback meter, the escape and avoidance conditioning succeeded for all the Group 4 members. More importantly they continued responding appropriately after the meter became unavailable. Apparently, after overt feedback of covert processes brings these processes under the control of explicit experimental contingencies, the overt feedback becomes unnecessary.

In summary, this study provides an early experimental example of escape and avoidance with humans. It also shows that once overt feedback helps bring covert processes under particular contingency controls, the overt feedback becomes unnecessary. Equally importantly, the study exemplifies how, for humans, stimuli affect responses without these stimuli or responses necessarily evoking human consciousness behaviors. In verbal shortcut terms, we need not be aware of responses or stimuli for them to occur, or to affect (as in evoke or consequate) our further behavior. Recall our example of daydreaming taking up the "single" neural consciousness channel while still driving safely. The driving responses correctly occur, not due to any evoked awareness of road features, but under the direct stimulus control that those road features provide.

Escape and avoidance, along with countercoercion, all remain likely outcomes of what Glenn Latham (1998, 1999) called the eight coercion traps. Let's look at these traps, as a prelude to covering *alternatives* to coercion.

Eight Coercion Contingency Traps

Dr. Glenn Latham spent decades working with parents at home and teachers at school. He was discovering the particulars of some parent and teacher behaviors that drive much of the misbehavior of children. He also helped design many of the best practices that help change these parent and teacher behaviors so that they lead to more appropriate child behavior at home and school (see Latham, 1994, 1998, 1999). Sadly, his discoveries are not yet nearly as widely applied as they could be. Perhaps your reading this book will help this happen.

When addressing the best steps parents and teachers can take to improve matters, Latham pointed to what he called "coercion traps" that ensnare people in coercive practices. He delineated what he found to be the eight most common of them. These eight coercion traps can befall anyone. Each trap increases the level of coercion, which then leads to more getting away, more staying away, and more getting even. While Latham cast these traps in terms of children, parents, and teachers, their description applies equally well to peers, friends, relatives, employees, managers, ministers, officers, soldiers, public servants, and diplomats. If I left *anyone* out, count them as included.

(1) The Criticism Trap

The criticism trap involves verbally berating people when their performance is inadequate. In focusing on negative performance aspects, the criticism usually fails to describe, let alone teach, what an adequate performance involves and how to reach it. For example, you may have heard (or said) "You have been cleaning this messy room for an hour but little has changed…" I doubt that any child has ever responded to such criticism by saying, "Oh, thank you for clarifying my inadequate effort. I will take more specific steps immediately, and quickly, to get this place cleaned up to your complete satisfaction." (Consider what you could instead say to the child that might be more helpful.)

Criticism also elicits negative emotions that exaggerate subsequent responses, and these responses are more likely to include various forms of escape, avoidance, and countercoercion. Indeed, *this applies to every one of these eight criticism traps*, so we will repeat it often.

(2) The Arguing Trap

Arguing is neither debating nor discussing, either of which can come to a reasonable, even helpful, outcome. The arguing trap often begins with two people verbally brawling in an un–winnable game of one–upmanship. But it easily escalates further by involving additional players. Tempers fray and flair up as the verbal weapons elicit negative emotional responses, all of which contributes to the usual, arousal–affected outcomes of leaving, avoiding, and hurting back. As coercion rises, positions polarize and nothing gets solved. While arguing provides each party with lots of attention–type added reinforcers—which is part of why it happens—it fails to build useful repertoires. Not only is arguing a most ineffective method for managing behavior, but arguing also takes a prize for counterproductivity. We need no specific example, because you surely have already heard too many arguments, and I doubt I need to suggest that you stay away from the arguing trap.

(3) The Logic Trap

When a question arises, and you are carefully staying away from arguing, logic becomes very tempting. And if the moment is calm, without the usual looming deadline hanging in the air and heightening emotions, then logic may help elucidate why something has happened or why it will happen or how you can improve it.

Instead, however, we often grasp at logic merely to make our own wisdom attractive to another person. We are likely only telling someone something that they already know, or we are giving logical instructions based on traditional lore, or common sense, or maybe reason. Unfortunately, these logic attempts likely provide attention for inappropriate behavior. And all these attempts to manage behavior through logic usually backfire by affecting others as aversive stimulation, which elicits negative emotions and their effects while evoking the three standard coercion and punishment outcomes, namely escape, avoidance, and counterattack.

For example if, in the middle of a school week, a young teen's friends arrive after dinner, and he is to go out with them, the question occurs regarding how late he may stay out. Avoiding an argument, you instead start listing a dozen various logical reasons why he should not stay out late, but before you get even half way, he—possibly politely—cuts you off saying, "We're in a hurry. Just what time?" Logic will not work here; other variables are at play (see Latham, 1994). For you to keep avoiding the aversiveness of an argument, you better not say too early a time, so instead you say a later time, a time later than the time you think you should have said. That is, the previously experienced aversiveness of an argument about the time leads you to stating a time later than the time that other variables had been compelling you to specify, and the removal of the threat of that argument reinforces stating that later time. Now, though, you might ask: How many months must pass before you would have considered your young teen to be even chronologically ready for the time that you stated? Look where the logic trap got you.

(4) The Questioning Trap

If you need information for problem solving, asking questions is not only appropriate but may also be necessary. If you arrive home to find the sitter taking an open bottle of window–washing solution away from your young child and saying that the child had been drinking it, asking the child *why* she was drinking it would only give attention to inappropriate behavior while also wasting time that would be better spent getting medical treatment. On the way to treatment, an appropriate question to ask would be "How much of the bottle did she drink?" That information might be of use to the ER physician.

The questioning trap is not about questions needed for problem solving. The questioning trap involves asking questions about inappropriate behavior, questions like the "why" question. You may get an answer, but it will likely displease you. For example, "Why is your homework incomplete?" "Because I hate the topic." Such answers usually evoke quite negative reactions, both emotional and operant, from the questioner, and a mutually coercive downward spiral begins, a spiral that includes the three standard outcomes of escape, avoidance, and countercoercion.

Prevent all that; never ask a child questions about misbehavior. Here are three more reasons not to ask questions about misbehavior. The questions are threatening, and so tend to evoke lying, evasion, and defensiveness. Asking such questions also provides lots of attention for the misbehavior, which makes it worse. And besides, usually those who ask such questions seldom want answers. They want appropriate behavior, and an answer does not equal appropriate behavior.

Anticipating our later discussion of coercion alternatives, while you should never ask questions about misbehavior, you should ask questions about future behavior. Extending our logic–trap example, and presuming that you have earlier clarified with your young teen that you expect him home by 10 P.M. on such nights, you should now ask "At what time do I expect you back home?" Ignoring all the verbal–behavioral "noise" that might occur, you calmly repeat your question as necessary and, when he finally answers "10 P.M.," you can say, "Thank you. I'm glad you understand that." Rather than you ordering your young teen home at a certain time, which is inherently aversive, *your young teen has told you* what time he should be home, which is far less aversive for everyone. Furthermore, if you have treated questions this way for the last decade, the exchange becomes rather routine and pleasantly lacks the otherwise usual verbal–behavioral noise. Is this not a desirable alternative to any coercive practice? It is a very predictable outcome (again, see Latham, 1994, 1998, 1999).

(5) The Sarcasm Trap

One of the most hurtful coercion traps is the sarcasm trap, which includes pointed teasing. Sarcasm occurs as a sort of shock treatment that supposedly helps manage behavior, but it fails miserably at this task. Sometimes disguised as humor, sarcasm seldom evokes laughs from the receiver, but its coercive effects are clear. These effects include the usual suite of negative emotions building walls between people, along with the standard outcomes of escape, avoidance, and countercoercion. We have all experienced sarcasm. We may have even indulged in it. A common and confusingly hurtful sarcastic comment on a student's poor grade is that, "At least they cannot accuse you of cheating." Imagine your feelings, and likely counterreactions, were someone to say that to or about you.

(6) The Despair/Pleading Trap

This "despair/pleading" trap features two names, because "despair" implies the emotional respondent component of the problem while "pleading" alludes to the operant component. This trap implies that the person whose behavior needs managing is at fault for these problems, and this implication begins the downward spiral of coercion. All the negative emotions and their effects ensue along with the three standard operant coercion outcomes of escape, avoidance, and counterattack. After a history of failed encounters regarding improvements, the circumstance of little or no progress on some often minor issue elicits the despair and evokes the pleading. These typically take forms such as saying, in strained emotional tones, "What am I *ever* going to do with you?… Everything I try gets nowhere!… After working my fingers to the bone day after day to pay your way, this is your best effort?…" And so on. I hope you have never heard such things except in comedy routines, although this trap is not very funny.

Now, what if someone, even the other person, actually responded with an answer? What if they offered to help, to tell you what to do to fix things? Then he or she had better be well out of reach, because the despairing and pleading person's behavior could quickly devolve from the verbal realm to the physical realm. This kind of offer to help is most unlikely to be helpful at this time, even if it is accurate. Just imagine the likely coercive reaction on the part of the despairing pleader if, in reply to "What am I going to do with you?" the other person said, "Well, I can make some suggestions. I have spent some time in the library and online studying the literature of behaviorology regarding alternatives to negative behavior management practices with children and others. With these in hand, I am sure I could tell you how to increase your competence in this area." If you are on the receiving end of the despair/pleading trap, I must recommend against taking this approach. It may allude to reasonable alternatives, but it is a polite yet still in–your–face form of subtle counterattack. It is not one of the reasonable alternatives, not in this form, and not with someone caught in the despair/pleading trap.

(7) The Threat Trap

We fall into the threat trap when we claim that something bad will happen if some performance fails to shape up. The "something bad" seldom involves something "reasonable" but instead is either extreme or absurd. Nevertheless the threat gets the negative emotions going, along with their effects. Often emotions are already running high, and the threat spikes them further. Escape, avoidance, and countercoercion are not far behind. For example, emotionally exaggerated threats commonly concern curfews: "If you are not back by midnight, I will ground you for five years!" Such threats are possibly more inappropriate than the behavior evoking the threat. And, as usual with any of these coercion traps, falling into the threat trap is counterproductive. The threat is useless verbal behavior that emotions like anger exaggerate, evoking subsequent regrets. Furthermore the threat is quite often unenforceable, which calls the threat–maker's competence into question. Even worse than these problems, threats are but a short step away from force. Indeed the abuse, inherent in some verbal threat behavior, pushes this behavior into the force–trap category.

(8) The Force Trap

The force trap involves using verbal force (e.g., shouting) or physical force (e.g., grabbing, hitting, kicking, shaking, shoving, squeezing, slapping) to "manage" another's behavior. None of these forms of coercion and punishment is appropriate, some exceed legality, and all fail to teach more desirable behavior. Furthermore, all of these coercion and punishment forms elicit a full range of long–lasting negative emotions and effects while also evoking extreme forms of escape, avoidance, and counterattack, including dropping out, escalating violence, and the ultimate escape, suicide.

Imagine what happens in and to society when our institutions and agencies institutionalize these or worse kinds of force as standard operating methods. Our challenge, which Sidman (2001) shows we have the means to meet, concerns applying appropriate scientific behavior controls in ways that reduce not only the use of interpersonal force, but also the use of institutionalized force, at all levels of social interaction. At the same time, we must institutionalize the many kinds of recognized and capable, positive, alternative controls already at our disposal through behaviorology (e.g., see Fraley, 2013, for some extensive coverage focused on the successful, scientific rehabilitation of people in the criminal justice system).

Trap Conclusion

In addition to all the other problems, all of those eight coercion traps undermine the relationships of the people involved, driving wedges between them. However, the world will not end if on occasion you fall into one or another of these traps. Indeed, the data that Latham and other researchers have collected shows that you can substantially reduce, perhaps even minimize, the negative effects of coercion if you maintain a positive–interactions to negative–interactions ratio of eight or more to one (e.g., see Latham, 1994, 1998). That is, react positively at least eight times for every one time that you react negatively. This may seem difficult at first, but gets easier, even automatic, with practice.

For example, after Latham trained teachers and staff at one school to implement positive practices, and to collect data tracking their progress, they reported moving from more negative interactions than positive interactions to a positives–to–negatives ratio of over 100 to one before the end of the term. What about those positive interactions? They range from a wink and a smile to some simple verbal praise or more elaborate acknowledgement or discussion of quality performance. These take little time, and they involve alternatives to aversive controls. Let's turn to that topic now.

Control by Alternatives to Aversive Controls

Recall that, just as an absence of control is not an option in all other natural–science subject matters, an absence of behavior control is also not an option. Natural variables, including the behavior of other people, control all behavior. However, while coercive controls like punishment leave us feeling constrained, the alternatives like added reinforcement leave us feeling free. The less coercion we experience, the more we experience that kind of freedom. Given the problems of coercion, including those we have discussed, alternatives to coercion are vital to our present well–being and future survival (see Sidman, 2001, and Skinner, 1971).

We will cover alternative controls to coercive control first in broad, general brush strokes. Then we will provide a sampling of some basic, specific alternative controls. To start these alternative controls, here is a story that addresses one non–coercive alternative. It exemplifies the rule that behavior usually responds better to positive (i.e., non–aversive) controls, like added reinforcement, than to negative (i.e., aversive) controls, like punishment or subtracted reinforcement.

A Story about Better Controls than Coercive Controls

This short children's story (which is adapted from Ledoux, 2002) introduces the positive–control alternatives to coercive control. The title is *Jamie's Lesson* and, under literary license, we need not take offense at the inner agents that the simple story phrasing might coincidently imply.

"Jamie's Lesson." *Let's start with some questions. You go to school? You see some kids behaving meanly? And you see others behaving nicely? Well, this is a story about Jamie, and about an early lesson that she had on helping others behave nicely.*

The middle of winter brought a cold sun to the bright blue sky, and a thin glaze of ice to the ground. Jamie's classmates played in the grounds surrounding the school. However their teacher, Mr. Glenn, saw Jamie off to one side, sniffling. Going over to her, he asked, "Jamie? Are you okay?"

"I don't like Freddy!" she replied adamantly. "He's so mean. He called me clumsy, just because I slipped on the ice."

"I can understand why you are upset," Mr. Glenn calmly said. "We feel unhappy when other people call us names."

"And everyone laughed, too," Jamie added.

"We feel even worse when others give attention to name calling," Mr. Glenn continued pleasantly. "We have talked in class about a helpful way to handle these things. Can you recall what we said?"

With a little hesitation, Jamie replied, "We said we should ignore bad behavior, and pay attention when people behave nicely." After a pause, she continued. "But Freddy never behaves nicely!"

"Well," Mr. Glenn said, "at times like these, seeing any good behaviors can be difficult. But tell me just one good behavior from Freddy recently."

"Well," Jamie said, deep in thought. Then, beaming, she said, "yesterday I saw him go right over to a little kid who fell off the slide, to see if he was okay. Oh, and this morning he helped pick up a box of spilled pencils—and he wasn't even the one who spilled them. That was nice of him."

"Wow!" said Mr. Glenn. "That's great. That's two good behaviors! Did you happen to tell him you thought that was nice of him?"

"...Oops," said Jamie.

"You can still tell him, if you want to," said Mr. Glenn. "That can still help him become better at nice behavior."

"That would be good," Jamie replied. "I will!" And off she went to tell him. *You too can catch people being good. Just once each day, notice someone's good behavior, and tell him or her that it was nice. Each of us paying such compliments every day makes a much better world.*

That story shows one alternative to coercive control. We will consider others, both general and specific. While we may describe many of these alternatives in terms of parents or teachers managing a child's behavior, all of these alternatives also apply, with appropriate adaptation, to friends, spouses, employers, employees, government agencies, international relations, and solutions to global problems.

Some General Alternatives to Coercion

The category of general alternatives to aversive control spans the range of non–coercive antecedent and postcedent processes and procedures. Think for a moment about those that we have already covered, and how the control that each exerts usually occurs non–aversively. They include evocative and function–altering stimuli, added reinforcement and differential reinforcement, shaping and fading, reinforcement schedules and backward chaining, certain uses of extinction, some respondent unconditioned and conditioned stimuli, and even some contingencies that we cover in later chapters (e.g., verbal "rules").

All those independent contingency variables control behavior in generally non–coercive ways. Let's look more closely at a few of the many specific ways to manage behavior without coercion.

Some Specific Alternatives to Coercion

Our specific alternatives to coercion focus on some different ways to deal with three categories of behavior. These categories cover appropriate behaviors, inconsequential behaviors, and inappropriate behaviors.

A new distinction for us separates *inconsequential* behaviors from *inappropriate* behaviors. We would label most of the annoying behaviors of children and others as inconsequential behavior. This is essentially harmless behavior. It continues, and becomes even more annoying, when it gets attention, especially adult attention, especially adult attention that it need not receive. On the other hand, inappropriate behavior consists of behavior that causes emotional harm or physical damage to the behaving person or others, or property damage. This is behavior that we cannot ignore. But we must manage it carefully, because some methods make it worse. Before looking at ways to manage each of these, let's consider better ways to generate and maintain appropriate behavior.

Appropriate behavior. Unfortunately, some inconsistent traditional cultural notions have led to the widespread practice of ignoring most appropriate behavior. These notions include (a) that behavior lacks causes, and (b) that only coercion affects behavior. However, words like "ignoring" merely provide a verbal shortcut for the process of *extinction.* "Ignoring" also implies an agent that "does" the ignoring, but since we know better, we can ignore this problem (all agential puns aside). Meanwhile, what happens to any behavior, including appropriate behavior, on extinction? Yes, its rate gradually reduces. For appropriate behavior this outcome ranks high in undesirability.

To reverse that outcome, *we must stop ignoring appropriate behavior.* We must stop saying, agentially, "We should not pay any attention to that behavior, because that is the way people should behave," as if their behaving appropriately, or not, merely depends on the type, good or bad, of mystical person agent inside them. Instead, we must *catch people being good* by following their good behavior with reinforcement. We must remember that stimuli evoke all behavior, including appropriate behavior, and these stimuli get their effectiveness from the consequences of the behavior, particularly attention reinforcers. The best way to generate appropriate behavior is to follow it with added reinforcers. Then use intermittent reinforcers to maintain the behavior.

For example, as a child grows up, most parents prefer reasonably uncluttered bedrooms. Some of the reasons for messy children's rooms concern incomplete or missing contingencies that otherwise ordinarily condition room–cleaning behavior. This could mean that, while some present contingencies would *maintain* room cleaning, no present contingencies *generate* room cleaning. Certainly, using coercive methods would solve nothing. They fail to teach how to clean rooms, but they do not fail to engender a host of negative effects.

Some simple contingencies that can generate room cleaning involve one or both parents modeling how to clean a room in a non–coercive manner that can even be fun. Such modeling, with its inherent added reinforcement for the child's imitation of the parent's responses, actually teaches children not only how to clean a room, but also how to keep rooms clean. And it can happen for any room (e.g., the parents' bedroom, a family room, the child's bedroom). Of course, the child's bedroom makes a good place to start.

For their toddler, parental modeling of room cleaning can begin soon after ordinary contingencies have generated walking. In line with the principle that behavior responds better to added reinforcers than to coercive consequences, parents can start by modeling "picking–up–after–themselves" behavior by first straightening up their own room with their child present. Then they can prompt this behavior by beginning to straighten up the child's room while paying attention to (i.e., reinforcing) the child's participation, even shaping it. This makes the participation fun for the child, even like a game, because the parents provide many reinforcing stimuli as enjoyable compliments and contingent praise. As a result, the quality of the room's condition improves, and that is easier to maintain even through adolescent years with the more occasional though still appreciated compliment and praise (in spite of the silly behavioral noise that such attention sometimes evokes). Some might even say that the child has a room–cleaning "habit," but by now you can easily say both why that is not accurate and what accounts for room cleaning.

Inconsequential behavior. The easiest behaviors to manage are those we call inconsequential behaviors. These include age–typical behaviors like dressing down, mild sibling rivalry, meaningless verbal blows, and jousting, or rough–housing that often leads to laughing, not crying. These behaviors threaten neither the basic quality of life nor the safety of limb and property. How should you respond to such junk behaviors? (Hint: The sometimes somewhat obnoxious verbal noise, which sometimes follows a parent's *compliments* of a child's good behavior, is inconsequential behavior, so *just ignore it.* That is, put it on extinction.)

You treat those kinds of inconsequential behaviors in the opposite way that you deal with appropriate behavior. While you attend to appropriate behavior, attending to junk behavior simply gets you more of it, which is counterproductive. Instead, basically, you *ignore* inconsequential behaviors. You simply put them, or leave them, on extinction. Also, rather than count instances in extinction, you can more easily time how long a current incident takes to stop when ignored. While later stimuli might re–evoke the annoying behavior, Dr. Latham's data show that about 80% of current inconsequential behaviors, if ignored, end within 90 seconds. Try it. Actively ignore the next annoying, inconsequential behavior. Turn away from it while timing it. Engage in something else during those 90 seconds. For example, the next time your child bothers you by tapping a fork against a plate at dinner, just surreptitiously time it while continuing other normal dinner–time behaviors. You will be pleased with the outcome. Behaviors that produce no results quickly extinguish.

Nevertheless, those may seem like the longest 90 seconds you have ever experienced. But that time interval is far shorter than the time required to undo the problems that coercion–based practices bring when they are used, in place of calm extinction, for managing inconsequential behavior. The calm extinction not only reduces the problem behavior but also models for children how they should deal with the problem behavior of others, a most valuable lesson for both right then, and adulthood later. Remember, indeed ponder the proverb, "Patience rather than anger escapes 100 days of sorrow." Inappropriate behavior, however, requires a more complex strategy. Let's consider that next.

Inappropriate behavior. We prefer appropriate behavior, which benefits from plenty of added reinforcement. We consider inconsequential behavior as relatively benign although annoying. It benefits from active ignoring. Inappropriate behavior, however, threatens the basic quality of life and the safety of limb and property. We cannot ignore it. We would not want to reinforce it. And responding to it coercively not only produces the usual negative effects but also still pays a sort of attention to it that only makes it worse. This is especially the case if the inappropriately behaving person is under attention deprivation, receiving little attention for better behaviors. So, what steps might mitigate inappropriate behavior?

One way to manage inappropriate behavior involves a multi–step practice that stops inappropriate behavior *and redirects it* toward reinforceable appropriate behavior. This practice, well–researched as the University of Kansas Achievement Place model, has long been implemented at Boys and Girls Town. We call this practice the *Teaching Interaction Strategy* (some call it the *Corrective Teaching Procedure)* because it stops the misbehavior and redirects it by teaching (i.e., conditioning) an appropriate, alternative behavior.

The Teaching Interaction Strategy has six steps that occur in a calm, empathetic and non–threatening manner. You can start by saying something positive. An easy and valuable enhancement involves first making eye contact with the misbehaving person, and holding that contact briefly, in silence if possible. This kind of contact often has a chilling effect on the misbehavior. In the next step, you describe the problem behavior in just a few words. For step three you describe a better, alternative, appropriate behavior, also in as few words as possible. Then, you state a clear reason why the new behavior is more desirable than the inappropriate behavior. In step five, you practice the desired behavior, at least verbally. Finally, you follow the appropriate practice behavior with positive feedback (i.e., added reinforcement). While the other steps occur in a matter–of–fact manner, you can and should show more emotional enthusiasm in reinforcing the appropriate practice behavior.

Let's list those six steps together, and then consider an example. Here are the six steps (from Latham, 1994 and 1998, which contain much more discussion of these steps, and many more examples, from both child care and education):

(1) After brief silent eye contact if possible, say something positive;
(2) Briefly describe the problem behavior;
(3) Describe a better, alternative behavior;
(4) Give a reason why the new behavior is more desirable;
(5) Practice the desired behavior, at least verbally;
(6) Provide positive feedback for the better behavior practice.

Here is an example of those six steps in action: Some of your child's friends stopped at your house on the way home from school one afternoon. You know from past experience that they sometimes seriously squabble. A commotion in the back of the kitchen induces directing your gaze that way. Just barely, you overhear Tom call Gerry a name unfit to repeat and, as you get close, you see Gerry punch Tom in the chest. Quickly putting your arms between them and separating them, you look each one in the eye for a second or two in turn while saying nothing. Then, you apply the Teaching Interaction Strategy with Gerry first because, even though both behaviors are inappropriate, Gerry's hitting behavior is more serious than Tom's name–calling behavior. Here is one way you could apply the strategy, with the steps numbered and summarized.

(1) Speak positively: "Gerry, you are usually quite pleasant to have around."

(2) State the problem behavior: "But when Tom called you that name just now, you reacted by hitting him."

(3) Give an alternative behavior: "One better way to respond to name calling is to ignore it and simply walk away."

(4) Say why the alternative is better: "By ignoring it, you stay out of trouble while also reducing that misbehavior by preventing it from getting further attention. Tom will probably forget it, and it won't affect your friendship."

(5) Practice the alternative: "So Gerry, when someone calls you a name, how will you deal with it next time?" (Some reasonable answers would be: "I will ignore it" or "I will just walk away.")

(6) Provide positive feedback: "That's good, Gerry. I know how hard that will be at the time, but in the long run, that will be easier on you too."

Of course you may need to ignore some behavioral noise during that strategy. You may even need calmly to repeat parts of it, possibly more than once. Still, you now have the six steps, and the example of dealing with Gerry's inappropriate behavior of punching Tom. You can see the importance of dealing with such problems *instructively*, without emotion but with direction. You may have thought of some other, more common—and coercive—ways to handle such a situation, but let's put those more damaging ways behind us. I cannot provide a laboratory or practicum setting in which you might practice the six steps. So, spend a few minutes outlining, for your own practice, what you might say when implementing the Teaching Interaction Strategy again, this time with Tom, because his name–calling behavior was also inappropriate.

Coercion Alternatives Summary and Extension

Again, independent variables in contingencies control all behavior at all times (e.g., even "my" writing behavior at this moment). But different types of control affect us in different ways. Emotionally, coercive control constrains us while added–reinforcement control frees us. The less coercion we experience, the less loss of freedom we experience. In other words the more we experience the alternatives to coercion, the greater is our freedom experience.

You can even find alternatives to coercion for some of the most coercive institutions in our culture. Lawrence Fraley has described coercion alternatives in detail for two of these institutions. In *Dignified Dying—A Behaviorological Thanatology* (2012) he presents the contingency engineering behind better ways to treat our terminally ill neighbors, friends, and loved ones, along with their survivors. And in *Behaviorological Rehabilitation and the Criminal Justice System* (2013) he elaborates on the contingency engineering behind improvements in managing penal incarceration in ways that help inmates as well as society.

Conclusion

Overall, however, we need far more detail to do even minimal justice to the topic of "aversive control, problems and alternatives." This chapter could not adequately introduce all the needed basics even if it was 50 pages longer. Happily, other books treat this topic thoroughly. You will find reading one or another of them quite advantageous. An excellent starting point is Murray Sidman's *Coercion and its Fallout* (Sidman, 2001). Note, however, that Sidman originally wrote his book before 1989, which was before the term *behaviorology* was in general use. So he refers to "behavior analysis" in his book, a name that psychology has recently claimed even though, back then, this term named the natural science of behavior. Sidman's book relays no kind of psychology. Readers can reduce confusion by substituting "behaviorology" when "behavior analysis" appears in this text. Instead, Sidman's book vastly expands this chapter. These topics deserve that kind of coverage, given our long–standing overuse and misunderstanding of coercion and the negative impact it has on efforts to solve global and local problems at all levels of society, from your household to the relations your country has with other nations. (You can find most of the books mentioned here on www.behaviorology.org or through the links listed on this website.)❧

References (with some annotations)

Dise, R. L. Jr. (2009). *Ancient Empires before Alexander* (Course 3150, Lecture 9). Chantilly, VA: The Teaching Company.

Fraley, L. E. (2012). *Dignified Dying—A Behaviorological Thanatology.* Canton, NY: ABCs.

Fraley, L. E. (2013). *Behaviorological Rehabilitation and the Criminal Justice System.* Canton, NY: ABCs.

Hefferline, R. F., Keenan, B., & Hartford, R. A. (1956). Escape and avoidance conditioning in human subjects without their observation of the responses. *Science, 130,* 1338–1339.

Latham, G. I. (1994). *The Power of Positive Parenting.* Logan, UT: P & T ink. This book is Dr. Latham's most comprehensive work applying behaviorology to child care (and education).

Latham, G. I. (1998). *Keys to Classroom Management.* Logan, UT: P & T ink. This book applies behaviorology to education.

Latham, G. I. (1999). *Parenting with Love.* Salt Lake City, UT: Bookcraft. This book, with the subtitle, *Making a Difference in a Day,* applies behaviorology to child care, and much of it parallels the author's two–part, two–hour video entitled "The Making of a Stable Family." Dr. Latham was a good friend and colleague, and one of the four founders of *The International Behaviorology Institute* (see "In Memoriam" at www.behaviorology.org).

Ledoux, S. F. (2002). A parable of past scribes and present possibilities. *Behaviorology Today,* 5 (1), 60–64. While opening with the "Jamie's Lesson" story, the principal point of this article concerned the educational abuse of children, and the financial abuse of taxpayers, that results from ignoring the "Project Follow Through" research outcomes (see Watkins, 1997).

Ledoux, S. F. (2014). *Running Out of Time—Introducing Behaviorology to Help Solve Global Problem.* Ottawa, CANADA: BehaveTech Publishing. Chapters 8, 9, and 6 contain details on single–subject experimental designs along with several extensive examples.

Sidman, M. (2001). *Coercion and its Fallout—Revised Edition.* Boston, MA: Authors Cooperative. Be aware that Sidman originally wrote this book before 1989, which was before the term *behaviorology* was in general use. Instead he refers to "behavior analysis" in his book, a name that psychology has recently claimed even though, back then, this was the name for the natural science of behavior. Since this book does not relay any kind of psychology, readers can avoid confusion by substituting "behaviorology" when "behavior analysis" appears in this text.

Skinner, B. F. (1955–56). Freedom and the control of men. In B. F. Skinner. (1972). *Cumulative Record: A Selection of Papers Third Edition* (pp. 3–18). New York: Appleton–Century–Crofts. The B. F. Skinner Foundation (www.bfskinner.org) in Cambridge MA, republished this book as "… *Definitive Edition,*" in 1999.

Skinner, B. F. (1971). *Beyond Freedom and Dignity.* New York: Knopf.

Watkins, C. L. (1997). *Project Follow Through: A Case Study of Contingencies Influencing Instructional Practices of the Educational Establishment.* Cambridge, MA: Cambridge Center for Behavioral Studies.

Whaley, D. L. & Malott, R. W. (1971). *Elementary Principles of Behavior.* Englewood Cliffs, NJ: Prentice–Hall.

Reader's Notes

Chapter 16
About a General Human Problem Behavior:

What can cause the occurrence of ideas and activities that contradict reality? Coincidental reinforcers cause superstitious behavior. ...

*T*o begin, recollect that both "ideas" and "activities" are behaviors. Thus, both come under the same natural contingency laws as any behavior. These not only include reinforcement in general, but also the particular kind of reinforcement that forms the focus of this chapter, *coincidental* reinforcement. We focus on coincidental reinforcers due to the role they play in superstitious behaviors, both some "idea" behaviors and some "activity" behaviors.

First, however, we need to deal with some even more basic misunderstandings than the misunderstanding over the origins of superstitions. This refers to the misunderstanding of reinforcers as rewards and the related question of bribery.

Rewards and Bribery

To start, let's clarify why we should avoid considering reinforcers as rewards. Then we can consider how the appropriate use of reinforcers differs from bribery in crucial ways.

Rewards. Interchanging the terms *reinforcers* and *rewards* presents problems, because *rewards are not necessarily reinforcers.* Rewards are stimuli that *others* think should reinforce your behavior, perhaps because these stimuli reinforce their behavior. However, rewards have often not yet met the definition of a reinforcer, at least with respect to the behavior of the organism of concern, possibly you. Remember, we have tested and observed stimulus consequences. They had to meet certain criteria before they could earn the title "reinforcers." We had to see the occurrence of these stimuli, immediately after a response, making the evocative stimulus for that kind of response more effective across subsequent occasions. *Rewards receive little such testing.* Besides, if and when a reward meets this definition after testing, then we should call it a reinforcer, not a reward. Also, the concept of rewards supports the false and scientifically irrelevant notion of personal agency. How? A reward is for "you" (as the inner agent inside the particular carbon unit that others call by your name). Furthermore a reward is *for you* rather than *for your behavior,* whereas reinforcers—as defined—do not reinforce *you*; they only reinforce behavior.

For example, someone likes not only dark–chocolate peanut–butter cups but also your behavior (i.e., both the candy cups and your behavior reinforce his behavior). If he follows your behavior with a bag of such candy cups, then you have received a reward. But if you are allergic to peanuts, then those cups definitely do not function as reinforcers for your behavior.

Bribery. One dictionary defines a bribe as "Something… offered or given to someone… to induce him [or her] to act dishonestly" (*The American Heritage Dictionary*, 1982, p. 207). We can broaden that somewhat. A bribe is anything that you offer or give, often beforehand, to induce someone to act in a manner that is immoral or illegal (Or fattening?) as well as "reinforcing" for the briber. Reinforcers, on the other hand, occur after desired behavior, and that behavior is usually in the behaver's best interest. So the occurrence of reinforcement, as part of the usual contingencies of life, is not bribery.

An example of bribery would be offering or giving your teenage son the latest, full–featured cell phone if he gives you a third of the illegal drugs he steals from others at school. On the other hand, if you give your toddler a cookie after she washes her hands when she finishes using the potty chair, then you are not bribing her. You are merely conditioning appropriate, life–long toilet and hygiene behavior with cookie reinforcers. Such cookies are neither "rewards" nor "bribes." Using them this way will not make your daughter a cookie junkie for the rest of her life, although other variables could lead to this outcome (e.g., cookie–manufacturer advertising schedules).

Coincidental Reinforcement

"Coincidental" rather than "accidental."

Historically, when a reinforcer occurs, and the response it followed did not produce it, we called it an "accidental" reinforcer. Now we describe such reinforcers as *coincidental* instead of accidental. Let me explain why with an example. Having avoided the outdoors for several months due to the winter cold, you are now walking across campus, as no indoor route is available, to a Monday morning meeting you must attend. On your way, you notice a couple inches of green sticking out from underneath a snow pile and, upon pulling it out, you find a $50 bill in your hand. For the rest of the week, you walk outside to your classes and meetings around campus, even between connected buildings, and even despite the persistent cold. (You may or may not be gazing strenuously at the bases of every snow pile you pass.)

Let's face it. Your walking outside did *not* put that $50–bill reinforcer under the snow pile. (If only finding petty cash was that easy, right?) In that sense your walking did not produce that $50 bill. But calling that reinforcer an "accidental" reinforcer has a problem. To some folks "accidental" implies some sort of chance, random, spontaneous or magical occurrence that lacks a natural functional history. But all of these are wrong, because such reinforcers do have their own natural functional history. Somehow that $50 bill got under the snow pile. It did not materialize mystically out of nothing. However, we often remain ignorant of the actual events in the functional history. Still, some other natural process led to the occurrence of the $50–bill reinforcer under the snow. But the response it followed did not put it there. Now, if an available term carries fewer such wrong implications, we benefit by using the new term to replace "accidental." *Coincidental* is such a term. It merely means "happened at the same time as…" *Coincidental* merely emphasizes the time relation.

In summary, "accidental" too easily implies no natural functional history for the stimulus of concern. On the other hand, "coincidental" implies fewer such problems. Thus we retain "coincidental."

Coincidental Reinforcers

We label a reinforcer as *coincidental* when the preceding, reinforced response did not produce the reinforcer. Some other natural process coincidentally led to the occurrence of the reinforcer at just that time, but the response it followed did not produce it. It happened coincidentally, hence the label *coincidental reinforcer.* For example, let's say that every day, on your way home from work, you stop at a favorite grocery for the day's supply of fresh vegetables. The grocery is on the northeast corner of the block while your office is on the southwest corner. Without any discernible pattern, on about half of the past occasions you went around the block clockwise while on the other half you went counterclockwise. About two weeks ago, however, you went counterclockwise every day for that whole week. Then last week you went that way three days out of five. And this week you are back to no discernible pattern. Why did that happen?

We could only discover one thing that happened. On the Monday of that counterclockwise week, as you neared the southeast corner, a $20 bill floated down in front of you. Of course you snatched it out of the air. You also looked up, looking for who might have lost it, or looking for more $20 bills floating down, or both. On the rather safe assumption that a $20 bill is a generalized reinforcer "for you," its occurrence, following your walking at that location, would have strengthened the behavior of walking along that route. We can rightly call the $20 bill a reinforcer from our observation that you walked counterclockwise every day the rest of that week. This way took you past that same location. On the other hand, no further $20 bills floated down over the next several days, or at any time since this episode occurred. This lack of reoccurrence of this reinforcer gradually led to the extinction of the extra counterclockwise walking that the occurrence of the reinforcer had conditioned. So, after a couple of weeks, you were back in the usual routine of walking clockwise half the time and counterclockwise the other half.

The *occurrence* of that $20–bill reinforcer, however, was not functionally related to your original counterclockwise walking behavior. That is, your walking behavior did not put the $20 bill in the air. And yet, while it only occurred coincidently with respect to your behavior, it did not occur magically. Some set of events comprised the natural functional history of its presence in the air. Perhaps someone left some $20 bills on a night stand near a third–floor window, and a gust of wind blew one of those bills out the window just a moment before you passed below that window. You got it, with the result that travelling counterclockwise occurred more frequently. By now, I think that *you* can improve the phrasing of this account in terms of the effect of the reinforcer *on the stimuli that evoke* travelling counterclockwise. What would you say?

Now, if the term *coincidental reinforcer* describes the $20 bill, what term describes the extra counterclockwise travelling behavior? The term we use is *superstitious behavior.* Superstitious behavior is the kind of behavior that coincidental reinforcers condition. Let's consider such behavior more closely.

Superstitious Behavior

Sometimes, using a non–technical term, we describe reinforcers as "fickle." After we have tested them and observed that they have earned the label reinforcer, we also note that they are not picky about what behavior they reinforce, because another principal property of reinforcers is merely temporal. They *follow* responses. Thus, if a stimulus *is* a reinforcer, it will reinforce *any* operant behavior that it follows, and we describe this as reinforcer fickleness. It almost sounds like a polite swear word: "Oh, fickleness!" Yet we may find cause to swear about it. Here is why.

Coincidental Reinforcers and Superstitious Behavior

A curious implication arises from the fickleness of reinforcers. Many people consider humans the smartest species on the planet, in part because, with humans, operant processes can regularly condition a new response in a single reinforcer occurrence. Perhaps they are right about human intellectual prowess. But consider that this reason for placing humans at the top of the intellectual heap ignores the fickleness of reinforcers. This fickleness makes humans a species most susceptible to the occurrence of superstitious behavior, which a single *coincidental* reinforcer occurrence can condition (e.g., recall the example of the single $20 bill in the air). Thus the implication of the fickleness of reinforcers is that people rather easily fall victim to coincidental reinforcers conditioning superstitious behaviors. The next time that the Halloween holiday rolls around, think about how the occurrence of coincidental reinforcers could have originally conditioned the superstitious behaviors that many well–known cultural superstitions describe, such as "walking under ladders brings bad luck." Sayings like this, and their origins, provide us with some examples of both "contingency–shaped" behavior, such as the original conditioning of a superstitious behavior, and "rule–governed" behavior, such as the behavior that a current superstition description controls (with details several chapters from now). Or, consider the origins and continued functioning of the many bothersome, even harmful, agential superstitions that permeate our culture… Can we swear now? Oh, fickleness!

Early Superstitious–Behavior Research

Summarizing some of the early research that analyzed superstitious behavior sets the superstition stage for us. B. F. Skinner reported some of this research on pp. 85–87 of his 1953 textbook. In one study the food hopper for each of several operant chambers, each containing a pigeon, provided momentary access to grain every 15 seconds *regardless* of the behavior of any of the birds. This is a *fixed–time* schedule, technically a fixed–time (FT)–15 seconds schedule, because reinforcement occurs solely by the clock without requiring any response (whereas fixed–interval schedules require both a time interval and one response). Of course, some response *was occurring* when the food hopper operated, and the food coincidentally reinforced that response, whatever it was, making it more likely to occur again. As a result, in only a few minutes with this type of schedule operating, observers saw every bird repeating one or another stereotyped response. Examples of stereotyped responses can include head bobbing, foot hopping, wing flapping, and turning in circles.

Since none of those various, stereotyped responses *produced* the reinforcers that followed them, we call the reinforcers coincidental, and we call all the responses *superstitious behavior*. The experimenters also tried longer times, such as FT–1 minute, but the effect took longer to appear, for reasons discussed in Skinner's textbook. Basically then, the experimenters discovered that *coincidental* reinforcers condition *superstitious behavior*.

Some Superstitious–Behavior Examples

Let's consider several more examples of superstitious behavior. These cover the coincidental but still reinforcing effects of a compliment, the coincidental but still reinforcing steps during an elevator ride, and the coincidental but still reinforcing points that follow various player rituals in some court sports.

The compliment. You found the morning temperatures for the first three weeks of October hovering around a cool but pleasant 16° C (i.e., 61° F). Then one day when you awoke in your dorm room, the outside temperature was at freezing (0° C / 32° F). Your first class was at 9 AM, and your winter wardrobe was still at home. To keep warm on your walk up to the classroom building, you had to put on a ratty old sweat shirt that you previously only used for house painting. On your way to class, you saw some friends from another dorm coming out of an 8 AM class on a side path. You waved and heard them say that they liked what you were wearing. Oh, really? Feeling good, being somewhat surprised, and thinking grunge must be back in style, you wore that sweat shirt nearly every day until taking it home for the Thanksgiving vacation. There, your mother disabused you of your misinformed fashion sense, and made sure you returned to campus with your proper winter wardrobe. What you never found out was that your friends had been addressing the person behind you who was sporting a new, stylish, cashmere cardigan. Your sweat shirt had not produced the compliment, but the coincidental compliment had nonetheless reinforced your sweat shirt–wearing behavior. It was a superstitious behavior (one that continued for several weeks after Halloween).

Elevator buttons. Another superstitious behavior example involves elevator buttons. Elevators get bad press in movies, and even in personal experience. Have you ever been stuck in one for any reason? If so, you would probably not be surprised to find that nearly every indicator, that a needed elevator ride is progressing toward a proper conclusion, turns out to be a conditioned reinforcer. I say "needed" because, if you are only going up or down a few floors, then taking the stairs is probably better for your health.

Now I presume that you have taken an elevator ride and so are familiar with the drill. You start by pressing the elevator call button, and it lights up, and the elevator is on its way. But what happens when other people also need to take the elevator? Invariably, one or another also pushes the already–lit call button, and some stand there pushing it several times, even continuously. Of course, none of those extra pushes has any effect on the elevator's arrival. Sooner or later (only a matter of a few seconds generally, and seldom longer than a minute) the elevator arrives. A light comes on and the door whooshes open while a chime sounds. All of those stimuli are conditioned reinforcers. So the button pushing of anyone pushing the call button unnecessarily, *when those reinforcers occur,* gets coincidently reinforced because, remember, those extra button pushes do not produce the reinforcers. They have no effect on the elevator's arrival. Those extra button pushes are superstitious behaviors, and past encounters with those coincidental reinforcers probably conditioned them in the first place. I suspect that you can now well describe what happens when such coincidental reinforcers follow the extra pushing of any of the many other buttons *inside* the elevator.

Sports rituals. Our last superstitious behavior example involves sports rituals. Perhaps you play one or another court sport (e.g., handball, racquet ball, squash). If so you may have witnessed, even participated in, various superstitious rituals that coincidental reinforcers condition during play.

Let's focus on squash, which I played often and thoroughly enjoyed during the four years that I taught in Australia. The squash ball has little bounce, so you must get to it quickly to hit it back to the wall. The long–handled racquet helps with this. But a good player can serve the ball such that it hugs the opposite side wall, forcing you to dig it out of the rear corner to return it. This puts your racquet at risk of damage—a good whack on the wall wrecks a racquet—and a good racquet is not cheap. In competition this might not matter, since the contingencies on winning trump economic concerns. Besides, you have several spare racquets. If you are playing, however, for exercise and enjoyment on a professor's meager state–college salary, then you might avoid that risk and lose the point. If you have never played squash, then this quick description may give you enough idea of what is going on. But where do superstitious rituals come into play?

Participating in a very active game, you are quickly ready, and eager to serve when your turn comes. But serving before your opponent is ready is really rather rude. So, while waiting, some of your energy goes into bouncing the ball, perhaps twice on the floor then once on the wall then once more on the floor. Then you serve. If your serve is a good one, right along the far–side wall and into the corner, your opponent may be unable to return the ball and you get a point.

All of the particular response components of your service produced that point, and that point further reinforced these components of your serve. However, that point also followed your little, pre–serve, wall and floor ball–bouncing ritual. So that point also coincidentally reinforced that ritual. We observe this in the repetition of that ritual before the next serve, and the one after that, perhaps until your opponent regains the serve. In any case, being a function of coincidental reinforcement, that ritual is an example of superstitious behavior.

We can even see such rituals drift in form. When you again get to serve, the ritual may change. After all, on that very last time you served, the loss of service rather than a point followed the ritual, so we experience no surprise if the ritual changes, drifting into another form (e.g., to bouncing the ball once on the floor then twice on the wall then once more on the floor). Whether that change is due to extinction (no reinforcer) or punishment (a lost point) remains unclear. What is clear is that if a point coincidently follows the new ritual form, this new form quickly becomes a new superstitious behavior ritual.

Can you think of other examples of superstitious sports rituals? Consider the antics of major–league players in various other sports. Generally one can observe pitchers and batters in baseball, and quarterbacks in American football, exhibiting coincidentally reinforced superstitious rituals. Take a break from this book before starting the next chapter, and watch for such rituals in a game or two of your preferred sport.

Before going out to the ball game, however, let's take a little time to broaden the picture. Let's expand from coincidental reinforcers and superstitious behavior, to mysticism and superstition.

Understanding and Managing Mysticism and Superstition

While mysticism and superstition are not among their teaching topics, behaviorologists will—for some time yet—have to teach what these are, as well as why and how to reduce their natural negative impact. Dealing with the world, day in and day out, compels plenty of interest in why events occur and how to make things better. You and I and everyone want answers, especially ones that seem to make us better at dealing with the world. The problem is, getting thorough and accurate answers seems to be directly proportional to the quality and amount of effort that we can expend in the process. In the remote past, staying alive, under minimum subsistence contingencies, occupied most if not all of peoples' energies. As the old saying goes, "Keeping body and soul together" was not easy. With current scientific knowledge, we would instead say, "Keeping body and behavior together" is easier today. In the millennia separating these two times, much has happened, even though too many folks are still today exposed to unhelpful accounts of behavior (e.g., astrology, theology, psychology).

In those earlier times, people had little extra energy for answering the questions that bothered them. The social pressures to provide some account or explanation, indeed *any* account or explanation, far outweighed competing contingencies for accuracy. Accounts simply got made up. If they were both plausible and not obviously contradictable, then these accounts gradually became intermixed, elaborated, and standardized as accounts for a culture. As these accounts accumulated such complexity over century upon century, some of them became institutionalized, often in the form of one or another religion.

Regardless of any contingencies compelling emotional support and intellectual rationalization for these theological accounts and explanations, they remained grounded in their mystical and superstitious beginnings. This is not to say that they were valueless. The rich and varied laws and codes of conduct that cultures derived from them were essentially the only game in town—better than simple bullying—for maintaining social organization during these periods. However, alternatives existed, but their discovery required the availability of energy for evaluation (i.e., testing, experimentation).

In the majority of major cases, those superstitious and mystical accounts went untested. Not only was testing energy–expensive, and thus unlikely to occur, but the accounts had grown layer upon layer of convoluted rationales and implications. Many included a range of non–natural agents and components, such that they became thoroughly untestable. With their institutionalized supporters, they simply endured, in spite of, or perhaps because of, being untestable. They became major sets of interrelated variables controlling peoples' behavior then, and on down through today, as traditional cultural lore.

Then, a couple of thousand years ago, social contingencies increasingly got complex enough to compel the occurrence of simpler accounts that people could, and did, test. The result was the beginnings of science. But let's make this long story short. After all, readers have quite likely already encountered one or another of the many authors of every stripe who have covered the historical developments of both religion and science over the last several thousand years. Some of the most readable ones address modern reoccurrences of this struggle between superstition/mysticism and science. See, for instance, Carl Sagan's *The Demon–Haunted World* (1995) and James Randi's *Flim–Flam* (1982). Authors have also focused on particular episodes of this cultural war zone. See, for instance, James Randi's *The Faith Healers* (1989) and William Ryan's and Walter Pitman's *Noah's Flood—The New Scientific Discoveries About the Event That Changed History* (1998).

One of the two possibly biggest problems with superstition and mysticism involves their continued competition with fully, and today experimentally, tested accounts, which we thus can apply more confidently. This competition relies not on any demonstrable validity to superstition and mysticism. Instead it relies on the weight, the contingent potency, of their thousands of years of history. This weight (another metaphor word would be *inertia*) induces the continuation of the cultural practice of conditioning (some call it indoctrinating) the young to respond early and often with emotional and (untestable) intellectual rationales in support of some particular version of superstition and mysticism (i.e., in support of some doctrine or dogma). For example, consider those whose religious conditioning leaves them refusing scientific medicine that would relieve their children's suffering and disease.

Probably the other biggest problem with superstition and mysticism involves their direct and invidious opposition to science itself. This occurs in spite of some religious denominations maintaining official involvement in scientific endeavors. The current oppositional episode began most obviously with the arrest of Galileo some 400 years ago. Opposition has continued in the us even into the twenty–first century with our repeated "modern" courtroom trials over whether or not evolution can or should be taught in the schools. And religious opposition will probably continue against behaviorology next. Why? Here are some reasons. This science can clarify the analysis of religion and religious behavior and religious faith and the explicit workings of religious contingencies throughout the culture. And, this science can describe how to mitigate the many damaging effects of some religious contingencies. It can also show how to reorganize and improve some other contingencies that still have beneficial effects even with religions controlling them.

Here is a bigger picture. Historically, physics showed that we can deal with the world—actually, the universe—better through science. Later, biology showed that we can deal with bodies better through science. And now, behaviorology shows that we can deal with behavior better through science. As contributions through science increase, any past need for contributions through superstition and mysticism decreases.

Religious details induce other superstitions. In theology one of the main mystical categories is the soul. Theologians and their faithful followers accept it as an active inner agent that supposedly exists and considers right and wrong and other relevant factors and subsequently directs feelings and actions and conduct. As such, many theologies attribute autonomous responsibility for human behavior to the soul. This prevalent philosophical position has continued since before Aristotle. It provides an age–old basis justifying beating up or killing a sinner whose soul directs transgressions against theologically couched societal laws, in addition to condemning that soul to everlasting hell.

Some would argue that god, not people, condemns to hell. But god and hell have the same questionable status. Just in which direction *does* the "spark of life" really travel in Michelangelo's Sistine Chapel ceiling painting? Who created whom? Meanwhile, people have always told, and regularly tell, each other in no uncertain terms that they are going to hell. They often then move to make the life circumstances of the other a veritable living hell. I hope those who would condemn me to hell for saying such things are not reading this book. The contingencies under which they live are likely so stable that contingency changes, like those that a mere book such as this could supply, are unlikely to have much effect on changing them, other than to make them uncomfortable and ticked off. Neither of these has ever been any part of my point in writing.

Conclusion

This book generally delays, until the last chapter, the contributions of behaviorology to helping solve global problems. Superstition and mysticism, however, carry such negative impacts on solving any problems, global or otherwise, that we must work to reduce their negative impacts at every opportunity. Our expanded understanding of the contingencies involving coincidental reinforcers producing superstitious behavior can help us avoid and reduce these negative impacts.❧

References (with some annotations)

The American Heritage Dictionary. (1982). Boston, MA: Houghton Mifflin.

Randi, J. (1982). *Flim—Flam.* Buffalo, NY: Prometheus Books.

Randi, J. (1989). *The Faith Healers.* Buffalo, NY: Prometheus Books.

Ryan, W. & Pitman, W. (1998). *Noah's Flood—The New Scientific Discoveries About the Event That Changed History.* New York: Touchstone.

Sagan, C. (1995). *The Demon Haunted World—Science as a Candle in the Dark.* New York: Random House.

Skinner, B. F. (1953). *Science and Human Behavior.* New York: Macmillan. The Free Press, New York, published a paperback edition in 1965.☜

Chapter 17
Accounting for Some Complex Behaviors:

How can contingencies produce behaviors separately that then create more complexity? Recombination of repertoires ...

$\mathcal{B}$eginning in the 1970s, behaviorological researchers discovered the process that we call *recombination of repertoires,* which increases the range of environmental–change options that lead to behavior changes. We discuss some further options before turning to recombination of repertoires.

Also during that time period, researchers recognized the misleading implications of the term "behavior modifier." They gradually moved through a series of more accurate terms, including the later ones that regularly appear in this book (e.g., contingency management, contingency engineering). They also acknowledged a wider range of environmental–change options, existing outside this science, that also lead to behavior changes. And yet, for some of these other environmental–change procedures, the term behavior modifier remains appropriate. Before considering recombination of repertoires, let's cover the residual "behavior modifier" procedures for their part in complex behaviors. To help put these procedures into perspective, we begin with a little review.

More Environmental–Change Options

Changes in accessible variables bring about changes in behavior. These accessible variables all reside in the environment (internal or external). The various postcedent processes, including reinforcement, extinction, punishment, and schedules, provide one source of accessible variables. A contingency modifier might access any part of this source when the need arises to bring about a change in behavior. The other sources...

Hang on. What is a "contingency modifier?" Is that simply some new term for "behavior modifier?" Are we now addressing "behavior modification?"

Let's answer those questions before returning to other sources of accessible variables that can change behavior. To answer them consider an important implication, one that stems from behavior–changing variables all residing *in the environment.* The implication is that the old term, *behavior modifier,* has more problems than merely early on evoking mad–scientist connotations in the popular culture. These connotations arose due to the long–standing and pervasive contingencies driving superstitious agential accounts, as we previously discussed. In addition the false notion, that any control must be of the punitive type, supports these connotations. And an even bigger problem with the term "behavior modifier" stems from it never really being all that accurate (e.g., behaviorologists modify environments, *not* behaviors). As a result the term fell from professional favor decades ago.

We generally leave the term "behavior modifier" out of favor, because it inaccurately emphasizes just the outcome, just the behavior change, just the modified behavior; hence *"behavior* modifier." This ignores, or at least de–emphasizes, the operating environmental variables in the natural functional history of the behavior change. In their turn these variables include any organism whose participation in the contingencies contributed to the behavior change. However, this old term placed such participants somehow outside of, or above, the laws of nature, including the laws of behavior, by labeling them the *modifiers* (of behavior) as if their part derived from having some agent inside the body (e.g., psyche, soul, mind, self, judge, decider, chooser, designer, evaluator, director, and so on). Supposedly this inner–agent modifier told the host therapist body what actions to take to modify the behavior of concern, without any traceable natural functional history. We critiqued this kind of causality pattern earlier, and necessarily set it aside. No inner agent inhabits the therapist body, which instead is a behaving organism immersed in contingencies like everyone else. And some of these contingencies drive changes to the client's environment that produce changes in behavior.

In more detail, therapist organisms, like all organisms, are limited to neuro–muscular and neural operantly and respondently conditioned responses that in one way or another *change the environmental contingencies* on another organism, human or other animal. *These contingency changes then bring about changes in the other organism's behavior.* So perhaps a more accurate label for such people—sans inner agents—could be *environment modifier* or *"contingency modifier."* But "modifier" continues to evoke and elicit the old, inaccurate and misleading reactions. As less problematic terms, what about *contingency manager,* or *contingency engineer,* with *contingency engineering* being a general procedural term? *Green contingency engineering* would then refer to the efforts of green contingency engineers to help solve global problems. After all, contingencies compel people to change contingencies that change, or "modify," behavior as detailed throughout this book. All people participate in contingency changes, in modifying contingencies, and so are *all* contingency modifiers. Perhaps, for those behaviorologists who work in applied settings outside the college classroom or laboratory, we should just stick to the label *contingency engineers* or, simply, *applied behaviorologists.*

Returning to our point, postcedent processes provide only one source of environmental–change variables. Contingencies might compel anyone to access other sources of variables when the need arises for a change in behavior. Some other sources include antecedent processes as well as processes that make more direct changes to body structures. Let's look at each of these in turn.

Antecedent Processes

Recall that we are always immersed in the present, immediate context, part of which is inside the skin, your skin and my skin. This context always contains numerous stimuli that could function to produce behavior. Antecedent stimuli, singly or more than one in some sequence or combination, function in the present to produce *every* behavior as a new behavior and response–class member. If that *"every* behavior" has not affected you previously, let it sink in a little. It covers all overt behaviors that we normally think of as behaviors, including language (i.e., verbal) behavior. And it also covers all covert behaviors, including thinking, and consciousness. It leaves *nothing* for magical or mystical accounts to explain. The naturalistic implications of that *"every* behavior" extend broadly to all our topics, including those we have yet to cover. As natural scientists we acknowledge that we cannot currently explain every behavior. Perhaps for some behaviors we may never have the access or resources to study and explain them. Nevertheless, we remain confident that only natural–science accounts will be able to interpret or explain behavior in ways that meet the highest scientific and practical standards, because magical or mystical accounts are not adequate explanations at all. This reminds us once again that, since psychology fundamentally adheres to magical or mystical accounts, mostly of the secular inner–agent variety, behaviorology is not any kind of psychology.

Of course, when we say "stimuli *produce* every behavior," we refer to both operant and respondent behaviors. For operant behavior we would say "stimuli *evoke* behavior." For respondent behavior we would say "stimuli *elicit* behavior." Using related terms in our usual manner, the antecedent stimuli that evoke and elicit behavior can be of either the *unconditioned* variety or the *conditioned* variety. These stimuli include eliciting stimuli, evocative stimuli and function–altering stimuli, among others. Also, other environmental–change processes can affect the functional operation of these stimuli. Examples include not only "equivalence relations" but also "establishing operations." The term *equivalence relations* covers a topic that remains far too technical for the scope of this book (see Ledoux, 2014, Chapter 18). The term *establishing operations* (e.g., deprivation and satiation) refers to processes with two outcomes. They increase the momentary effectiveness of a reinforcer, and they increase the momentary effectiveness of any stimulus that in the past evoked any behavior that resulted in the occurrence of the reinforcer. For instance, let's extend our past water fountain example. In it the establishing operation of water deprivation not only momentarily made water a more potent reinforcer, but it also momentarily made the water fountain a more effective stimulus at evoking the behavior of drinking. By this point you can supply the accuracy expansion that connects *unchanging* stimuli with changing nervous system structures and functions.

Any of those antecedent conditions, or processes, or stimuli provide potential access points for making environmental changes that would produce behavior changes. While the operations of these antecedents are the focus of other chapters, here is a relevant example. When an open package of cookies on the kitchen counter evokes a child's taking–and–eating–cookies behaviors, a parent is certainly not limited to altering the consequential postcedents of these behaviors as a means of reducing excess or untimely cookie consumption. A much easier intervention involves merely moving the cookie package onto an upper shelf in a cupboard and closing the door. Indeed, keeping the cookie package there in the first place could have prevented that first inappropriate cookie consumption. Many parents have long been under contingencies that generate and maintain top–cupboard storage for cookies. Culturally, some old and misconstrued but witty proverbs describe such contingencies, like "out of sight, out of mind." We could, of course, repeat such a saying using "mind" not as an inner agent but as some sort of standard verbal shortcut for behavior. I think, however, that we would be even better off instituting a new, more accurate saying, something like "out of sight, out of repertoire."

Will the prompt from that kind of antecedent–related saying induce new sayings on your part that relate to other antecedents, *or to any postcedents?* Perhaps a very short postcedent review will help. Meanwhile, please send me any new behaviorologically sound sayings that contingencies make occur to you, along with your name so that I can credit them to you.

Postcedent Mini Review

Postcedent variables include all the kinds of reinforcers and punishers, and any other postcedent processes (e.g., extinction) and patterns (e.g., schedules). A further point about postcedent variables also applies to other variables. This includes all the variables that contingencies might compel a contingency engineer, or anyone, to access when the need arises to bring about a change in behavior. The accessing can occur two ways. Contingencies can induce access "knowingly" (i.e., for contingency engineers, access can occur with the language–based verbal–behavior supplementation that later we will call "rules"). Or contingencies can induce access directly (i.e., for anyone, access can occur "intuitively," without verbal–behavior supplementation).

Sources from Altering Body Structures

When the need arises to bring about a change in behavior, some other sources of variables, that a behavior modifier might access, go way beyond our usual antecedent or postcedent variables. In this case, the "behavior modifier" label *is* more appropriate than the "contingency engineer" label. It is even more appropriate than the "contingency modifier" label. These other sources of variables address processes that we generally treat independently of the usual contingencies controlling behavior, because these processes are more invasive than the contingency management or contingency engineering from behaviorology. The five processes of this sort that we mention here are surgery, drugs, nutrition, disease, and restraints.

Surgery. One kind of body alteration, generally both large scale and permanent, involves surgical practices of different kinds. These practices affect the occurrence of behavior in multiple ways, and have occurred around the world and across the millennia. Several even occur in our time. Some physiological processes gradually constrain some behaviors. These processes and their affected behavior often respond rather well to surgical interventions. For example, when an accumulation of certain kinds of injuries or stresses reduces walking to a painful gait, limited in both extent and duration, knee surgery often returns the behavioral walking functions to nearly normal.

However, not all surgical practices have such beneficial effects. The practice of severing some portion of the frontal lobes of a brain (e.g., a lobotomy) virtually always yields some detrimental effects. On the other hand, it so seldom yields solely beneficial effects that such an occurrence gets widely reported as news. Some insist that such practices fit the category of unethical experiments. Among other reasons for this claim, no surgeon can know for certain what function the removed brain portion really serves. Removing it might make things better, but more likely—according to outcome data—will make things worse. Is taking that risk ethical? (We discuss ethics in a later chapter.) Others suggest that such practices should be consigned to museums of medical history.

Perhaps both of those views are right, at least until behaviorology expands to the point when we have developed, tried, and evaluated appropriate but less–invasive procedures. When we can finally say that we really have tried everything and nothing works, can we then try practices like lobotomy? Maybe we will never reach that point, or maybe the contingencies will simply not allow such practices. One such contingency is that "try" usually implies reversible effects, which lobotomy, along with most surgery, disallows.

Some practices of some past cultures inform us about the extremes to which "surgical" practices once carried in their role of producing permanent and large scale changes in behavioral function in multiple ways. In times and places past, a common practice was to cut off the hand of a thief immediately. This practice certainly consequated the earlier theft response. Also, if the thief survived the "surgery," the wrist stump probably functioned to evoke responses alternative to, perhaps even incompatible with, further thieving responses that could cost the thief's other hand if not the thief's life. Very likely the wrist stump also regularly elicited emotional responses, as well as various responses that we would label countercoercion (i.e., getting back at those who were related to the "surgery") which the emotional responses could exaggerate, all usually to the thief's detriment. The wrist stump also likely served as a function–altering stimulus (an S^{FA}) that altered the function that the *body with the missing hand* served in evoking responses. The missing hand evoked extra–watchful responses from merchants, regarding the one–handed shopper, compared to the usual friendly responses that two–handed, non–thief shoppers evoked.

Drugs. Drugs change behavior by affecting physiological function and structure, including the functioning of the structures that mediate behavior. That is, drugs produce a set of behavior–related changes similar to those that surgery produces in that different drugs change physiological structures in different ways. Different drugs can alter the mediation of responses. Or they can alter the functioning of antecedent stimuli and processes. Or they can alter the functioning of postcedent stimuli and processes. They can even alter some combination of these contingency components. And any of these outcomes can occur as direct effects or side effects of single drugs or of multiple drug interactions. However, compared to the large scale, usually permanent bodily changes of surgery, drugs operate on the smaller scale of chemical changes that in most cases are temporary.

Since many sources report the effects of drugs, both beneficial and detrimental, to both physiology and behavior, we make the point here by addressing only a couple of the possibilities (and not the most common ones, like the effects that alcohol consumption has on behavior). For example, the complicated motions present when traveling, especially by flying, often interfere with behavioral functions by making travelers dizzy or queasy. In that situation, given a body about to fall over, or on the verge of vomiting, stimuli succeed poorly in evoking the usual responses. A common drug remedy that restores some affected behavior–controlling functional relations for a few hours, at some cost of increased drowsiness, involves taking a 50 mg dose of dimenhydrinate (which is available in various over–the–counter forms).

The caffeine in coffee provides another example. You are probably among those motor vehicle drivers for whom caffeine consumption, in coffee or pills, has extended your daily driving range, or at least has increased the safety of your driving, by decreasing the drowsiness that fatigue or repetitive, unreinforcing scenery otherwise induces. Alternatives that avoid drugs, while helping solve other problems, include accessing increased public transport systems.

Nutrition. Another accessible variable that affects behavior concerns nutrition, which is a variable in maintaining both physical and behavioral health. A body that consumes a balanced, nutritionally sound diet is a body better prepared to meet more of the demands that some sets of daily contingencies can impose. For example, consider the nutritional needs to support the stressful contingency demands on the engineer managing the shop floor where workers are crafting and assembling fuel–system components for commercial aircraft. Those nutritional needs differ, possibly substantially, from the nutritional needs to support the contingency demands on a retired engineer engaged in trout fishing several days each week at a beautiful stream located a few hundred yards from the front door of her retirement home. To support their differing daily contingency demands, the shop engineer and the retired engineer each need a different daily nutritional balance to maintain both behavioral and physical health.

Sometimes the need to bring about a change in behavior arises from problems with nutrition. For instance students who arrive at an 8 A.M. class, without first eating breakfast, quickly fall asleep, mostly from low blood sugar rather than from the professor's performance.

That can apply more broadly. If you are taking in the wrong daily nutritional balance to support the contingency demands on your full day's behavior, then your behavior with respect to nutritional intake needs to change. For example, consider that what looks like a healthy, behavior–supporting diet can be an unhealthy behavior–wrecking diet in some circumstances. Let's say you work a desk job that requires only 3,000 calories per day of food intake. Meanwhile, to stay healthy, your physiology needs the nutrients (e.g., vitamins and minerals) that only a 5,000 calorie per day diet could supply. This is the kind of diet that a healthy family farmer eats, without gaining excess weight, to support hard work in the fields every day from dawn to dusk. However, if you, with your desk job, eat the 5,000–calorie diet, then over time you would gain quite a bit of excess weight. Such unhealthy weight interferes in numerous ways with normal contingency functioning, particularly by limiting behavioral ranges. Typical daily–life contingencies produce behavior more effectively when the mediating body is not overweight.

On the other hand, eating the 3,000 calorie diet prevents you from receiving all the nutrients you require. This too is unhealthy and in numerous ways interferes with contingency effectiveness. Your physiology can better support behavior mediation when it receives the full complement of nutrients that it needs. In such a case, you would likely benefit from getting the needed nutrients through appropriate supplements—to support health, not to cure disease (although much research misses this point)—without gaining excess weight. This would help maintain behavioral as well as physical health. Prior to dietary changes, however, your contingencies should compel consultations with your physician and nutritionist.

Disease. In a manner rather the opposite of nutrition, disease proves detrimental to behavior, sometimes permanently so. Before reaching that stage, however, a common problem that disease presents to behavior and contingencies is that the physiological fight against disease uses up bodily resources that are then not available for mediating behavior. The body is run down and fatigued in spite of hopefully plentiful rest and relaxation, and we observe a general reduction in the effectiveness of behavior–affecting contingencies. The more specific ways that this happens are as numerous as the number of diseases.

Restraints. The last type of direct changes to body structures concerns restraints. These remain among the variables that some folks outside behaviorology might access for changing behavior. Restraints *temporarily* constrain behavior in some of the same ways that surgery *permanently* constrains behavior. Restraints, which we also call bondage, simulate surgical bodily changes that prevent contingencies from being effective. For example, while wearing handcuffs, contingencies that might otherwise induce upper–limb aggressive behavior against others remain temporarily ineffective in a manner similar to the manner in which surgical amputation of the arms makes such contingencies permanently ineffective. Similarly, the coercive contingencies of being "in custody" induce escape behaviors, but those contingencies are far less effective when wearing leg irons, a temporary effect that simulates some of the permanent effects of leg or foot amputation.

Recombination of Repertoires

The time has come to look at another option for expanding access to contingency variables that change behavior. Let's consider some examples of the experimental research concerning *recombination of repertoires*. This research carries important implications particularly for scientific, engineering, and educational problem–solving behavior. For over a decade starting in the late 1970s, Robert Epstein, with B. F. Skinner, coordinated some studies at Harvard called the *Columban Simulation Project* (see Epstein, 1996). In this project, pigeon behaviors were functionally related to explicit variables in ways that simulated complex human behaviors. Some of these complex behaviors concerned novel behavior, symbolic communication, and the use of memoranda and tools. Others were traditionally thought to arise from various mentalistic notions such as "insight" or "self concept." The result of this research was a more objective explication of complex human behaviors. The same kinds of common contingencies that people observed producing the pigeon simulation of human behaviors were at work with the human behaviors.

The pigeon simulations began with analysis of the complex human behavior of concern. This enabled surmising the minimum repertoire components likely needed for that complex behavior to occur when a challenge situation confronts the organism. Then, for each pigeon subject, after conditioning each required repertoire component (in isolation from other components, to avoid confounded results) the experimenters placed each pigeon in the challenge situation. The researchers found that, for different problematic tasks, if the conditioning of all necessary component behaviors had occurred, then (and only then) the challenge situations evoked successful responses appropriately combining the conditioned repertoire components. From this came the term *recombination of repertoires.*

"Self concept." Our first recombination of repertoire example concerns the test for a supposed "self concept." In this test a young child faces a mirror with a rouge spot on his or her forehead, a location that makes the spot visible only in a mirror. If the image of the child–*cum*–spot in a mirror only evokes responses typical of the presence of another child, we are supposed to accept that the child lacks a concept of self. However, if the image of the child–*cum*–spot in a mirror evokes responses of touching the spot, then we are supposed to accept that the child's having a "self concept" caused those responses.

In experimenting to discover the actual variables involved in the mirror test, the researchers came upon two classes of responses. These they needed to condition in their pigeon subjects. They began by conditioning the birds *with no mirror present* to peck blue stick–on dots placed on virtually every part of the bird's body that it could reach. They could easily have used red dots the color of rouge. Separately, they also conditioned effective responding in a mirrored space, *with no blue dots on the bird's body.* This conditioning involved reinforcing each bird's pecking at each of the varying, correct locations on the unmirrored wall of the chamber upon which a blue light had briefly flashed. The flashes occurred while the bird faced a mirror and could only see the flash locations in the mirror image. Finally they placed a blue stick–on dot on the bird's breast. They also placed a bib around the bird's neck that prevented the bird from seeing the dot directly, because any even slight lowering of its head moved the bib downward thereby covering the dot. When in the chamber with the mirror covered, none of the birds tried to peck the dot, which was possible and likely if they could detect it in any way. With the mirror uncovered, however, every bird began bobbing its head as the dot, repeatedly visible in the mirror whenever standing erect, evoked repeated attempts to peck the dot, which each time disappeared from view due to downward bib movement.

Does this mean that these birds had a self concept? Parsimony requires accepting that spot images in mirrors evoke spot–touching. This touching occurs not as a function of self concepts, for pigeons or humans. Rather, it occurs as a function of a relevant conditioning history and current evocative circumstances. The contingencies of this history and these circumstances, for these birds, is an explicit matter of record.

"Insight." The other recombination of repertoire example concerns testing the "insight" account of some complex human behaviors. Consider that many proud parents have watched as their child, too short to get a cookie from a jar atop a table and having never faced this situation before, looks around and, spotting a chair, moves it over to the table, climbs on it, and retrieves a cookie from the jar, supposedly due to something called *insight*.

In experimenting to discover the variables involved in this situation, the researchers came upon three necessary pigeon response classes using boxes and toy bananas. They conditioned each of these response classes in turn. They conditioned the birds *with no banana present* to push a box around the chamber toward a target spot. Separately, they conditioned the birds to climb on a stationary box. Lastly, and again separately, they also conditioned the birds *with no box or target spot present* to peck a toy banana within normal reach. These three conditioned response classes (i.e., pushing boxes, climbing on boxes, and pecking toy bananas) approximated the components of the child's complex cookie retrieval behavior (i.e., pushing a chair, climbing on a chair, and getting a cookie) which contingencies had likely also separately conditioned. Finally, they placed each bird in a chamber with a box to one side and a toy banana *suspended from the ceiling*, a challenge situation that had never confronted the birds before. With some apparent confusion and sighting, like the child's behavior, the birds pushed the box under the banana, climbed on the box, and pecked the banana. Birds that lacked any one of these three response classes failed the challenge situation.

Does this mean that the successful birds showed "insight?" Was the child's behavior due to "insight?" Or, like the pigeon's behavior, was the child's behavior also an example of previously conditioned repertoires combining under novel circumstances? We seldom observe children closely enough to track the conditioning of various repertoire components. Still, parsimony requires accepting that the occurrence of the challenge–meeting responses is not a function of supposed higher mental processes like insight, for pigeons or humans. Rather these challenge–meeting responses are a function both of the organism's history having included the conditioning of all relevant repertoire parts, and of the current evocative control in the new pattern of related multiple stimuli in the challenge situation. For organisms with the necessary brain structures, evoked neural responses of consciousness, like thinking, may also naturally be *supplementing* the principal and more obvious sources of control. More research may clarify this situation, which we address in a later chapter.

Recombination of Repertoires Conclusion

The recombination of repertoires line of research not only benefits the analysis of problem solving but also enhances the justifications for multi–disciplinary education in scientist/engineer training curricula. As the range of an individual's conditioned repertoire of behavior expands, so does the likelihood that needed parts will be available to combine successfully and achieve solutions in new problem circumstances for which no previous conditioning has provided *explicit* solution responses. We welcome physiological research showing how nervous systems mediate the combining of behavioral elements. Meanwhile the recombination of repertoire contingency accounts successfully replace the unnecessary and counterproductive traditional mentalistic accounts of these complex behaviors.❧

References (with some annotations)

Epstein, R. (1996). *Cognition, Creativity, and Behavior.* Westport, CT: Praeger. This title of Epstein's book repeats the title of a half–hour long video that remains well worth viewing if you can find it. It shows many more of the recombination of repertoires examples that the research covered.

Ledoux, S. F. (2014). *Running Out of Time—Introducing Behaviorology to Help Solve Global Problem.* Ottawa, CANADA: BehaveTech Publishing.℘

Chapter 18
A Comprehensive
Intervention Example:

Can humans engineer contingencies that produce
healthier behavior? A way to end smoking ...

*I*n a previous chapter, we described what we might see as the "dark side" of the behavior–control reality. This dark side holds the coercive parts of the laws of behavior, the aversive controls that humans used against each other long before our natural science of behavior experimentally discovered any behavior controls.

Now, for this chapter, we emphasize the positive side of contingency control over behavior. We turn our attention to the contingency controls in a singular but comprehensive example that covers a natural–science solution to a continuing problem behavior for humans, namely smoking. This example portends impacts and implications well beyond its own borders. *By describing how to end smoking, we actually portray, by an example, how contingency engineers develop behaviorological practices from behaviorological principles.*

Developing Behaviorological Therapies

Therapies and interventions stemming from different sources address problems with different and unequal methods. Interventions with behaviorological science as the source thoroughly analyze and directly control, change, and eliminate the independent variables that cause the problems. However, other therapy approaches, originating in disciplines not fully respectful of natural science, deal with problems without thorough analysis of, and direct control of, causal variables. In many cases these therapies instead attend to mystical or other unreal variables. Or, these other approaches to therapy focus on variables that may relate, if at all, only indirectly to the relevant functional variables, although sometimes with a measure of adventitious (i.e., coincidental) success (see Fraley & Ledoux, 2015, Chapter 3). By countercontrolling problem–producing variables, the benefits of the behaviorological approach include an inherently greater likelihood of clinical success.

Scientific Therapy Steps
Designing a scientifically informed therapy, such as a behaviorological intervention, takes several steps to keep it consistent with the functional realities

of both the problem behavior and the desirable alternatives to this behavior. Of the two most fundamental of these steps, the first involves thoroughly analyzing the behavior of concern to discover and rank as many as possible of the variables—historical and present, antecedent and postcedent—of which it is currently a function. The second step involves arranging environmental changes as helpful, separate, reproducible independent–variable techniques that address the problem–causing independent variables in ways that countercontrol as many of them as possible. This begins with those that contribute the most to the occurrence of the problem behavior. In other steps the therapy builds in extra supports that the laws of behavior can provide, such as processes that gradually shape up the alternative desired behavior while the therapist introduces the techniques that countercontrol the problematic independent variables. As success builds, a good design includes adding processes that fade out client reliance on the therapist and the therapy process. A good design also includes conditioning the client skills that take over the therapeutic role. Depending on the nature of the problem, the therapy may include other steps, perhaps ones that build mutual supports, through scheduled and unscheduled group interactions, among clients who share similar problems and solutions.

Those are some of the general steps that behaviorologists take in designing any therapy or intervention. Keep thinking about them so that you can spot their appearance and application as we go through our extended example on smoking behavior. Why examine smoking behavior as the example?

Many therapies to combat smoking exist. Some simply take an intuitive approach. These provide a contrasting background against which you can more easily see what makes a behaviorologically designed therapy different, and why the differences are important. The most basic contrast concerns the failure of scientifically uninformed "therapies" to analyze smoking as a behavior functionally related to accessible environmental independent variables. Some such therapies completely ignore the behavior status of the smoking problem. Others address important but secondary issues such as nicotine addiction. I call this a secondary issue, because the data from our smoking–control example (see Morrow, Gmiender, Sachs, & Burgess, 1973) showed that 90% of those who completed the full therapeutic program were still not smoking one year later, even though no special pharmaceutical techniques to address nicotine addiction were even available at the time. We will elaborate on these results later. Meanwhile some folks push currently available pharmaceutical aids for nicotine addiction as sole or primary solutions for smoking problems. How vital these aids might be in a scientifically grounded stop–smoking program remains an unanswered experimental question. I suspect that they will prove helpful but not vital, because smoking is mostly a function of many other variables. All these other variables must be addressed, because they build the smoking problem both *before* and *after* the addiction occurs.

In any case we also emphasize smoking control as our example due to the tremendous pressures that smoking puts on both public–health concerns and health–care costs. After you see, by way of the stop–smoking example, how the behaviorological steps to therapy design work, and succeed, you will be able to conceive how the steps can apply to any number of problematic behaviors.

A Comprehensive Stop–Smoking Therapy Example

Anti–smoking therapies come in two varieties. In the more common variety, they essentially involve guessing what steps to have clients take to try to stop smoking while also adding *new* variables to control smoking behavior. The most common *new* variable concerns the "Record" of time since the last smoked cigarette. The other variety involves behaviorologically analyzing the variables explicitly responsible for the smoking behavior and then designing steps to countercontrol those variables directly. Joseph Morrow, Susan Gmiender, Lewis Sachs, and Helene Burgess took the *behaviorological* approach, although this term was not yet available for them to use. In 1973 they reported their evaluation of such a therapy (i.e., Morrow *et al*, 1973; henceforth we will refer to their procedures and report simply as the "Morrow Study").

The Morrow–Study therapy differs from most non–behavioral smoking–control therapies in several ways. Primarily this therapy treats smoking as a behavior rather than, as was then the fashion, as a pathological symptom perhaps indicative of problems with the client's putative inner agential self. Also, as we already noted, this therapy demonstrates a high clinical success rate. And this therapy focuses on directly countercontrolling the explicitly analyzed functional variables producing smoking. It also refuses to use the added variable of the client's Record of how long they have gone without ever again smoking any cigarettes, a variable on which many other smoking therapies tend ultimately to rely to prevent smoking.

Today, decades later, savvy therapists continue using versions of this smoking–control therapy. We mostly feature the original Morrow–Study therapy here—as old as it may be—because it was the earliest *comprehensive* behaviorological–science smoking intervention. This early comprehensive and successful therapy helps us appreciate the value of deriving scientific practices from scientific principles, a general process that fundamentally enhances the success of most efforts to improve the human condition.

Let's cover the Morrow–Study therapy by discussing its three main aspects. These are (a) the therapy success rate, (b) the therapy process, and (c) the therapy procedures.

The Therapy Success Rate

The phrase "smoking–control therapy" covers many different intervention types. Many if not most of them initially invoke aversive techniques as the main method to bring the client's cigarette smoking down to a low daily rate. These aversive techniques directly address only one or two of the variables functionally related to smoking They include using nausea–inducing drugs, horror stories about the diseases to which smokers are more susceptible than non–smokers, gory visual comparisons of healthy and smoke–damaged lung tissues, ash trays shaped like open lungs, or emotional appeals about children wanting to imitate smoking parents. These therapies typically schedule the aversive techniques, individually or in groups, for the first week or two of the therapy program, which may constitute the full extent of the therapy.

Regardless of all that variety, nearly all of these therapies initially succeed. By the scheduled end of the aversive–technique period, the daily smoking rate of nearly all clients has dropped to zero or nearly zero.

However, at the end of the scheduled therapy intervention (i.e., after that initial success) clients are on their own. To prepare clients to stay off cigarettes after their initial success, most therapies leave clients with only one fundamental controlling variable. This added variable, which does not directly address variables producing smoking, takes the form of their "Record" of how long they have gone without smoking a cigarette. Unfortunately, these therapies condition clients to see a single post–quitting cigarette as breaking, and thus wrecking, the Record. So the smoking problem resumes.

That Record of smoking–free days provides a powerful but fragile variable. Clients say: "How long have I gone without smoking? Six days!" Later: "Six weeks!" Later still (maybe): "Six months!" And even later (rarely): "Six years!" The longer one goes without smoking, the more control this Record variable seems to exert. The problem with this variable, however, is its fragility. Sooner or later, for most clients, too many stresses inevitably pile up on a particular day (e.g., your spouse gets upset over undone household chores, and your boss increases your work, and a broken pipe floods the basement, or the family's teenager crashes the car). This results in the otherwise successful quitter smoking a single cigarette, and then, usually, more cigarettes. "How long since my last cigarette? Just six minutes? My Record [of weeks or months or years] is wrecked. I might as well have another cigarette!" The unbroken Record, so powerful while unbroken, is erased.

Behaviorologically designed therapy arrangements, however, discard the notion that clients must never have another cigarette. In the Morrow Study, the "quitter's procedure" allows occasional cigarettes but only under conditions that lead to even less smoking, not more. Besides, in terms of health, any single cigarette probably causes less damage than taking a long walk on a summer afternoon in a smog–polluted city. Without the quitter's procedure, however, a single post–quitting cigarette erases the Record and so can induce a return to smoking. So relying on the Record as a therapy technique actually undercuts continued success, keeping clinical success rates relatively low.

The stress–produced breaking of clients' Records also produces other adverse effects. Clients' self–esteem aside, not only have they returned to smoking but also their *efforts* to quit have ultimately gone unreinforced. This makes quitting again even more difficult to attempt. The behavior of quitting is on extinction. The joke that "it's easy to quit smoking; I've done it hundreds of times" is not funny to smokers who have tried, and failed, to quit.

In contrast to those problems, the behaviorological smoking–control therapy featured in the Morrow Study shows a high, long–term clinical success rate. Their study involved 55 clients (25 women and 30 men). Ages ranged from 28 to 65 years. Initial smoking age ranged from 13 to 28 years. The number of years that these clients had smoked ranged from 9 to 45 years. The daily rate of cigarette smoking at the start of the study ranged from 15 to 60 (three–quarters of a pack per day to three packs per day). As with clients in other smoking–control therapies, these clients all ceased smoking, as scheduled, by the end of the initial week of therapy. Overall, of the 55 clients starting the program, 34 (or 62%) were not smoking one year later. However, the therapy consisted of two sequential components. One was individual–based (during the first week of daily meetings) and the other was group–based (during three months of weekly meetings). The results also showed that 20 clients (36% of the 55 starting clients) completed *both* of the therapy components. Of these, 18 clients *(90%)* were still not smoking one year after completing the full program.

Therapies seldom achieve that kind of clinical success rate, regardless of the presenting problems they handle. To what might the success rate of this therapy be due? The Morrow Study implicitly suggested, and here we explicitly suggest, that the success is due to the comprehensive, designed connection, based on a behaviorological analysis, between the therapy techniques and the variables functionally related to smoking. We cover these connections by detailing the behaviorological process and procedures of this therapy, including how to get more clients to complete both therapy components and so increase their chances for long–term success.

Based on their results, the Morrow group made two recommendations for adjustments to their therapy design. Later therapists adopted these adjustments (e.g., see Ledoux, 2015a). Both adjustments could reduce the number of clients who, having returned to smoking, needed to return to therapy.

One adjustment concerned the number of meetings after the daily sessions. The Morrow group recommended that clients continue attendance at group meetings on a monthly basis for nine more months after the three months of weekly meetings. This adjustment would extend therapy duration to a full year by increasing the interval between meetings through an additional step (daily to weekly *to monthly*). Such a schedule more gradually reduces—compared to the original Morrow study—the therapist's part in the contingencies generating and maintaining efforts to quit smoking. This gradual reduction in reliance on the therapist (akin to a fading procedure) also contributes to clients' continued success when the therapy ends and they proceed independently.

The other adjustment concerned special fees. The Morrow team suggested that therapists require each client to pay a special fee that would be refunded if, and only if, the client had attended virtually all therapy meetings *regardless* of whether or not the client had quit smoking. They suggested that making the refund contingent on attendance, rather than on success, might reduce the chance of clients quitting therapy. This should increase the likelihood of their quitting smoking. Given the money that quitting smoking saves (even after paying the therapist's regular fees) they suggested setting this special fee at $50 (in 1970 dollars, which might be $500 in 2015). In addition, some clients see the refund as a nice bonus for their successful efforts. In agreement with subsequent therapists who have used this therapy, we include these adjustments in the therapy process and procedures that we report here.

The Therapy Process

This intervention process contains two sequential components. The first involves daily *individual* sessions and the second involves weekly (and later, monthly) *group* sessions. Detailed descriptions of all the "self–control" techniques, which we merely mention by name in this section, appear in the subsequent "Therapy Procedures" section.

Individual daily sessions. In this therapy the first component emphasizes contingencies on the individual. The individual client and therapist meet Monday through Friday for five daily sessions, each one hour long. Each of these daily sessions has some different activities. These focus especially on the introduction of various "self–control" techniques (i.e., techniques that, after the therapist verbally conditions them, the client's environment evokes without the therapist present; no inner self–agent is involved).

The daily sessions also have some activities in common, including these seven, which occur each day: (a) The therapist collects the record of the number of cigarettes the client smoked in the previous 24 hours and praises drops in smoking rate. (b) The therapist supervises while the client engages in the technique called *satiation smoking* (a maximum of three "satiation cigarettes" per session). (c) The therapist and client discuss the previously introduced techniques and any difficulties the client may have experienced using them. (d) The therapist teaches the client the new techniques scheduled for the session. (e) The therapist provides any instructions related to the client's efforts between then and the next session (e.g., any restrictions on smoking between sessions, and the need to keep an accurate record of how many cigarettes get smoked between sessions). (f) The therapist responds with empathy to a client's doubts about his or her ability to succeed in quitting. And (g) regarding any difficulty the client is experiencing, the therapist again with empathy expresses confidence, draws comparisons with similar clients who have successfully quit smoking, and engages in other helpful verbal exchanges. Clients may smoke between some daily sessions. However, they must always engage in all relevant techniques that the therapist had previously introduced, and continue accurately counting all the cigarettes that they smoke.

On **Monday** the therapist first stresses the importance of client cooperation if the client is to reach zero smoking by the end of the first week. The therapist's role is to aid the client in reducing smoking so that the smoking rate is zero between the Thursday and Friday sessions. After that the client will go without smoking from Friday's individual session to Monday's first weekly group session, which is the first step of fading out the therapist. Then the therapist teaches the client the technique that we call *satiation smoking*. (Again, we cover the details for all of the techniques later.) The client smokes three satiation cigarettes during the session, using the client's normal brand of cigarettes. The therapist intersperses these satiation cigarettes with teaching the three techniques that we call *pure activity, anti–social chair,* and *difficult to obtain.* The therapist also explains the point of these and later–introduced techniques in countercontrolling common causes of smoking and therefore in ultimately eliminating smoking. Using all the techniques that the therapist has introduced, clients may smoke between the Monday and Tuesday sessions.

On **Tuesday** the client again smokes three satiation cigarettes, the first two of the client's normal brand. The third is a strong, domestic, unfiltered brand provided by the therapist. If the client normally smokes this brand, the therapist provides an even stronger tasting imported brand. Between these satiation cigarettes, the therapist reviews previously introduced techniques, and introduces four more techniques. These new techniques include those we call *alternative behaviors, review emotional responses,* the filter–related part of *change cigarettes* (which is a two–step technique; the therapist will introduce the second step, the brand–related part, at the next session), and *damage cigarettes.* Again, using *all* the techniques that the therapist has introduced so far, clients may smoke between the Tuesday and Wednesday sessions.

The **Wednesday** session again includes three satiation cigarettes, but only one is of the client's normal brand; the other two are the strong, unfiltered brand. Between these, the therapist reviews previously introduced techniques, and introduces several more. The first two are for immediate use, as usual. These include the technique that we call *rehearsal of difficult times,* and the second, brand–related part of the two–step *change–cigarettes* technique. The therapist also introduces, for *later* use, the technique that we call the *quitter's procedure.* The therapist stresses the importance of getting the smoking rate down to the lowest possible level between this session and the next so that the client can more easily go without any smoking between the Thursday and Friday sessions. To help reduce the rate to the lowest possible level, the client should postpone smoking as long as possible, and as usual use *all* the techniques that the therapist has introduced while also smoking any cigarettes as satiation cigarettes. Finally, the client is to destroy or give away all of her or his remaining cigarettes just before the Thursday session.

On **Thursday** the client smokes only two strong, unfiltered–brand satiation cigarettes. For most of this session, the therapist and the client discuss the importance of going without cigarettes for the next 24 hours, and again review all the techniques. As part of this review, the therapist advises the client on managing any problems that she or he has experienced in the application of these techniques. As an aid to abstaining, the client can look forward to smoking a satiation cigarette at the Friday session. And from now on the client is to use the *quitter's procedure* technique for any cigarette smoked.

On **Friday,** the last individual session, the therapist allows the client to smoke a strong, unfiltered–brand satiation cigarette. Having abstained for 24 hours already, most clients experience great discomfort over this cigarette, and they need not complete it. The client and therapist review what the client has been through, and discuss what the client might experience in the adjustment to no more smoking except through the quitter's procedure. In the rest of this session, the therapist describes to the client the scheduling, purposes, and benefits of the upcoming weekly—and later, monthly—group sessions.

Virtually all clients achieve zero smoking between the Thursday and Friday sessions. Something akin to a shaping process usually occurs as they extend this one–day success. It extends first to three days, by abstaining from the Friday individual session across the weekend to the first group session on the following Monday. Then it extends across a week between group sessions. Later it extends across a month between group sessions.

Group sessions. The second therapy component emphasizes group contingencies. The therapy continues with eleven *weekly* group meetings spread over three months, followed by nine monthly group meetings, which extends the complete therapy program to one year. Each group meeting lasts about two hours in the evening. Clients who have completed the individual therapy sessions make up the group. At any particular meeting, some clients may be attending for the first time, others for the last time, and most somewhere between these points.

At each group meeting, similar activities take place in an informal atmosphere of healthy beverages and snacks. (No smoking!) Generally clients introduce themselves, state how long they have been "on" the quitter's procedure, and describe any problems they are encountering or what topics they would like to see the group discuss. Together they discuss events in their lives related to no longer smoking, and the difficult situations they face and how to handle them. Thus each client benefits from the experience of the others. When they exhaust therapy–related subjects, they are free to discuss any other topics of mutual interest. In these ways they bond together into a mutual support group of non–smoking friends who have had similar quit–smoking experiences. For some clients these constitute their first, and often their only, such friends. These friendships and social contacts often endure on their own after the end of formally scheduled meetings.

The therapist keeps the group focused and provides verbal reinforcement when applicable. Should a client need additional help, the therapist works out an appropriate individual program with that client.

What's the point? All those process details behind that successful smoking therapy, and the procedure details that we cover next, can readily distract us from the bigger picture. This therapy exemplifies the general behaviorological approach to developing therapeutic interventions. In this approach we locate and analyze the independent variables responsible for a behavior of concern. Then we arrange change techniques that directly counter those variables, using established principles as the foundation for successful intervention practices.

A summary table. Table 18–1 tracks the daily sessions at which the therapist introduces the various anti–smoking procedures. This table also specifies the general type of variables that each anti–smoking procedure addresses.

The Therapy Procedures

The Morrow–Study therapy employs a particular set of procedures to produce appropriate responses. Most of these procedures involve self–control techniques that can be found in the literature (e.g., Skinner, 1953, Chapter 15, gives a general discussion while Stuart & Davis, 1972, focuses specifically on overeating). Our coverage here acknowledges some inherent technique overlaps, and includes some refinements to the techniques and their names.

The Morrow Study specifically designed each technique to counter one or another of the variables, or variable components, that explicitly control smoking. These variables fall under four categories. (a) *Antecedent stimulus considerations* include respondent unconditioned and conditioned emotions as well as operant evocative stimuli. (b) *Response considerations* cover the behavior vacuum caused by quitting. (c) *Postcedent stimulus considerations* include unconditioned and conditioned reinforcement. And (d) *the combination of variables* applies when clients are "on their own" during and after the group–based component of the therapy (i.e., the "quitter's procedure").

Antecedent stimulus variables. Four techniques address antecedent stimulus variables. One addresses unconditioned emotional respondents. Another addresses conditioned emotional respondents. And two address two different sources of evocative stimuli. While ignoring inner agents as unneeded, we invoke descriptions here containing verbal shortcuts, of an agential style, that typify the non–technical terms and phrases we might use with clients.

We use the term *satiation smoking* for the technique that addresses unconditioned respondent emotional variables. This is the initial aversive respondent procedure in this therapy. In this technique the client *inhales on a cigarette every six seconds.* Satiation smoking offsets the unconditioned pleasant respondent effects that smoking elicits. As the rate of smoking outside the sessions decreases, satiation smoking becomes increasingly uncomfortable. The aversive effects are temporary, and subtracted reinforcement occurs when clients discard an uncomfortable satiation cigarette.

We use the term *review emotional responses* for the technique that addresses conditioned respondent emotional variables. In this technique clients list, on a card that they carry with them, all their main quitting–smoking reasons that carry an emotional impact. This technique offsets the conditioned pleasant respondent effects that smoking elicits. Therapists instruct clients to review the list several times each day during the week of individual sessions, and as part of the quitter's procedure. In reviewing the list, they should focus on each reason until it elicits the relevant emotion. A list item could involve feeling the pleasure that a positive reason for ending smoking elicits (e.g., pleasure at being able to smell flowers again) or feeling the fear that a negative reason for ending smoking elicits (e.g., fear that the client's toddler might play with the cigarette lighter). As time passes clients remove from the list any reason that fails to elicit an appropriate, strong emotional reaction. Clients also add to the list newly discovered reasons that elicit appropriate emotional reactions (e.g., hearing the infamous saying, "kiss a smoker, lick an ashtray").

Two other techniques address sources of evocative stimuli. Both techniques help bring about the initial cessation of smoking and then decrease in relevance.

In one evocative–stimulus focused technique, *difficult to obtain,* clients put their cigarettes and matches in places that are separated in space as much as reasonably possible. This technique offsets the evocative power of *client–owned* smoking materials. The new locations necessitate additional response effort such as bending over to retrieve an item. This makes satisfying an urge to smoke more difficult (see Ledoux, 1973, for research support of the response status of urges to smoke). For example, at home the client keeps cigarettes in a suitcase on an upstairs bedroom closet shelf (with the TV on the main floor) while the matches are in the basement. Should a TV program or commercial (or other S^{ev}) evoke an urge to smoke, the increase in time and effort, that obtaining the needed items requires, is likely to prevent the occurrence of smoking at that time. The difficult–to–obtain technique helps because, when matches or, especially, cigarettes are nearby, they all too easily evoke smoking. For example a smoker reaching into a pocket or purse for a cell phone to record a telephone number would, upon encountering a pack of cigarettes, likely light up.

In the other evocative–stimulus focused technique, *rehearsal of difficult times,* the client makes special arrangements with everyone from whom he or she might receive cigarettes. This technique offsets the evocative power of smoking materials *owned by others.* For one day (the day immediately after this technique is introduced) all these people are to offer cigarettes unexpectedly to the client. The client is never to accept an offered cigarette but instead is to practice saying "no" to the offers. Clients who experience an excessive urge to smoke are to get their own cigarettes under already introduced techniques. Clients need to rehearse difficult times, due to the minimal likelihood of saying "no," when someone offers a cigarette, especially given the price of cigarettes. Some clients report *never* having said "no" under such circumstances. Under conditions of deprivation during and after stop–smoking therapy, saying "no" is likely to be all the more difficult. Rehearsing difficult times enables the client, under controlled if theatrical conditions, to practice saying "no." Then, having already said "no" at least under some relevant stimulus circumstances, clients find saying "no" easier when they are later faced with a real need to say "no."

Response considerations. We use the term *alternative behaviors* for the technique that addresses the problems associated with the response status of smoking. In this technique the therapist helps the client construct a list of alternative behaviors in which the client is increasingly to engage as smoking decreases. This technique offsets the response vacuum created when smoking behavior stops. For this list many clients emphasize past activities, such as hobbies or skills, in which they no longer engage but to which they would like to return. The cessation of smoking may even make some preferred activities affordable. The Morrow Study provided these details:

> [We ask clients] to include at least three activities in each of three categories: physical activities (such as jogging, bike riding, gardening, walking, etc.); quiet activities (i.e., crossword or jigsaw puzzles, sewing, arts and crafts, carpentry, reading, etc.); and oral activities (such as sucking, chewing or biting on lemon drops, life–savers, gum, celery, fruits, various non–toxic objects, etc.) (Appendix A, item IV).

Table 18–1

Names of techniques by the session/day of their introduction and under the general type of variables that they countercontrol

SESSION	VARIABLES		
	Antecedents	Responses	Postcedents
Monday	Satiation smoking		Pure activity
	Difficult to obtain		Anti–social chair
Tuesday	Review emotional responses	Alternative behaviors	Change cigarettes (filter part)
			Damage cigarettes
Wednesday	Rehearsal of difficult times		Change cigarettes (brand part)
	Quitter's procedure: introduction only		
Thursday	(Quitter's procedure: begin use)		

The alternative–behaviors technique is needed, because behavior does not occur in a vacuum; it occurs in time. For smokers, smoking behavior fills part of each day. As therapy produces less smoking, more of each day becomes behaviorally empty. If we make no plans to fill this time by design with beneficial activities, then other, often troublesome, activities fill it through other normal behavioral processes like generalization. Extra eating is a common but undesirable replacement, as generalization occurs from one consumption response class to another. Many people who have quit smoking have reported an increase in eating, and consequently weight. The alternative–behaviors technique fills this otherwise empty time with behaviors that are less likely to cause problems like weight gain.

Postcedent stimulus variables. Four overlapping therapy techniques address *the reinforcing capacity of smoking*. All work to weaken that capacity. The first two focus on countercontrolling *conditioned*–reinforcement value. The other two focus on countercontrolling **un**conditioned–reinforcement value. We need these four techniques to help counter any continued generating and maintaining of smoking due to its reinforcement value.

We use the term *pure activity* for one of the two techniques that offset the *conditioned*–reinforcement value of smoking. Normally a smoker engages in other reinforcing activities while smoking. This pairs the reinforcing aspects of these activities with smoking, which establishes some of the conditioned–reinforcement value of smoking. The pure–activity technique countercontrols much of this otherwise normal pairing of various physical–activity reinforcers with smoking. In this technique, when clients smoke, they *only* smoke. They engage in no other activities while they smoke. This means, for instance, no television viewing, no stereo listening, no eating or drinking, no reading, no driving, no studying, and so on. This technique thus reduces some of the conditioned–reinforcement value of smoking, because it stops the pairing of smoking with the reinforcers from various physical activities.

We use the term *anti–social chair* (which we sometimes call the "smoking spot") for the other technique that offsets the *conditioned*–reinforcement value of smoking. Engaging simultaneously in smoking and *social* activities, a common occurrence, also pairs the *social*–reinforcer components of many activities with smoking. This again increases and maintains some of the conditioned–reinforcement value of smoking. The anti–social chair technique extends the pure–activity technique by countercontrolling much of this normal, automatic pairing of social reinforcers with smoking. In this technique the therapist and client discuss locations, suitably isolated from normal social environments, that the client can use as smoking spots where she or he will be *alone* when smoking. Consistent with the pure–activity technique, these are locations where the client can keep a chair to be used exclusively for smoking. The overlap of the pure–activity technique and the anti–social chair technique presents no problems. The client selects one spot for use at home and one spot for use at work. The client is not to smoke anywhere else. When away from these places, a public toilet is a suitable anti–social smoking spot (where legal). Some clients put an uncomfortable chair in spots like a stairwell, bathroom, basement, attic, garage, or spare room. This technique thus further reduces the conditioned–reinforcement value of smoking, because it stops the pairing of smoking with social reinforcers.

The Morrow Study kept the two techniques that are focused on the *unconditioned*–reinforcement value together. They referred to the combination as *reduction of reinforcement value,* because both shared in reducing this reinforcement value. Here we describe these two techniques separately.

We use the term *change cigarettes* for one of the two techniques that offsets the *unconditioned*–reinforcement value of smoking. In this technique clients change their cigarettes in two steps. The first step is filter–related. At the scheduled session, clients who smoke filtered cigarettes remove the filters, and use only unfiltered cigarettes thereafter, making the cigarettes somewhat aversive (or "distasteful," as clients describe them). Clients who smoke unfiltered cigarettes change to a filtered brand. The second step is brand–related. At the next session all clients change to a strong, even imported, unfiltered brand, making the cigarettes even more distasteful. These changes in the cigarettes reduce some of the unconditioned–reinforcement value of smoking.

We use the term *damage cigarettes* for the other technique that offsets the *unconditioned*–reinforcement value of smoking. Clients damage their cigarettes either by poking half–a–dozen pin holes in them, or by waterlogging them, or both. Damaging the cigarettes in any of these ways before smoking them makes them distasteful. Thus such damage to the cigarettes also reduces some of the unconditioned–reinforcement value of smoking.

The combination of variables. The *quitter's procedure* addresses the combination of variables that we invoke near the end of the week of individual therapy sessions. During this week some previously introduced techniques stand alone (e.g., *alternative–behaviors,* which continues whether or not other techniques are in use). Other techniques become irrelevant to the client's efforts (e.g., *difficult–to–obtain,* since the client no longer owns cigarettes). The quitter's procedure combines all remaining, relevant techniques in a way that avoids clients having to rely on an "it's been x amount of time since the last cigarette" Record. Such a Record verbally implies falsely to them that they can never again have another cigarette. Instead the quitter's procedure allows future cigarettes but only under specific, rate–reducing controls.

The Morrow Study designed the quitter's procedure so that successful clients need not fear falling for the temptation to have a post–quitting cigarette.

> [We developed this technique for clients] in response to the expressed fear… that they would again return to smoking if and when they ever had a cigarette after they quit, and their expressed inability to stop smoking if that meant they never, under any circumstances, could have another cigarette. [We told them] that they could have a cigarette, after a period of zero smoking, which would satisfy their

> urge to smoke while at the same time making the smoking response
> so aversive that the probability of emitting the smoking response
> would not be increased… (Appendix A, item VIII)

Therapists describe the quitter's procedure to clients as a series of steps. These become successively more difficult so as to counter increasingly strong urges to smoke. In this procedure, when stimuli evoke an urge to smoke, clients face a series of what, in lay language, we may conveniently (i.e., without the usual scientific objections to inner agents) call choice or decision points.

In the presence of some likely unrecognized stimulus that is evoking smoking, clients initially observe and report to themselves (as a "public of one"; see Ledoux, 2015b) an urge to smoke. Knowing that some environmental stimuli are evoking smoking, clients leave the immediate environment (i.e., the room or other location) in which the urge to smoke occurred. The first step in weakening such control is to move away from that environment. Once in another setting, they consider—the first decision point—whether or not to smoke. Simply changing environments is often enough to render ineffective whatever stimulus started evoking smoking.

If that evocative effect still favors smoking, then clients next review each item on their emotional responses list until it elicits the relevant emotion. Having reviewed the list, they again consider—the second decision point— whether or not to smoke. Reviewing this list often renders ineffective whatever stimulus is still evoking smoking.

If the evocative effect on smoking continues, then clients review the final sequence of the quitter's procedure. They review their required actions *if* the urge to smoke persists: (a) They must go buy an expensive pack of specified, strong tasting, unfiltered cigarettes. (b) They must trash 19 of the cigarettes from the pack right then at the store, because the urge to smoke involves only one cigarette, and keeping the other 19 would only provide them with 19 more evocative–stimulus "temptations." And (c) they must take the remaining cigarette to an agreed upon smoking spot where they must put pin holes into the cigarette and/or waterlog it before, as a pure activity, smoking that cigarette as a satiation cigarette! Having reviewed the required actions of the final sequence of the quitter's procedure, they again consider—the third decision point—whether or not to smoke. Reviewing this final sequence, and experiencing the typical aversive emotional reaction to it, often renders ineffective whatever strong stimuli are still evoking smoking.

However, if the contingencies are still compelling smoking, then clients walk, if possible, to a store where they *might* buy a pack of cigarettes, since they no longer own or "bum" any. Just prior to paying for a pack, however, they review the quitter's–procedure steps one last time and consider—the fourth decision point—whether or not contingencies compelling smoking are really that strong. This second full review of the remaining quitter's–procedure steps, just before actually paying for a pack, often succeeds at the last moment to deter smoking. If this proves insufficient against the current compelling contingencies, then clients pay for the pack and follow through with the rest of the quitter's–procedure steps, finishing with a satiation cigarette.

By completing the quitter's procedure on such rare, necessary occasions, the client has satisfied the urge to smoke but has not reinforced smoking. We soon discuss *why* clients follow these stringent steps.

One other factor bears on the continued successful application of the quitter's procedure. As part of the discussions on this procedure, therapists also repeat two points to clients. They *can* smoke a rare post–quitting cigarette. But smoking a post–quitting cigarette, in any way other than through the use of the quitter's procedure, carries a high probability of producing a return to smoking.

Clients need the quitter's procedure, because virtually everyone who has smoked and quit can expect to experience an occasional urge to smoke. Presumably one or another as yet inadequately addressed smoking–evoking variable has momentarily raised the probability of smoking. Some of these variables are essentially unaddressable. For example, only a rare client, such as the president of a college campus, would have the authority to ban, say, cigarette vending machines from the work place. Yet such a measure would be required to minimize the smoking–evoking effects of these machines in this environment. Nevertheless, such measures are usually beyond a client's reach. Under these circumstances the quitter's procedure remains vital to continued success in therapy compliance, because it provides the client a way to "satisfy" urges to smoke while minimizing the reinforcers from smoking.

The quitter's procedure also increases the understanding of those who have never smoked regarding the extent of the problems faced by smokers who are under contingencies (i.e., who "want") to quit. The rigorous rate–reducing controls of the quitter's procedure surprise many of these non–smokers who see such controls as extreme. They ask "How could anyone ever follow that procedure? How could they throw away the money tied up in 19 cigarettes, or take those other steps? Why would they put themselves through all that?" Smokers who quit through using the quitter's procedure provide the answer. They experience, in various ways, that *the quitter's procedure works,* so they follow it, finding this procedure to be an invaluable aid to successfully quitting smoking. Actually most smokers find the prospect of *never again* having another cigarette not only far worse than the quitter's procedure but also a virtually impassible barrier to successful quitting. Compared to quitters tied to a Record, those who have instead used the quitter's procedure to stop smoking have shown, *through their compliance with it,* that using this procedure is indeed easier than never ever again having another cigarette.

Conclusion

This smoking–control therapy *provides an example* of a therapy that addresses its concerns comprehensively and is theory based in that it derives its practices from established behaviorological principles. Its substantial clinical success rate exemplifies the value of behaviorologically analyzing explicit functionally controlling variables *and designing intervention steps that countercontrol them.* This benefits clients, in this case clients with smoking problems. Extension of these design steps to many other kinds of concerns will likely enable people with those concerns to attain similar success rates. Potential candidate problems for behaviorologically designed interventions range from over–indulging in food or drink to wrecking planetary resources. By the way, this therapy still works, and research–grounded therapists apply it. Today, however, advertising contingencies instead compel clients toward therapies that feature drugs and nicotine patches or gum as components even though the relative importance of these to therapy remains inadequately researched, as we described earlier.

References (with some annotations)

Fraley, L. E. & Ledoux, S. F. (2015). Origins, status, and mission of behaviorology. In S. F. Ledoux. *Origins and Components of Behaviorology—Third Edition* (pp. 33–169). Ottawa, CANADA: BehaveTech Publishing. This multi–chapter paper also appeared across 2006–2008 in these five parts in *Behaviorology Today* (the precursor to *Journal of Behaviorology):* Chapters 1 & 2: *9* (2), 13–32. Chapter 3: *10* (1), 15–25. Chapter 4: *10* (2), 9–33. Chapter 5: *11* (1), 3–30. Chapters 6 & 7: *11* (2), 3–17.

Ledoux, S. F. (1973). *The Experimental Analysis of Coverants.* M.A. thesis, California State University, Sacramento.

Ledoux, S. F. (2015a). Successful smoking control as an example of a comprehensive behaviorological therapy. In S. F. Ledoux. *Origins and Components of Behaviorology—Third Edition* (pp. 243–258). Ottawa, CANADA: BehaveTech Publishing. This paper documents and extends most of the contents in the original 1973 Morrow Study (see Morrow, Gmiender, Sachs, & Burgess, 1973). This paper also appeared (2011) in *Behaviorology Today, 14* (1), 3–13.

Ledoux, S. F. (2015b). An introduction to the philosophy called radical behaviorism. In S. F. Ledoux. *Origins and Components of Behaviorology—Third Edition* (pp. 25–32). Ottawa, CANADA: BehaveTech Publishing. This paper also appeared (2004) in *Behaviorology Today, 7* (2), 37–41.

Morrow, J., Gmiender, S., Sachs, L., & Burgess, H. (1973, April). Elimination of cigarette smoking behavior by stimulus satiation, self–control techniques, and group therapy. Paper presented at the annual convention of the Western Psychological Association.

Skinner, B. F. (1953). *Science and Human Behavior.* New York: Macmillan. The Free Press, New York, published a paperback edition in 1965.

Stuart, R. B. & Davis, B. (1972). *Slim Chance in a Fat World: Behavioral Control of Obesity.* Champaign, IL: Research Press.

Chapter 19
More Types of Behavior:

In what other ways can we consider behavior besides respondent and operant?
Overt (i.e., "neuro–muscular" or simply "muscular") behavior and *covert* (i.e., just "neural") behavior ...

$\mathcal{W}$e discussed functional classifications of behavior (i.e., operant and respondent) in earlier chapters. With them we also discussed some descriptive classifications of behavior (e.g., motor and emotional). Since then we have mentioned other, often overlapping, descriptive classifications, including "overt" behavior and "covert" behavior. Having covered a range of supporting topics in the intervening chapters, we can now effectively elaborate on covert and overt behavior as neural behavior and muscular (i.e., actually neuro–muscular) behavior respectively.

We also consider a couple of related topics. One concerns contingency accounts for behavior from the common "public of others" perspective as well as accounts for behavior from the important but often ignored "public of one" perspective. The *public of others* perspective supports our understanding of, and dealing with, overt behavior while the *public of one* perspective supports our understanding and interpretation of covert behavior. The other related topic concerns "behavior passivity," the nature of all behaviors as natural events—dependent variables (i.e., effects)—occurring in natural reaction to the independent (i.e., causal) variables in contingencies. *All these possibilities pertain to both operant behavior and respondent behavior.*

Overt/Covert Behavior

The distinction between overt behavior and covert behavior helps us separate response classes that have neuro–muscular members from other response classes that have strictly neural members. The *overt/covert* distinction refers not to another type of behavior but rather to another way to talk about various behavior types. The distinction combines and emphasizes both the general location in which behavior occurs (i.e., the behaving organ, like a hand, mouth, or brain) and the amount of accessibility others have to the behavior. We consider some behaviors as public events, because they occur in the external environment and so other people have ready access to them. These we call *overt* behaviors. Other behaviors, however, we consider as private events, because they occur in the internal environment and so other people have perhaps severely limited access to them. These we call *covert* behaviors. Overt behaviors allow easy access for anyone within observing range of the behavior (i.e., a public of others). Covert behaviors, however, allow only limited access to a public of others while providing access most easily to a public of one. The body, the physiology, that eliciting or evocative stimulation causes to mediate covert behavior, comprises the only "public" that has access to it. That is, the covert behavior, as a real event, serves to affect the public of one by evoking chains of additional covert neural consciousness behaviors such as observing and reporting the behavior that the public of one observes (with more details in a later chapter). As an overt/covert example, we consider some verbal (i.e., language) behaviors as overt and other verbal behaviors as covert. We consider neuro–muscular verbal behavior (e.g., speaking, writing, signing) as overt and easily accessible to virtually anyone. And we consider strictly neural verbal behavior (e.g., much thinking and knowing) as covert and accessible mostly to the public of one in whom it occurs and whom it affects.

Our more extended concern with the overt/covert distinction involves avoiding private internal events in general in our analyses due to the accessibility issues. But simply calling behaviors public or private works poorly, because behavior is not the only kind of event that fits a public/private distinction. The overt/covert distinction separates behavior from those other events.

Regardless of whether we are dealing with stimuli or behaviors or something else as private events, we acknowledge them, rather than exclude them, if they are real events. Except for interpretative analyses, we use the same method to avoid all the private events in our experimental or applied analyses. This includes covert behaviors that start as real *dependent variables* —that real *independent* variables cause to occur—*and that then become, as real events, real independent variables* causing subsequent covert or overt behavior. The main method, to avoid private events in our analyses, involves moving back the necessary steps in the functional chain of events. It involves moving away from the interpretive analysis of private independent variables or covert behavior dependent variables. It involves moving back in the functional chain of events until we are dealing with *public* independent variables or *overt* behavior dependent variables. Since the functional chain usually travels from the external environment to the internal environment and then back to the external environment, we can keep our analysis in the external environment by jumping the internal environment. This takes the general analytical form of "if A causes B, and B causes C, then A causes C," where A and C share the external environment. Furthermore this move often allows us to account adequately, if imperfectly, for the external, internal, and subsequent external events.

Let's consider an example, which also shows additional aspects of the analytical method. Your checking account statement arrives in the mail (or online), and you face the challenge of checking your figures and balancing your checkbook. (The monthly arrival of your checking account statement does evoke these checking and balancing reactions, right? If not, the danger of financial loss, both legal and illegal, increases substantially.) Speaking less agentially, the account statement evokes a little operant number crunching behavior (which can be overt or covert). The reinforcing outcome occurs when the statement's final figure and the checkbook record's final figure agree precisely. Presuming you wrote but a few checks this cycle, outsiders only see you sitting at your desk glancing back and forth between the statement and the records, both on the desk before you, and occasionally placing a tick mark somewhere on the statement or on the records. After a final pause, the observers see you write "ok" on both the statement and the records, and leave the desk, the reconciliation between the two satisfactorily complete.

Now, those observers could independently review the documents and math, and thereby confirm that assessment. What they cannot as easily confirm is any claims you report, even as a legitimate public of one, about verbal behavior computations you *thought* through privately while at the table. We take your claims seriously, though, as we have no reason to disbelieve you, particularly when, as in this case, no contingencies exist that compel lying about your thinking math behaviors. After all, some contingencies compel people to tell the truth and other contingencies compel people to tell lies. Neither activity occurs due to good or bad inner agents.

While we as outside observers can interpretatively accept that those covert behaviors occurred, we cannot easily confirm that they occurred, nor can we easily confirm that they did not occur. Accordingly, we instead let stand the external functional chain of events showing the account statement evoking document review behavior (of the statement and the records) leading to the reinforcing outcome of the statement's final figure and the checkbook record's final figure agreeing precisely.

Still, we *interpretively* accept that the "document review behavior" legitimately involved covert behavior chains as scientifically acceptable real events. At least theoretically these are of a type that may become more accessible in the future, perhaps through greater collaboration with our physiology colleagues. By the way, one of the reasons for accepting the reality of your covert math behavior, which you reported as a public of one, is that the outside reviewers, when they reviewed the documents and math for correctness, could have completed this task either overtly *or covertly.* In the latter case, they would accept their own covert behavior part of the reviewing chain, which they observed as individual publics of one. So why would they deny the occurrence of your public–of–one observed covert behavior? They would not. No good reason exists to accept the covert behavior in one case and deny it in another when all else is equal.

Much of our interpretative analysis relates to facts that our physiology colleagues handle. This includes evocative or eliciting stimulus–caused *effects* on parts of the *nervous system* that produce interactive stimulus–response chains of *neuro–muscular* responses *and neural* responses. As real events, any of these responses can serve a stimulus function for the next response in the chain. This can even involve emotions and feelings. Once some stimulus elicits emotional behavior, then the detection of feelings, and even some of the changes in behavior mediation, occur on the level of neural behavior. So let's review that kind of behavior a little.

Neural Behavior

While motor behavior involves innervated muscle movement, which we more accurately call neuro–muscular behavior, when we speak of *neural behavior,* we refer to behavior comprised only of neurons firing. Recall that the skin is not a boundary to the laws of the universe. These laws operate on both sides of the skin. However, the skin may represent a boundary that reduces human access to what goes on inside the skin. Even then, some human activities, like physiology, have more access to inside the skin than other human activities, like behaviorology. We manage these access limitations through our interpretive analyses. Including internal events like neural behaviors in our analyses helps dramatically in extending our understanding of human nature and human behavior. Indeed, we must include internal events if our analyses are to be complete. And gradually, the work of our physiology colleagues will turn the interpretative analyses into accurate functional analyses.

What makes that so important? Perhaps it is not so important. But our past conditioning has convinced us that we cannot understand human nature and human behavior thoroughly enough unless we embrace all real, related topics, including the really difficult and limited–access topics like neural behavior, even if only at the level of interpretation. We cannot leave these topics to magic or mysticism, theological or secular. The well supported assumptions of our philosophy of science, radical behaviorism, enable us to reach some scientific grasp of various phenomena. These include the status and function of consciousness (to which we devote a full chapter later) because we can begin by considering consciousness as real neural behavior. We can then track the extent to which the same variables that control more accessible behaviors also control the behaviors of consciousness and the full range of other neural behaviors.

Neural behavior summary. As with emotional behavior, neural behavior occurs as a dependent variable in functional relationships with various independent variables. Our physiology colleagues are getting ever better at directly tracking neural behavior, because it is real. And once some stimulation produces it, neural behavior can serve stimulus functions producing either other directly observable (i.e., neuro–muscular) behaviors, or other neural behaviors. The neural behaviors often occur in the kinds of chains of cascading neural firings that seem to make up the various behaviors of consciousness or awareness, and so forth. Neural behavior is not the expression of an inner agent's existence. The brain, a real neurally behaving organ, is not "mind" which, as an inner agent, is not real. Neural behavior occurs neither spontaneously nor because some mystical inner agent directed its occurrence. And neural behavior remains accessible to the public of one in whom contingencies compel its occurrence. Let's look more closely at the public of one and the public of others.

Accounts from a Public of One *and from a* Public of Others

Contingencies produce people's overt and covert responses. People's responses produce overt and covert stimuli. Some stimuli affect us as members of the public at large, a *public of others*. These stimuli, mostly from the overt responses of others, participate in contingencies driving "analysis responding," that is, our responses in which we analyze observed behavior. These analyses describe the most likely contingencies responsible for the observed overt behavior. They may also provide access that enables helpful interventions when the overt behavior is in some way problematic.

Yet at the same time, *each* of us is also a *public of one,* a one–member public, privy to the occurrence of our own covert responses and stimuli (with no inner agents implied). Covert stimuli and responses, as real events, occur inside our body where others cannot be privy. With appropriate physiological measurement equipment, these others might gain some access.

As real events either respondent or operant behaviors—either of these as covert or overt behaviors—can evoke *public–of–one* or *public–of–others* observational responses. We can make some valid analyses both regarding our own behavior, as a public of one, and regarding our own behavior and others' behavior as members of the pubic of others. But these two types of analyses differ, particularly in terms of accessibility. Being able to delineate clearly the *public–of–others* version provides a better foundation for interventions.

Let's look at multiple sides of an extended example that will show these differences and their implications. Consider an all–to–common scene. The denial of access to a demanded snack, for a child in a grocery cart at the store, evokes the start of a tantrum. In short order the embarrassed mom or dad calmly wheels the cart out the door where gradually the tantrum subsides. Yes, I know, this is not what usually happens. The reinforcing value of relief from the screaming usually induces the behavior of giving in. Unfortunately this leads to the predictable result that tantrums increase across later store visits (and, by generalization, across other occasions).

In that example, however, we are portraying scientifically (i.e., behaviorologically) informed parents whose education has conditioned them about how to deal with tantrums. Furthermore, once in a developing tantrum circumstance, this example shows one of the recommended steps for handling the situation. Without delivering whatever the tantrum was about, calmly move from the current setting to a different one. Glenn Latham points out that a parent should never give in before the child gives out (see Latham, 1994, and especially 1999). After tantrums start, giving children what they want inadvertently reinforces tantrum behavior, something quite definitely opposite to what would reinforce parental behavior. This clarifies a little about the child's behavior as the behavior of concern. But the parents' behavior better serves our comparative analysis of *public–of–one* and *public–of–others* perspectives.

So, consider that example again. The denial of access to a demanded snack, for a child in a grocery cart at the store, evokes the start of a tantrum and, in short order, the embarrassed parent calmly wheels the cart out the door where gradually the tantrum fades. This time, however, we consider *leaving the store* as the behavior of concern.

From the parent's public–of–one perspective, what happened? The parent might report, "The tantrum made me feel embarrassed. This evoked my leaving the store. Past experience taught me that leaving the store would make me feel better, which it did. I will leave the store again the next time a tantrum starts." Here is how we might symbolically diagram this *public–of–one* reported contingency. Note that since this is a specific activity, we use the "Response" label rather than the more general "Behavior" label:

S^{Ev}		Response		Consequence		[Result]
Aversive feelings of embarrassment	→	Leave store	→	Aversive feelings diminish	→?	[Leave store in future when tantrum starts]

We start with the aversive feelings of embarrassment. Of course, only the width of the page prevents us from including additional prior evocative stimuli (e.g., even prior to the "No cookies" that evoked the tantrum, and the tantrum plus the store's social environment that elicited the emotional response that the parent reported as those feelings of embarrassment). Also, notice that the reduction in the aversive feelings of embarrassment occurs upon leaving (i.e., escaping) the social environment of the store regardless of how long the tantrum persists before trailing off. So we cannot say that a reducing tantrum contributes to reduced embarrassment. What if the parent had reported feelings of anxiety that the tantrum evokes. A reducing tantrum, whenever that occurs, leads to a decrease in the anxiety. Consider this by writing such a diagram.

However, while this situation seems to lack any contingency that would induce a false parental report, being unable to be more confident about the accuracy of the report leaves a certain residual discomfort. We also often lack ready access to observing emotions or feelings, except one's own. The same holds for access to changes in, or relief from, the reported emotion. Here, we lack ready access to the embarrassment (or anxiety) feelings, and the subsequent change in them, that the involved parent reported. We are stuck with this kind of interpretive difficulty with all public–of–one analyses.

To improve this situation, we must move to a public–of–others analysis. So, from our public–of–others perspective, what happened? Other observers, upon witnessing the events, might report, "Shortly after the tantrum began, the parent pushed the cart holding the child out the store door, after which the tantrum subsided." A prediction consistent with these events would be that "the parent will leave the store again the next time a tantrum starts." Here is how we might diagram this public–of–others reported contingency:

S^{Ev}		Response		Consequence		[Result]
Tantrum begins	→	Leave store	→	Tantrum subsides	→?	[Leave store in future when tantrum starts]

While these overt events are related to the very real covert events that the participants experienced, far fewer accessibility problems inhere in this more public analysis. Moving to a public–of–others perspective renders our analysis practical. We can consider additional helpful intervention steps when the overt behavior turns out to be in some way problematic. But all this rests on a certain well–supported presumption about behavior, the presumption of *behavior passivity*, a natural–science presumption to which we now turn.

Behavior Passivity

The benefit of public–of–others analyses will be evident in our continuing consideration of contingencies to cover more complex human behaviors. However, before expanding into such areas, we should more explicitly visit a principle that we have mostly taken for granted thus far. This is the principle of *behavior passivity*.

Behavior passivity refers to the nature of behavior, all behavior, including all human behavior of any type. Like all real events in nature, on this planet, in this universe, behavior is natural, that is, it comprises natural events. It is an inevitable reaction. Operant behaviors as well as respondent behaviors are all reactions to variables in contingencies. As such, behavior merely "happens." It happens as a function of other also real variables. When these variables happen, as a result of their own traceable natural functional history, behavior follows. If a behavior happens, then it had to happen. Conversely, if a behavior does not happen, then it could not have happened. Various authors (e.g., Skinner, 1953, p. 112) have made this same point in similar ways.

As real events, behaviors participate in the accumulating natural functional history of yet other real events, some of which are also behavior. Due to stimulation, neural behavior happens mostly as the firing of brain neurons while neuro–muscular behavior happens as innervated muscle contractions making limbs, or other behaving organs, move in various ways. Since, in an appropriate time frame, we cannot really separate the observing of behavior and the firing of neurons along with the muscle contractions, we describe the situation in terms of the physiology *mediating* the behavior.

"Physiology mediating behavior" provides a direct description of those events that excludes traditional, pre–scientific (yet still all too current in some quarters) mystical notions about the origins of behavior. Some of these excluded notions involve physiology somehow originating or spontaneously initiating behavior, which it does not. Other excluded notions involve a wide range of physiologically independent self agents that the cultural–level, even educational, conditioning of some people has led them to see as responsible for behavior essentially in scientifically untraceable or untestable ways. These mystical notions can be as specific as the notions that the agent does the behavior or tells the body to do the behavior, or decides which behavior will occur, or chooses the next behavior. Alternatively, these notions can be as general as the terms mind or psyche or self or soul causing behavior. Yet no evidence can exist for an inner agent's doing, telling, deciding, or choosing, due to the untestable mystical status of inner agents.

According to the evidence of our scientific experience with behavior over the last century (and our more general scientific experience over the last four centuries) all those mystical notions—specific or general—treat behavior as a magical event outside scientific reality. But allowing any part of those mystical notions to surface in discussions about understanding, predicting, controlling, or interpreting behavior, which is of necessity our scientific agenda, greatly reduces the possibility of dealing with the behavior part of our world effectively. This is because these prescientific, mystical notions stand in opposition to a century's worth of data from our so far successful natural behavior science, including its behavior–passivity foundation.

If we try to work outside behavior passivity, then we sacrifice our best chance to help each other, or to improve our world, successfully. Across some later chapters, we look at contingencies in the context of discovering more of the properties of complex behavior.✣

References (with some annotations)

Latham, G. I. (1994). *The Power of Positive Parenting*. Logan, UT: P & T ink.

Latham, G. I. (1999). *Parenting with Love*. Salt Lake City, UT: Bookcraft. This short, inexpensive book makes, I believe, the best first book for persons interested in supporting their parenting skills with scientific information.

Skinner, B. F. (1953). *Science and Human Behavior*. New York: Macmillan. The Free Press, New York, published a paperback edition in 1965.✑

Chapter 20
Linguistic Behavior, Also Known As Verbal Behavior:

How does language relate to behavior?

Verbal behavior and some of its parameters ...

A Large portion of this chapter's topic, *verbal behavior,* makes up another kind of neural behavior. This adds an expansive aspect to our considerations. However, having gotten a little "geeky" about neural behavior in the last chapter, we avoid geeky in this chapter. The details on verbal operant behaviors, even just the overt variety, stand out as both interesting and applicable. But even providing formal definitions for discussing them gets rather technical. So here let's just name these verbal operants, and leave you to follow up through other books for the details according to the interest your contingencies generate. The verbal operants include *mands, tacts, intraverbals, codics* (i.e., textuals, taking dictation, etc.), and *duplics* (i.e., echoic, copying text, mimetic, etc.). Do not worry if you have no idea what these things are, for further discoveries always await all of us. To satisfy any residual curiosity after this chapter about verbal behavior, I recommend Chapter 20 of my *Running Out of Time...* book (Ledoux, 2014), which contains implications and applications to teaching non–native languages as well. Another book, *An Introduction to Verbal Behavior—Second Edition,* which is a self–study book that I co–authored with Norm Peterson (Peterson & Ledoux, 2015), may also satisfy some of your curiosity.

In this chapter, however, we only consider some general points about verbal behavior. Even discussing these general points may require some small reference to a couple of geeky technical terms, but here these will be harmless.

A Partial Verbal Behavior Overview

Language is perhaps the single most noticeably complex behavior repertoire that conditioning processes can produce. We use the term *verbal behavior,* however, to encompass more than just what we usually call language. Groups of people speaking the same language (i.e., a verbal community) initially condition—a far more complex process than this single word conveys—much of the basic verbal behavior of new community members during the several years of their early childhood. The verbal community then continues, among its members, to condition refinements and extensions throughout life, aiding the quality and continuation of the community.

Verbal behavior is not involved in *directly* producing the reinforcing consequences that shape and maintain it (i.e., the way turning and pushing or pulling a door knob *directly* produces an open door). Rather, verbal behavior *indirectly* produces its reinforcing consequences through the behavior of one or more other verbal community members. For example, "Open!" or, in a more refined, later iteration, "Please open the door," in the presence of another verbal community member and a closed door, indirectly produces an open door through the other's behavior of door opening.

While interpreting verbal behavior in light of the basic laws of behavior, Skinner (1957) differentiated several types of verbal behavior based on increasingly complex functional differences among them. More than three decades of published experimental research on verbal behavior relations and applications has accumulated. This research validates Skinner's original interpretive analysis. Most of this research appears in the journal, *The Analysis of Verbal Behavior* (TAVB).

Since we mention one or another verbal operant at different points, here are some quick descriptions. We also add a point or two about how they work.

Mands are verbal responses that occur due to, among other things, deprivations. The occurrence of the deprived items reinforces this verbal response. For example, under conditions of going without food, the mand "Let's eat!" occurs upon arrival home after a day's work.

Tacts are verbal responses that occur due to the evocative functions of *non–verbal stimuli.* The reactions of listeners reinforce this verbal response. For example, say you are camping in the forest. Then, in the middle of the night, the noises of a big dark animal rummaging through your food stores evokes your whispered tact "Bear." Your companions provide reinforcers when they whisper their thanks to you as you all creep out the other side of the tent to safety. Note that your yelling "BEAR!" instead of whispering it, is also a tact, but perhaps less likely to produce such reinforcement.

Intraverbals are verbal responses that occur due to the evocative functions of *verbal stimuli.* Again, the reactions of listeners reinforces the response. For example, under conditions of an American teacher saying, "Red, white and ____?" in the presence of a student, the student's intraverbal response "blue" occurs. The teacher's response, "That's right" (based, in this case, on a flag–based thematically related set of colors) provides reinforcement.

And *textuals* are verbal responses that also occur due to the evocative functions of verbal stimuli. But in this case the stimuli are printed words, perhaps on a page, that share "point–to–point correspondence" (more later) with your responses. Additional variables also apply, leading to the comprehension that we tact as *reading*, which can have covert and overt components. Your current responses, that the words on this page are evoking, provide the example, and your increasing understanding provides one kind of reinforcement.

Those four verbal operants represent not quite half of the elementary verbal operants. For the others, and details, see the recommended books.

Much verbal behavior is clearly neuro–muscular behavior. However, verbal behavior also comprises a substantial segment of the strictly neural behaviors of consciousness, including thinking, hearing, and comprehending.

Language is Verbal Behavior

In this and some later chapters, we begin to supply some current scientific answers to some of humanity's ancient questions. Humans have always had questions about "language." Here we cover some basics about language, which is also the medium in which these questions and answers occur. However, we set aside the ancient, cultural traditions that have language coming from a mind— or any other mystical inner agent—as its supposed psychic powers magically take over after any real stimuli and direct the body to produce real responses of its mere choosing. Instead, we take language and analyze it scientifically, under the label *verbal behavior,* because contingencies induce both the ancient questions and their answers to manifest in the form of this kind of behavior.

We approach verbal behavior in two steps. After providing some general considerations about verbal behavior, we cover nine characteristics of verbal–behavior analysis.

General Verbal Behavior Considerations

Both terms, language and verbal behavior, describe a topic that many see as among the most, if not *the* most, complex, omnipresent and vital phenomena of our world. The stimuli that we call words, and their effects, literally surround us, in a myriad of spoken, written, gestured, signed and symbolized forms. Functional word recombinations, which contingencies continuously vary in form, comprise large components of every human's everyday behavior. We go to great lengths to assure this verbal extent and variation (e.g., through education) partly because words are the vehicles that contingencies drive into "rules" (the topic of a later chapter). Of major significance, the further behaviors that rules evoke, such as education, build accumulations of knowledge and practices that last beyond individual lifetimes. We describe these accumulations as cultures, accumulations that drastically reduce the need for conditioning to build each individual's repertoire from "scratch" through direct contingency contact. The further behaviors that rules evoke include words combining into larger units, from phrases and sentences to fiction and non–fiction books, and from rhymes and couplets to ballads, songs, and epic poetry. All of these units occur in relation to nearly every imaginable subject matter. Many of these units occur in both the usual overt forms as well as the even more intimately familiar covert varieties that we call thinking. And all these units occur under the ordinary, natural control of generally non–coercive contingencies.

Let's begin our minimal coverage of this vast verbal–behavior topic with some of its recent history, and then with some ancient history regarding the relation of evolution and physiology to verbal behavior. These provide a foundation for defining not only verbal behavior but also the verbal community. Then we will describe some characteristics of our verbal–behavior analysis.

Recent history

The recent history of verbal–behavior analysis sets the stage for us. The term *language,* arising out of traditional, mystical cultural lore, has always been, and continues to be, fraught with agential implications. B F. Skinner (1957) adopted the term *verbal behavior,* rather than struggle with the term *language,* out of concern to stress only the real, measurable, naturalistic aspects of linguistic phenomena, which include the operation of contingencies in the generation of language repertoires and the production of language products.

Skinner began considering complex human linguistic behavior early in his career. Some circumstances in 1934 induced his particular interest in this topic. His efforts over the next 20–plus years, including various lectures and courses, culminated in his 1957 book, *Verbal Behavior.* Late in this book, he described the circumstances that induced his increased interest in verbal behavior:

> In 1934, while dining at the Harvard Society of Fellows, I found myself seated next to Professor Alfred North Whitehead. We dropped into a discussion of behaviorism… and I began to set forth the principal arguments… with enthusiasm. Professor Whitehead was equally in earnest—not in defending his own position, but in trying to understand what I was saying and (I suppose) to discover how I could possibly bring myself to say it. Eventually we took the following stand. He agreed that science might be successful in accounting for human behavior provided one made an exception of *verbal* behavior. Here, he insisted, something else must be at work. He brought the discussion to a close with a friendly challenge: "Let me see you," he said, "account for my behavior as I sit here saying, 'No black scorpion is falling upon this table.'" The next morning I drew up the outline of the present study (pp. 456–457).

The answer to Whitehead's challenge, which runs several pages, provides interesting and fun material. I recommend it to you, but only after completing this chapter, of course, and perhaps also after completing some further resources as well, before tackling Skinner's book (e.g., Skinner, 1953).

The appearance of Skinner's *Verbal Behavior,* which many consider his most important work, began to focus attention on how far a natural science of human behavior could reach, particularly in discounting agential explanations of behavior such as those common in traditional linguistics. Traditional linguistics represents a stronghold of agential habitation wherein the superstitious side of the culture takes language as directly communicating the expressions of the mind, psyche, or self. Following this kind of lead, in 1959 Noam Chomsky published a paper highly critical of Skinner's book, and purporting to be a review of it. By seeming to discredit Skinner's book, especially in the eyes of those who had not, and now would not, read it, Chomsky's paper misled many into further support for fundamentally mystical accounts of language. It also misled them into ignoring Skinner's account, an account that is strictly a natural–science account of linguistic phenomena. Even though Skinner's book accounts for verbal behavior in a largely interpretive manner, those who read Skinner's book held that Chomsky had missed the point. He had argued effectively against things that Skinner had not said, and against viewpoints that Skinner had not held. In 1970, a widely respected paper by Ken MacCorquodale reviewed Chomsky's paper and largely set the record straight. Data supporting the natural–science account that Skinner had offered began to appear in the 1970s and continues to pile up (e.g., the still published journal, *The Analysis of Verbal Behavior,* began publication in 1982).

Before moving on, however, we should note that, after setting aside the inherent inner agents, the field of traditional linguistic analyses still has much to offer. Its structural–analysis sometimes dovetails smoothly with the functional analysis that characterizes our coverage of verbal–behavior. We should also recognize a simple convention regarding verbal behavior. Exposure to verbal–behavior material quickly leads to the occasional use of the abbreviation, "vb," in place of "verbal behavior." Occasionally we also invoke this convention.

VB, Evolution, and Physiology

Moving from recent history to ancient history, let's consider the origins of language and the origins of anatomically modern humans. Across several sciences some researchers say, interpretatively, that our species "began" when the brain and vocal musculature evolved in a way that happened to enable contingencies to induce verbal behavior. Let's unpack that a bit.

Until more data say otherwise, researchers generally accept that our species gradually appeared as natural selection happened to produce certain evolutionary changes in the anatomy and physiology of some proto–species members. These changes accumulated to the point where our remote ancestors had neural structures and vocal–musculature structures functioning together as a system that operant contingencies increasingly affected. These contingencies produced expanding verbal effects during the lifetimes of numerous socially interacting individuals. Then, across the multiple generations of groups of individuals, the contingencies established more and more complex varieties of verbal behavior. These continued to occur—they did not stop for others when someone died—because the lifetimes of community members overlapped each other. The resulting beneficial effects for community members maintained the group contingencies, which we now call cultural practices, that drove further verbal behavior developments.

While that description gives you a broad view, here are some specifics. The vocal neuro–musculature increasingly came under control of operant contingencies due to the resulting contributions to survival. These contingencies altered various vocal sounds during the individual's lifetime such that particular stimuli came to evoke particular sounds. Postcedent processes would shape and maintain the consistency of these sounds across *groups* of individuals that shared contingencies in common. Each group member would then say the "right" thing (i.e., the thing others would reinforce) under the appropriate conditions (e.g., the appropriate evocative stimuli).

As contingencies expanded an individual's verbal repertoire, the *meaning* of a particular response always stemmed from the particular controlling variables. Meaning resides in the contingencies. Meaning never resided in agential characteristics or components. And similar contingencies operated on different sounds making them "mean" the same thing across different groups that were separated from, and thus not in contact with, one another. This led to the development of different languages and language groups with a variety of language similarities and differences. The contingencies of physical reality that produce verbal behavior *remain similar across all the different verbal communities* around the globe. These similar contingencies result in common linguistic characteristics in different forms worldwide. One example pertains to the presence of verb tenses in various languages that result from similar contingencies regarding past, present, and future events. You will find that a thorough study of the contingencies related to any verbal behavior results in eliminating any relevance for fictional explanations or inner agents in accounting for verbal behavior.

All those recent and ancient historical considerations give us some overall perspective regarding verbal behavior. With this in place, let's turn to verbal–behavior analysis definitions and characteristics.

VB Definitions

Verbal behavior has proven to be of vital importance to humans and their cultures and survival. So let's look a little more closely at the definition of verbal behavior. Let's also look more closely at the definition of verbal community, because it continuously keeps us focused on the inherent social and interactive character of verbal behavior.

Verbal behavior. We basically define behavior as *verbal behavior* when the consequences, particularly the reinforcers, of the behavior occur through the activity (i.e., the mediation) of another organism. In other words, verbal behavior is behavior that produces reinforcers that occur through another organism's behavior. Conversely, non–verbal behavior is behavior that produces reinforcers that occur through the direct effects of the behavior on the environment. For example, a wasp flying into the room through a window serves as an establishing operation that increases the reinforcing value of an open door through which the further response of exiting can occur. While alone in the room, if the wasp stimulus evokes a grasp–the–door–knob–and–turn response, the resulting open–door reinforcer occurs through the *direct* effects of the grasp–and–turn response, so this response is not verbal behavior. However, with another living human body present and closer to the door, if the wasp stimulus evokes frantic pointing–at–the–wasp and pointing–at–the–door responses, and this gesturing evokes the other body's door–opening response, then for the gesturing body, the open–door reinforcer occurs *indirectly*, through the response of the other body. The reinforcer that gesturing produces occurs through the mediation of the other body, making the gesturing meet our basic definition of verbal behavior. The wasp could also evoke the vocal response "Open the door!" This response could also produce an open door through another body's physical door–opening response, so it too would fulfill the basic definition of verbal behavior.

That verbal–behavior definition takes verbal behavior beyond the *vocal* behavior that, for many thousands of years, held sway not only over earlier gestural or sign languages but also before contingencies extended verbal behavior into written and other forms. Indeed, this basic definition takes verbal behavior well beyond the very notion of linguistics. Under this basic definition, if the reinforcers that a behavior produces occur through the responses of another organism, then we can call the behavior *verbal behavior*. For example, when the pet dog scratches at the back door and the dog's owner opens the door thereby mediating the reinforcing consequence of the dog's door–scratching behavior, that door scratching *technically* meets the basic definition of verbal behavior.

Nevertheless, we generally construe behavior like that door–scratching as on the periphery of the kind of behavior that more commonly evokes our response of "That is verbal behavior." Along these lines, we usually add some qualifiers to our definition: *Verbal behavior* is behavior that (as a real stimulus event) evokes another organism's responses that mediate—as in provide—the reinforcers for the first organism's behavior, *after verbal–community contingencies have conditioned such mediating behavior, and where typically both organisms are members of the same verbal community.* That conditioning of mediating behavior is part of what makes people members of the verbal community.

Verbal community. We define the *verbal community* simply as the group of people whose mutual mediating reinforcements condition the verbal and mediating behaviors of the group members as a result of the benefits that accrue to the group from generating and maintaining these verbal behaviors. Many of these benefits become obvious when you consider that this definition broadly refers to all those people, again with overlapping lifetimes, conditioned in the set of verbal responses and practices that constitute a particular language (e.g., English). For instance, the majority of objects and events in our urban surroundings result from responses that verbal behaviors make possible, including, and perhaps emphasizing, verbal behaviors that stem from, and result in applications of, the written verbal reports of scientific research and engineering outcomes. In a more limited sense, the verbal community for any individual consists of those with whom the individual shares verbal and mediating behaviors.

Nine VB–Analysis Characteristics

Here we consider many characteristics of verbal behavior. The first five apply more to verbal behavior in general while the rest also remain pertinent to categorizing the variety of elementary verbal behaviors.

(1) Function rather than structure. The long standing tradition of approaching speaking and writing as language has produced some worthwhile results. From this approach linguists know much about the *structures* of many languages, such as their vocabulary, grammar, and syntax. The facts that they have discovered help us compare languages, trace language families, teach about languages, and even improve our teaching about how to write well in a particular language. This structural approach to language, however, usually begins with the damagingly common traditional agential presumption. It presumes that language, especially as language behavior, flows as communication from immaterial, including mentalistic, agents residing inside a body, the body from which the speaking or writing appears merely to emanate magically. Yet linguistic analyses of grammar, syntax, and so on, lack any need for this assumption, an assumption which in any case starts out and remains mystical and thus scientifically inappropriate.

But what if speaking or writing "appears merely to emanate magically" from a body only because no one has scientifically analyzed *why* the speaking and writing occur? What if we scientifically analyze *why*? What if we analyze *why* in terms of real, measurable independent variables? Such *why* questions approach language in terms of *function* rather than structure. Skinner's verbal–behavior analysis began scientifically to explore the extent to which language behavior might be a function of the same kinds of contingencies, with the same kinds of independent variables, of which other behaviors are a function.

In continuing Skinner's analysis, we analyze verbal behavior as a function of mostly evocative and consequential stimuli. This not only gives us details about the variables of which verbal behaviors are a function, but it also tells us about the function of verbal behaviors in affecting other events. By revealing these functions of verbal behavior, behaviorological analysis shows many more of the ways through which people affect the world around them. This makes particularly inescapable the relevance of verbal behavior, and the natural–science behind its analysis, to solving individual, local, and global problems.

(2) No new fundamental principles. Throughout its historical development, verbal–behavior analysis has succeeded while working only with the same set of independent variables, the same set of concepts and principles and processes, that control non–verbal behavior. Verbal–behavior analysis has not needed any new fundamental independent variables that only apply to verbal behavior. Of course research continues to expand our understanding of behavior, both verbal and non–verbal, and sooner or later the data may require additional fundamental concepts or principles or processes. At this time, however, all the independent variables that make up our concept of contingencies of reinforcement are the only variables needed in accounting for

any behavior of humans and other animals, including verbal behavior. These concepts and principles and processes include all those that we have covered in past chapters, such as conditioning, generalization, establishing operations, function–altering stimuli, evocation, reinforcement, schedules, punishment, extinction, shaping, chaining, fading (plus others from later chapters).

(3) VB sense modes. When people first hear about verbal behavior, many think of making sounds with the vocal musculature. But verbal behavior occurs in other sense modes as well. While we *hear* vocal verbal behavior, we *see* written verbal behavior. Furthermore, we see the verbal behavior of gesturing and signing, and we *feel* the written verbal behavior of Braille. Any behavior in any of these modes that is reinforced through another person's behavior is verbal behavior. However, sometimes a bodily sound, such as a sneeze, evokes another person's behavior but without any reinforcing effect. Another's subsequent polite comment after a respondent sneeze has little to no effect in making you sneeze more often. Thus behavior such as sneezing would qualify as non–verbal behavior.

That is one source of confusion about which you should be wary. This source of confusion comes from mixing stimulus and response words with respect to various sense modes. For example, the terms *auditory* and *visual* describe stimuli while the terms *vocal* and *writing* describe responses.

(4) "Speaker" and "listener." The terms *speaker* and *listener* refer to different behavior repertoires. The vocal verbal community conditions both speaker and listener repertoires in its members, while the signing verbal community conditions both signer and viewer repertoires in its members.

In verbal–behavior analysis, we tend to de–emphasize the "listener," along with the "viewer" of signs and gestures. This is because the stimuli from a speaker or signer affect a listener's or viewer's behavior in ways essentially similar to the ways that *any* non–verbal stimulation affects a listener or viewer. Indeed, the very notion of verbal behavior focuses attention on the behavior of speakers, along with the behavior of gesturers and signers, and writers. Basically, the speaker's behavior is verbal while the listener's behavior often is not verbal. Listener behavior responds to another's verbal behavior but otherwise need not itself be verbal. Note that for convenience, we often simply speak of the "speaker." Our VB analysis, however, usually applies equally well to writers, signers, and gesturers. Of course, none of these words imply inner agents. These terms *only refer to the body that mediates* (i.e., to the physiology, the working of which mediates) *the responses* due to current contingencies.

You may have even noticed that the term, *mediate*, can cause a little confusion, because we use it two ways. Many chapters ago, and just now, we described bodies (i.e., nervous systems) as mediating behaviors. In this chapter *we also describe listeners as mediating the reinforcers* of a speaker's behavior, which is what makes the latter verbal behavior. For the key to unlock any confusion, consider that while *bodies mediate behavior, listeners mediate reinforcers* through the behaviors that "their" bodies mediate.

Furthermore, the contingencies on speakers and listeners often involve interlocking interactions, sometimes producing speaker behavior and at other times producing listener behavior. For example, a conversation might proceed this way: Circumstances evoke a question from party A (a speaker) which evokes an answer from party B (a listener, then a speaker). This answer in turn evokes a comment from party A (now a listener, then a speaker) which evokes a question from party B (a listener, then a speaker) which evokes an answer from party A (a listener, then a speaker) and so forth.

Such interactions show the speaker and listener repertoires coexisting as functionally different and analytically separate repertoires. Still, they become virtually inseparable *when stimuli evoke* the neural behavior of verbal thinking, which we can describe as covert *simultaneous* speaking–listening.

On a terminological note, Julie and Ernie Vargas (Skinner's daughter and her spouse), in their verbal behavior courses at West Virginia University in the 1990s, replaced "speaker" with "verbalizer," which better addresses all the non–vocal forms of verbal behavior. They also replaced "listener" with "mediator," which better addresses the functional role of mediating reinforcement delivery following the verbalizer's behavior. Given the traditions in which the "speaker" and "listener" terms arose, these newer terms may carry fewer agential implications. While "speaker" and "listener" remain the common terms at present, verbalizer and mediator may become the common terms in the future.

(5) Audience controls. In terms of overt responding, while particular variables control the form of a given verbal response, a general contextual variable controls whether or not any verbal behavior occurs at all. We call this general variable the *audience*. Without an audience, without a listener present, overt verbal responses seldom occur. For example, if you are the last to leave the classroom but cannot open the door due to carrying a large stack of books, no verbal behavior occurs that requests someone to open the door. Instead, the current circumstances evoke some other door–opening solution. However, the presence of someone else, the presence of an audience, leads to a verbal door–opening solution. In this sense, the presence of an audience serves as a function–altering stimulus. An audience alters the function of the other relevant stimuli from neutral to evocative of a *verbal* response. The form of the verbal response depends on the particular controlling stimuli. With an audience present, a closed door that you cannot get open evokes a "Please open the door" request. With an audience present, for example a parent at the zoo, a tiger emerging from a den evokes a child's "Tiger!" statement.

Of course, the automatic establishment of a listener repertoire during the conditioning of a speaker repertoire leaves people, at the covert level of verbal thinking responses, as their own reinforcing listener. Thus, we consider that an audience essentially remains ever present. The general lack of mediation of substantive reinforcers, which requires a separate audience/listener, appears to balance the lower level of energy expenditure associated with covert responses.

The control that an audience exerts also affects other aspects of verbal responding. For instance, the audience also controls which parts of a repertoire a particular stimulus evokes. For example, if a physician is the audience, then the discoloration on a patient's leg evokes a "contusion" response from the nurse. But if the patient is the audience, then the discoloration evokes a "bruise" response from the nurse.

We also find that *places* exert an audience–control kind of stimulus control. For example, along with other variables, the local classroom produces a different part of a student's verbal repertoire than the local theater produces or the local pub produces. Students talk about different things in those places. They also talk in different ways in those places because, along with other variables, places exert control over other response characteristics. Generally classrooms produce medium sound levels of verbal behavior while theaters produce whispered levels and pubs produce boisterous levels.

Those first five VB–analysis characteristics apply generally to verbal behavior. The remaining four pertain both generally and to categorizing various kinds of elementary verbal operants, a topic you can find in those recommended books mentioned at the start of the chapter.

(6) Responses and response products. An analysis of elementary verbal operants calls for a level of detail that requires a distinction between a *response* and what we call the *response product.* As we have pointed out before, the occurrence of behavior constitutes a real event that can function as a stimulus. In many cases the response produces direct and immediate changes in the environment. The term *response product* applies both to such changes and to the stimulus status of responses. Response products are the stimuli that the responses produce.

For example any response may leave one or more response products. When stimuli induce innervated changes in the vocal musculature (i.e., the response) a result (i.e., a response product) involves changes in air–wave patterns that function as auditory stimuli (i.e., "words"), especially for others. These auditory stimuli are the response products of speaking. Longhand writing involves arm and hand movements that not only produce motion visual stimuli, especially for others—we see someone writing—but the movements also produce marks on paper. Both the visible movement stimuli, and the visible marks stimuli on the paper, are response products of the writing responses. Not all of these response–product stimuli leave stimulus records that can affect others later. The movement stimuli, once complete, are gone, while the marks stimuli, once complete, remain and so can affect others later.

This distinction between responses and response products helps us categorize some formally controlled verbal operants. Formally...? What? "Formal control" refers to one of two general kinds of verbal operant controls. The other is "thematic control," and next we will consider both of them.

(7) Thematic and formal controls. Analyzing and categorizing verbal operant behaviors first requires sorting them as either under a *thematic* kind of

control or under a *formal* kind of control. While we take "thematic" directly from *themes*, we use "formal" in its *structural* connotation. Thus we call one kind of control *thematic*, because we can generally trace some sort of theme connecting the verbal response with either its controlling variable or its consequence. Meanwhile we call the other kind of control *formal*, because we can trace some fairly explicit structural similarities between the whole or parts of the controlling stimulus and the verbal response or response product. In this usage, "formal" is a fancy word for "structural."

Note, however, that themes can seem arbitrary or ambiguous. So we use the presence or absence of a basic structural feature, the feature that we call point–to–point correspondence (which we discuss next) as the dividing line between thematic and formal controls. We consider verbal operants that *lack* point–to–point correspondence as under thematic control and verbal operants that *have* point–to–point correspondence as under formal control.

Yes. These last four vb characteristics sound awfully "theoretical." They get very concrete, however, in the sorting of verbal behaviors into their elementary categories. But that is a very technically challenging activity that we have left for those recommended books. Covering these last four vb characteristics here makes that activity easier for you when you go for it.

Let's return to thematic and formal controls. Consistency across some dimension of stimulus and response establishes a theme, which we name for that to which it pertains. Here are some examples of verbal operants under *thematic* controls. After playing soccer for half an hour you say "water," and receive some, which is a theme of getting what you said. You say "dog" when you see a dog, which is a theme of the same thing seen and said. And you say "lights" when you hear someone else say "phone, gas, and…," which is a theme of home utilities. The next section contains *formal* control examples.

(8) Point–to–point correspondence. When we examine verbal stimuli and verbal responses, we find that verbal stimuli contain parts that by themselves can control parts of verbal responses in a manner that we describe as *point–to–point correspondence.* This involves formal (i.e., structural) control. Each part of the stimulus controls a corresponding part of the response. Here is an example of this controlling correspondence. The three parts of the written–word stimulus *dog* (which is the response product of someone's writing behavior) can control other responses including your saying "dog," or its letters "dee" and "oh" and "gee," or its phoneme sounds. Also, when someone says "dog," that auditory stimulus (which is the response product of someone's vocal musculature movements) can control other responses including your writing *dog* (i.e., *d* and *o* and *g*). In all these cases, you see or hear the parts of the stimulus (written *dog* or spoken "dog") corresponding to, and controlling, the parts of your responses (spoken "dog" or written *dog*). Again, we call this correspondence *point–to–point correspondence,* and it describes some details of the independent and dependent variables of verbal operants, details that help us better organize and categorize verbal operants, and thereby deal more effectively with them.

Here are additional examples of verbal operants showing *formal* controls, with at least point–to–point correspondence between the stimulus and the response. You say "bus" when you see the written or printed word *bus*. You say "apple" when you hear the word "apple." And you make the ASL (American Sign Language) sign for cat when you see another signer make this sign.

(9) Formal similarity. Many of those example stimuli and responses show point–to–point correspondence in *different* sense modes (e.g., seeing *dog* and saying "dog"). That is, they are either (a) spoken *then written* or (b) written *then spoken.* However, when stimuli and response products that have point–to–point correspondence are *also in the same sense mode* (e.g., they are *both* spoken, or *both* written, or *both* signed) they share something more. They share *formal similarity.* They share a kind of physical, structural similarity on a part by part basis that we call *formal similarity.* The parts of the stimulus not only correspond point–to–point with the parts of the response, but each part of the stimulus physically, structurally (i.e., formally) resembles the corresponding part of the response product. Here is an example of this similarity that could also use dogs, as in our point–to–point correspondence example, but instead uses cats. The *written*–word stimulus *cat* controls your *writing*–response parts that produce the written word *cat* again. This control occurs through the three parts of the written–word stimulus *cat* (i.e., the written letters *c, a,* and *t*) not only having point–to–point correspondence with, but also having a structural, formal similarity to, the three parts of your writing response *c–a–t.* Similarly the three parts of the auditory stimulus "cat," (i.e., the sounds from the "c," "a," and "t") control your saying "cat," not only through point–to–point correspondence with, but also formal similarity to, the three parts of your vocal response "c–a–t." The same analysis applies to the arm–and–hand movements in the ASL sign for cat when it then controls the same arm–and–hand movements of another signer signing cat. When a stimulus and response product not only show point–to–point correspondence but also are in the same sense mode, they have *formal similarity.* This describes some further details of the independent and dependent variables of verbal operants, details that also help us better organize and categorize verbal operants and thereby deal more effectively with them.

Conclusion

With those nine verbal–behavior characteristics in hand, particularly the last four, you are ready to cover some major categories of verbal behavior, *using the recommended materials mentioned at the start of the chapter.* These include the elementary relations that we call *mands, tacts, intraverbals, codics,* and *duplics,* along with various subtypes in the *codic* and *duplic* categories.

With those basics in hand, we can also start considering applications of our verbal–behavior analysis to the task of building verbal repertoires. The analysis contains practical implications for improving language conditioning in at least three areas. (a) We can improve the conditioning of first–language repertoires for newborn verbal–community members over the first five years of life (e.g., see Hart & Risley, 1995, 1999). (b) We can also improve the conditioning of first–language repertoires for developmentally delayed and autistic verbal–community members, often by starting with the conditioning of a sign–language repertoire (e.g., see Maurice, 1993; also, see Maurice, Green, & Luce, 1996). And (c) we can improve the conditioning of subsequent, non-native language repertoires, particularly in regular educational settings (e.g., in elementary school, high school, and college; see Ledoux, 2014, Chapter 20).

References (with some annotations)

Chomsky, N. (1959). Review of B. F. Skinner's *Verbal Behavior. Language, 35,* 26–58.

Hart, B. & Risley, T. R. (1995). *Meaningful Differences in the Everyday Experience of Young American Children.* Baltimore, MD: Paul H. Brookes.

Hart, B. & Risley, T. R. (1999). *The Social World of Children Learning to Talk.* Baltimore, MD: Paul H. Brookes.

Ledoux, S. F. (2014). *Running Out of Time—Introducing Behaviorology to Help Solve Global Problem.* Ottawa, CANADA: BehaveTech Publishing.

MacCorquodale, K. (1970). On Chomsky's review of Skinner's *Verbal Behavior. Journal of the Experimental Analysis of Behavior, 13,* 83–99.

Maurice, C. (1993). *Let Me Hear Your Voice—A Family's Triumph over Autism.* New York: Ballantine Books.

Maurice, C., Green, G., & Luce, S. (Eds.). (1996). *Behavioral Intervention for Young Children with Autism.* Austin, TX: Pro–Ed.

Peterson, N. & Ledoux, S. F. (2014). *An Introduction to Verbal Behavior— Second Edition.* Canton, NY: ABCs.

Skinner, B. F. (1953). *Science and Human Behavior.* New York: Macmillan. The Free Press, New York, published a paperback edition in 1965.

Skinner, B. F. (1957). *Verbal Behavior.* New York: Appleton–Century–Crofts. The B. F. Skinner Foundation (www.bfskinner.org) in Cambridge, MA, republished this book in 1992.&

Chapter 21
Linguistic (i.e., Verbal) Behavior—II:

What difference can verbal behavior make?

Contingencies can govern behavior through "rules"

that are *statements* of contingencies. ...

*W*ith the background in verbal behavior from the last chapter, we can now turn in this chapter to one of the most valuable benefits of verbal behavior. Language enables contingencies to drive verbal descriptions (i.e., statements) of contingencies. These statements of contingencies can then serve stimulus functions in controlling behavior without every human body having to undergo the slow, extensive, and sometimes dangerous shaping of behavior by the original contingencies that the statements describe. We summarize these "statements of contingencies" with the term *rules*.

As a brief example, consider the danger of walking too close to the edge of a cliff. What is the contingency if loose stones cause you to slip when you are close to the edge versus when you are well back from the edge? If you slip when you are close, and you are only injured, then you are less likely to walk close to the edge in the future. But you may not be only injured. If you die, then you never walk the cliff edge again. On the other hand, slipping when you are well back from the edge involves far less danger. Must everyone experience slipping off the edge for the contingency to produce the behavior for everyone of walking well back from the edge? No. Verbal–behavior contingencies drive a verbal description (i.e., statement) of the danger, as a rule. "Avoid danger by staying well back from the edge of cliffs." Note that this statement of the contingency (i.e., this rule) contains not only the consequence (danger) but also the response (staying well back) as well as the evocative stimulus (cliff edges). This rule then produces walking well back from the edge without having to personally experience the danger. Such is the basic benefit of such rules.

Those kinds of contingency developments lead us to differentiate between *contingency–shaped behavior* and *rule–governed behavior*. While some researchers may prefer other names, these developments open an extensive expansion of non–agential complexity for human nature and human behavior. In this expansion we can see that the notion of human nature revolves around the contingency relations that so comprehensively involve human activity at the individual and cultural levels. Behaviorologists have worked with this expansion since Skinner first described it back in 1966 (see Skinner, 1969).

Rules as Contingency Statements

The rules that we discuss here address a range of phenomena that not only subsume the usual notion of rules, such as school rules, but also enable us to deal with more of the more complicated aspects of human behavior. Again, the term *rules* refers to *statements of contingencies*. These statements explicitly or implicitly include at least a behavior and its contingent consequences, and often its evocative stimuli as well. These statements are verbal behavior, behavior that arises in humans through conditioning that other humans mediate.

Rules are present verbal stimuli that *supplement* other stimuli in ongoing contingencies. They sometimes serve this function by bridging the gap between behavior and the delayed consequences of defective contingencies (i.e., contingencies in which the consequences are too far removed in time to be directly effective). For example, good grades are contingent upon thorough studying. But grades occur far too long after the studying to function as reinforcers of that studying. A rule like, "More thorough study produces better grades," repeated *in the present,* can supplement other contingencies on study behavior, leading to the kind of studying that gets good grades.

Through the conditioning of verbal behavior, such rules occur when circumstances evoke verbal descriptions of contingencies that someone has experienced. The rules then function *as stimuli* for others. They become a part of other contingencies as function–altering stimuli or evocative stimuli. But these stimuli (i.e., rules) that specify contingencies (i.e., "contingency–specifying stimuli") lead us to differentiate between two kinds of operant behavior, "contingency–shaped behavior" and "rule–governed behavior." As we consider each of these in turn, note that contingencies control *both* of these kinds of behavior. However, only the "rule–governed behavior" involves contingencies that include rules. Again, we differentiate between them, because treating them separately helps us better deal with complex human behavior.

Contingency–Shaped Behavior

We call behavior *contingency shaped* when contingencies condition that behavior without involving stimuli that equate with statements of contingencies. In common language we might say that the behavior occurs through direct personal experience. For instance, we keep a safe distance from bee hives because, in the past, getting close resulted in getting stung. Such contingencies thus directly determine the distance–keeping movements, so we call that behavior contingency–shaped. Let's trace relevant variations of *this* example beyond contingency–shaped behavior and through (a) rule–governed behavior, (b) observational contingencies, (c) instructions, and (d) other verbal supplemental stimuli.

Rule–Governed Behavior

The behavior that some contingencies compel can be dangerous for our health (e.g., honey reinforces behavior requiring more than mere closeness to hives). Must we all experience such contingencies directly for those contingencies to produce appropriate behavior? Do the natural laws of behavior include any process to save at least some, perhaps most, of us from having to experience the bad effects of dangerous outcomes? Yes. We need not each experience all contingencies directly, because the laws of behavior encompass *rule–governed behavior.* The process involves a contingency evoking a verbal–behavior description of another contingency, and that description, which we call a rule, then affects (i.e., supplements the controls on) other behavior. We call the result *rule–governed behavior.* This process reduces the need to experience directly every contingency of daily life, including the dangerous ones. Indeed, this is a major contingency cause for cultural practices, for example the general cultural practice of formal education.

We call behavior *rule–governed* when the contingencies governing the behavior *include* rules that function as either function–altering stimuli or as evocative stimuli. That is, the behavior occurs in part due to rules rather than solely through direct conditioning. For instance, while simple contingencies can directly condition keeping a safe distance from bee hives, you probably must get stung as part of that conditioning. However, observing someone else, even another animal, get stung when close to a hive can evoke a rule, a verbal–description behavior of the contingencies. This rule might state "getting close to bees can get you stung," or "staying away from bee hives reduces getting stung." These rules can generalize to stimuli in future settings (i.e., stimuli in future settings might evoke the rule). For example the later buzzing of bees may evoke restating the rule, which then helps compel safe–distance–keeping movements, movements that we then call rule–governed behavior.

Observational Contingencies

Those kinds of rules can become *instructions* for others who have neither experienced the contingency that the rule describes nor seen someone else experience it… Hang on. How can the stimuli from *seeing someone else* experience a contingency make our behavior change as if we had experienced it? At first glance, this scenario looks magical. The consequence in the contingency follows another organism's overt behavior. You only observe it. The consequence does not follow your own response. So how can it affect you?

Good questions. The consequence affects you, because basically it does follow your response. It follows your observing response, which the other organism's behavior evoked. The real problematic magic here is that the question arises from the notion that "you" refers to an inner agent. We already know, however, that the "you" (and the "we") are merely verbal shortcuts that simplify our discussion. Leaving out the inner agents leaves real energy exchanges (as evocative stimuli) interacting with real physiologies that mediate real energy exchanges (as behaviors) that produce further real energy exchanges (as consequences). Stimuli evoke another organism's response, and that response produces a consequence. The stimuli, from both the other organism's response and its produced consequence, change the neural structures of the "observing" organism. These changes are somewhat similar to the way that the "observing" organism's neural structures would change *if* the evocative stimuli had caused the "observing" organism to respond in the same way, thus directly producing the same consequence. This similarity is enough such that, for the "observing" organism later, similar stimuli evoke similar responses that produce similar consequences. Outsiders see this as a generalization effect. No mystical magic is involved.

Instructions

Again, rules can become instructions for *others* who have neither experienced the contingency that the rule describes nor seen someone else experience it. For these others the instructions become part of contingencies that lead to rule–governed behavior.

While we usually phrase rules in a rather general manner, we often phrase instructions in a more situation–specific fashion, although an instruction can simply be someone else stating or restating a rule in the presence of others for their benefit. Still, instructions are rules, statements of contingencies, that the contingencies of someone else have induced. They or others then later pass the rules on, as instructions, to still more others whose behavior is then appropriate without their coming directly under the control of the original contingencies. For instance perhaps your grandfather had once got too close to bees and got stung. When his daughter (your mother) was getting too close to bees as a child, your grandfather instructed her with the rule, "getting close to bees can get you stung." Or perhaps he phrased the rule as the instruction, "Stay away from *those* bees or you will get stung." Either way, the important point is that his statement controlled the distance–keeping behavior of your mother. Note that each of these forms of the rule specifies the evocative stimuli (bees), the appropriate (distance–related) response, and the consequence (stings).

Now, your grandfather stayed away from bees, and never *again* got stung, due to the contingencies of his personal experience perhaps with rules serving supplementary functions. Your mother, on the other hand, *has never been stung,* but that is due to her distance–keeping behavior being rule–governed, that is, being under contingencies that include a rule about keeping a safe distance from bees, a rule that possibly occurred as an instruction from her father. And when you were little, and getting too close to bees, she passed this instruction on to you. Perhaps you also have never been stung and will never get stung.

A Role for Supplemental Verbal Stimuli

The verbal stimuli of rules and instructions, as statements of contingencies, often supplement the other variables in contingencies, making the contingencies more successful in various ways. While rules can certainly function this way on the overt level of some stimulus that evokes writing down the rule or speaking or reading it out loud, rules commonly function this way through the covert, purely neural, verbal behavior of consciousness that we call thinking. Such behaviors are real events and their occurrence then participates in the current contingency as a supplemental stimulus altering the function of some other stimuli. In this case the behavior occurs under additional variables besides direct conditioning. For instance, the buzzing of bees may evoke the covert thinking response, "getting close to bees can get you stung," a response the elements of which your mother first conditioned decades ago when she saw you about to touch a bee on a garden flower, and so instructed you. This thought, from a past instruction, then supplements, as a supplemental verbal stimulus, the current contingency in which the annoying bee buzzing was already arresting forward motion toward a hive such that the now combined stimuli instead induce motion that puts greater distance between you and the hive, preventing you from getting stung.

Concerns about Rules

A Rule–Governed Behavior Shortcoming
In spite of the help that rules provide, rule–governed behavior has a major shortcoming. Once conditioned, the rules can last longer than their benefits, because the contingencies can change faster than the rules that we derive from the contingencies. For example, what if the safe distance from bee hives in *current contingencies* has increased, over what the rule implies? This could happen due to the hive harboring genetically altered bees. The genetic alteration could arise from interbreeding with certain bee strains for whom stimuli (e.g., people or other animals) evoke (or elicit) attack responses at greater distances than the previously safe distance that *current rules* help evoke. This previously safe distance is now an unsafe distance, and behavior that the rule still governs will get you stung. The contingencies have changed faster than the original rule. Perhaps we must come under the control of a rule about rules: *Keep comparing rules with the actual contingencies they purport to describe,* or get stung (i.e., get hurt in whatever way the contingency involves).

A Residual Confusion about Contingencies and Rules
Rules and contingencies share a residual confusion. Rules are *not* separate phenomena from contingencies. *Rules are parts of some contingencies.* As evoked statements of contingency components, rules constitute natural phenomena that are part of various contingencies. A rule can even be a part of a contingency that involves contingency–shaped behavior. Again, the differentiation between contingency–shaped behavior and rule–governed behavior occurs, because it helps us deal with some complicated aspects of human behavior. Rules are not

magical, or agential or "cognitive." Contingencies generate rules. And rules become natural parts of other contingencies that also generate, maintain, and reduce as well as shape behavior. Let's consider another example of a rule, a simple statement of a contingency. In this case *the rule is simply a part of another contingency.* In this example the rule serves as a function–altering stimulus that changes an evocative stimulus for one response into an evocative stimulus for another response. (Of course, you surely recall that the change is not in the stimuli but in the effects that the stimuli have on physiology, on neural structures.) This new example describes something that has never happened to you before, and you have not discussed such events with anyone before, because you come from a nice small town where events like those in this example are extremely rare. But you are in the big city now, enjoying some spare days before presenting data as a scientist at a sustainable–living convention. Thus, this example is an example of a contingency–shaped behavior that encompasses a "rule." Note that some authors prefer the term "contingency–conditioned" over "contingency–shaped," especially with examples like this, because no clearly separate shaping steps occur in the example.

Imagine that you are walking on a poorly lighted street late on a cold winter–holiday night, and you see someone else walking from the other direction. This other walker is fighting the biting breeze with a cute ski mask that looks like it has white rings around the eye openings making the mask look like one of those cute old–time cartoon dogs or racoons. But just as you are about to pay a compliment on the ski mask, the other walker waves an illegal, black–market gun at you and says, "Give me your wallet or I'll shoot you." Now, just because every non–criminal citizen like you has a right to keep and bear arms for many legitimate reasons, including self defense as in this case, *not everyone exercises their right,* and for this example that includes you. Where people are allowed to exercise this right, even if few actually exercise it, the allowance itself serves as an S^{FA} that, for a potential robber, changes many potential victims from S^{Ev}s to S^As for robbery. Think about that (and check the FBI statistics for verification). For our example, however, in this particular big city, exercising that right is currently (and probably illegally) disallowed.

So, those words, "Give me your wallet or I'll shoot you," occurring under those circumstances, constitutes a contingency–specifying stimulus (i.e., a rule) that serves as a function–altering stimulus that changes the "cute" ski mask from an evocative stimulus for a compliment into an evocative stimulus for compliance. While at first you hesitate, which evokes some further abusive and potentially very dangerous (for you) posturing responses on the part of the robber, you finally comply by handing over your wallet. You may or may not also notice that the "white rings" were just overly large eye openings. Your compliance reduces the threat as the thief removes your cash, throws your wallet behind you, and disappears up an alley. Again, this is an example of a rule, as a contingency–specifying stimulus, participating in a contingency as a function–altering stimulus. (Due to its situation–specific nature, we could call the robber's statement an instruction, but we call it a rule here, because its implication of severe consequences matters more in our example than its specificity.) Here is the diagram for this contingency.

$S^{FA\,(CS)}$	S^{Ev}_1 (for compliment) S^{Ev}_2 (for compliance)	R	S^{r-}
"Give wallet or else" →	[Cute] Ski mask →	Compliance (rather than compliment) →	Threat reduced

Note that in this diagram, the S^{FA} appears as $S^{FA\,(CS)}$. The S^{FA} in this example "specifies a contingency." It is a function–altering stimulus of the contingency-specifying type, hence $S^{FA\,(CS)}$. However, only rare circumstances, like an example in a book, evoke inclusion of the "^{CS}."

Now, returning to our concern over a residual confusion, is your behavior in that example an example of contingency–shaped behavior, or an example of rule–governed behavior? If, as described, this is the first robbery you experienced, we would describe your compliance as an example of contingency–shaped behavior, even though a rule was involved in the contingency controlling your behavior. Consider, though, that generalization from this instance could lead to a verbal description of the contingency that leads to more effective behavior in these circumstances. This rule could be, "When a robber confronts you in that city, comply right away." Whether the rule arises from having survived a robbery, or from talking about your—or someone else's—experiencing a robbery, is immaterial. The rule is now a verbal S^{CS} supplementing other contingencies. Now, if a robber again confronts you and you comply more quickly, due to this rule, we call *this* compliance *rule–governed behavior*.

Conclusion

Rules carry grand implications for much of complex human behavior. So take a little time to consider how the various parts of stimulus control, contingency-shaped behavior, and rule–governed behavior all figure into a range of cultural practices. These include education, as well as our share of making timely solutions to the behavior–related components of global as well as local and individual problems.

In even broader terms, scientific research involves scientists coming under the contingencies that exist in their research efforts. The discoveries (i.e., the results, the outcomes) from their scientific–research efforts then produce rules regarding the independent and dependent variables in the research contingencies. These rules enable others, who are not involved in the research (i.e., all those—everyone else—who are not under the contingencies inhering in the research effort) to benefit from the research by applying the rules. They need not experience the contingencies that produced the rules, yet they can apply the rules and reap the benefits. In addition the rules accumulate as a culture's scientific knowledge which, through the education of later generations, lasts well beyond the lifetimes of the original researchers.

As a simple example, consider the scientific research that led to the discovery of the surgical technique for safely removing inflamed appendixes. This discovery led to rules about the technique, rules that medical schools teach to new surgeons who were never under the research contingencies. Yet these new surgeons can then apply the technique, first in contrived, practice situations that build the necessary skill, and then in real–life situations where the technique has saved many lives. Similar scientific research efforts have led to many surgery techniques. Have such surgery techniques saved your life?

The interrelations between rule–governed behavior and the development of some needed, green cultural practices, including their connection with education, will likely prove particularly pertinent to avoiding early human species extinction. (We delve into some details in a later chapter.)☙

References (with some annotations)

Skinner, B. F. (1969). *Contingencies of Reinforcement: A Theoretical Analysis.* New York: Appleton–Century–Crofts. The B. F. Skinner Foundation (www.bfskinner.org) in Cambridge, MA, republished this book in 2013.☙

Chapter 22
Some Scientific Answers to Ancient Human Questions:

Can science inform how to live?

Consider values, rights, ethics, and morals. ...

*H*umans have many more long–standing questions than just the ones about verbal behavior, which our last two chapters touched on. Our most common ancient questions concern topics like consciousness, life, personhood, death, reality and the interrelated–topic series of values, rights, ethics, and morals, which is the topic of this chapter. We humans have asked questions and sought answers about such topics for a very long time. But until recently we have had to manage with answers that always seemed somewhat unsatisfactory. Finally we can begin to consider the kind of answers that would differ from those of the last several thousand years, the kind of scientific answers that would make a difference in our dealings with ourselves and the world around us, the kind of answers that behaviorology, as the natural science of behavior, has begun to provide. This works, because so many of your questions and mine so thoroughly involve human behavior. Over this and the next several chapters, we consider some initial scientific answers to some of our ancient questions.

Some Historical Context

Past answers to ancient questions have of course been helpful in various ways. But human historical records—going back, for example, to ancient China and ancient Egypt—show that until quite recently these questions and answers have changed little over at least the last 5,000 years. Perhaps little change has occurred since as far back as the development of verbal behavior, which we currently consider to have occurred at least 50,000 years ago. Across most of that time frame, relatively little else regarding living conditions changed much as well. Change began to accelerate about 400 years ago with the rise of current science. Until then the extremely gradual changes accruing in cultural contingencies remained relatively insufficient to induce much of the behavior patterns that we call science. Being thus unavailable, science could shed little light on anything, including these ancient questions and answers.

Then, around 1600 CE, the accumulating cultural–contingency changes began to produce scientific behavior patterns (e.g., behaviors that respected evidence from solely natural events as independent and dependent variables that people can observe or measure). Across the four intervening centuries since then, developments in science and supportive cultural contingencies increasingly accelerated. Thus, over the last 150 years—relative to the preceding thousands of years—science brought generally helpful and quite extensive changes to most people around the globe (e.g., changes in communication, food preservation, construction, sanitation, medicine, and transportation).

However, scientific answers to ancient, as well as new, questions about human nature and human behavior (i.e., the general behavioral subject matter) remained for several centuries outside the purview of the developing sciences, the traditional natural sciences that became physics, chemistry, and biology. In part this happened because, as part of surviving in the still theologically controlled culture around the 1600s, the science of the time had to go along with a dangerous compromise with religion. Science could deal with physical events, so long as it left human nature and human behavior to the mystically grounded theological and philosophical disciplines of the time, and ever since.

As church authority over the culture later gradually decreased, some changes have occurred. In the 1800s, mystically grounded *secular* disciplines also laid claims to the general behavioral subject matter of human nature and human behavior that encompasses many of those ancient questions. Then, in the early 1900s, natural behavior science began to develop and subsequently address these questions (e.g., see Skinner, 1938). Nevertheless, natural behavior science is still opposed by religion. In addition, due to residual effects of that early compromise, the traditional natural sciences provide little appropriate support for natural behavior science. No wonder many say we cannot afford that compromise. In his 1997 novel, *The Millennium Man,* Joseph Wyatt provides an educational and entertaining account that summarizes the range of these changes and developments, around the globe, in traditional and behavioral natural sciences, and makes comparisons with mystical disciplines.

No law of nature, however, says that science cannot address ancient questions about human–nature and human–behavior. Biology has begun to address parts of them, particularly some parts pertaining to the natural origins of the human body. Behaviorology extends this effort to the natural origin of human behavior. You can read B. F. Skinner's 1948 novel, *Walden Two,* for an educational and entertaining account of some of the changes and developments that might stem from a behaviorological approach to answering some of the ancient questions confronting humanity.

The topics of this chapter directly involve behavior and ancient questions. We already introduced many basics that a natural science of behavior has discovered in the last 100 years about behavior and the variables of which it is a function. So let's see if we can now begin to get any closer to some satisfactory, difference–making, answers to some of humanity's long–standing questions.

Recognize, though, that behaviorologists are not the only natural scientists concerned with these questions. Various well–known and deserving traditional natural scientists (e.g., Richard Dawkins and Sam Harris) have made efforts to tackle some of these questions (e.g., ethics and morality). However, lacking a minimal background in behaviorology, the kind of background that would provide the natural behavior science that dovetails with their own specializations, they have ended up trying to shoehorn accounts of behavioral phenomena soley under headings like brain physiology, evolution, and genetics. Each of these clearly plays a role, but that role is more along the lines of *how* the phenomena work. What behaviorology addresses tracks more of *why* these phenomena happen. Humanity needs this line of inquiry, because it can delineate the *accessible independent variables* involved in these phenomena. These accessible variables then lead to developing beneficial interventions.

In this chapter, we can see how those accessible independent variables can evoke more active human participation in the contingencies that compel the development of values, rights, ethics, and morals in particular directions— some of which may have a distinct bearing on our continued survival—rather than leaving these directions to *coincidence*. Not a new theme, this is the same dichotomy that Skinner and others addressed multiple times in the past in terms of "accident" (i.e., coincidence) versus design.

One example of our current topic, having a potentially sizable bearing on our continued survival, can show the interconnections of our chapter concepts. This example concerns the value of sustainable living. The conditioning history that turns components of sustainable living into conditioned *reinforcers* provides substantial *value* for our survival. We claim this value as a *right* that deserves *ethical* respect and may even be *morally* correct at this point. Such possibilities, of course, threaten the mystical and superstitious assumptions about human nature and human behavior that ancient cultural institutions and lore induce as they come to us through generations of cultural conditioning. This cultural conditioning arose from our 50,000 years—documented for the last 5,000 years—of accumulated verbal conditioning in circumstances that disallowed much thorough reality testing. So now this conditioning includes lots of untestable mystical and superstitious accounts. Perhaps that conditioning still affects your behavior in some ways, possibly even inducing a negative emotional reaction to discussions of these kinds of topics. I can only hope that the counterconditioning that the last 21 chapters have provided for you will prove to be an adequate intellectual and emotional counterbalance, altering any negative emotional reactions to positive emotional reactions.

With that hint about some of the interrelationships of our chapter concepts, let's take, in turn, our beginning look at values, rights, ethics, and morals.

Values, Rights, Ethics, and Morals

Many of the concepts that we consider in this chapter relate to each other in a crescendo of complexity, namely the concepts of values, rights, ethics, and morals. For thousands of years now, these topics have evoked questions and discussions among humans. While the answers we consider here are not extensive, they are at least scientifically informed. We start with values because they tie these concepts directly to measurable variables in the prevailing contingencies (e.g., see Skinner, 1971).

Values

The things that we value are the things that we need, appreciate, hold dear, maintain access to, and so on. Examination shows that these things function as reinforcing stimuli. For example we often confront a relatively simple stimulus circumstance, such as needing to discover a stimulus that will serve as a reinforcer to improve the behavior of a poor and hungry student. This circumstance evokes our asking this seemingly small question: "What reinforces this student's behavior?" With some observation we may discover an accessible and affordable stimulus type or two that serves this function and so answers this question. The answer also tells us something about what the student values, although as values, reinforcers can become rather complex. Let's say that a broader set of stimuli (which we need not specify here) evokes our asking this seemingly bigger question: "What are this student's values?" With some observations, we may discover and make a sizeable list of her values. When we examine the list, though, we find that it contains the names of numerous stimuli (e.g., as objects, events, circumstances, processes) that function as reinforcers for her behavior. The list starts with food and money. But at the other end, the list might include love, world peace, and sustainable living for all. For simplicity, we focus on the start of the list. For what poor and hungry student would money and food not serve as reinforcers? (Perhaps you thought that by "poor" I meant that the student was incompetent or lazy, rather than "impecunious.") A similar comment applies to the more complex values at the other end of the list.

Between food and sustainable living, a wide range of additional stimuli could appear on this values list, including current and historically based stimuli, all of which could also function as reinforcers for the student's behavior. For example on the list we could find comfortable living quarters, honest friends, fair and capable professors, a quiet place for study, a sophisticated computer, a good sound system, and lots of music to play. We might also find various opportunities for the student on the list, such as opportunities to craft, participate in, or attend entertaining events (e.g., concerts, operas, plays, films) and opportunities to practice fun or practical skills. These could include her musical instrument playing skills that band–related contingencies originally conditioned in middle school, or her target–shooting skills that team–participation contingencies originally conditioned in high school, or hunting skills that regular field trips, with extended family members, originally conditioned, trips that put food on the table and in the freezer. We might even find on the list a big, strong, excessively safe vehicle that runs reliably although with poor gas mileage; apparently for this student the necessary social conditioning has not yet made a reinforcer out of a personally smaller carbon footprint. All these things could comprise a portion of the student's values, a portion of her reinforcers.

Now look over that list again. The seemingly bigger question (i.e., "What are the student's values?") is essentially the same as the supposedly smaller question (i.e., "What reinforces the student's behavior?"). Both questions concern both the student's reinforcers and the student's values. These are the same, which applies for everyone. How many or how few appear on such a list is of little importance. The reinforcers are the things that they and we value and, conversely, the values are our and their reinforcers. You can see this by making some more lists. Recalling that behaviors, as real events, function as stimuli, make a list of the stimuli that reinforce your behavior, and a list of the stimuli that reinforce the behavior of a friend that you know well, and a list about someone you know poorly, and a list about the members of some thematically related group of people (e.g., people connected by shared contingencies concerning conservation). In every case, you will find that a list of what they value repeats a list of their reinforcers. *Values are reinforcers.* Follow behavior with the things people value, and you will find their behavior occurring more often. Their reinforcers are their values.

As an observation, you may also note that the length of those lists gets shorter if you make them in the order that we described. You may be intimately familiar with a long list of your values, and the values of your friends, values that you likely share. However, you can spot only a few of the values of persons you know poorly, and possibly you can recognize only the main, and shared, value of a thematically specified group of people. This main value, the main thing that reinforces the behavior of all of the members of the group as a group, likely appears in the group's name. For example, what is the main value—the main thing that reinforces the behavior of all the members of the group as a group—for the group that calls itself the *Death with Dignity Alliance?*

Alternatively some people define a value as the *behavior* that produces a reinforcer. This makes one's values the behaviors that produce one's reinforcers. By our first definition, if a small carbon footprint is among the stimuli that reinforce your behavior, then one of the things you value is a small carbon footprint. By the second definition, one of your values would instead be the behaviors that produce a small carbon footprint.

Values as *behavior,* however, carries a danger. The causes of the behavior remain neither specified not implied. What if the same scientifically uninformed cultural conditioning that includes agential behavior explanations causes the "values as behavior" statement? Then the term *value* ends up referring to something that an inner agent possesses. In the usual agential accounts for behavior, the agent (of whatever sort) then directs the body to behavior in ways comporting with the possessed value or value characterizations. We of course exorcize the inner agent as scientifically unworthy, and reject the fictitious accounts for behavior. Hence my preference for our definition of values as reinforcers. This definition applies as we move our discussion on to rights and ethics and morals.

Before moving on, however, recognize that we only consider the values that are *unconditioned* reinforcers, which comprise necessary stimuli for individual and species survival (e.g., food, water, even sex) as inherently valuable, in the sense of *absolute* values (although even these have at least partial exceptions, as we will see). *Absolute,* then, merely refers to the unconditined origin of the value. Other values gain their status as values through the conditioning process, the pairing that conditions the reinforcing function of otherwise non–reinforcing (i.e., initially neutral) stimuli. This process makes these stimuli function as *conditioned* reinforcers, and thereby also makes them values. But they are values in the sense of conditional or *relative* values, because without the pairing, they function neither as reinforcers nor as values. This difference will soon show up in the dichotomy between *absolute* and *relative* rights, ethics, and morals as well. Consider rights.

Rights

While the term *values* refers to reinforcers, the term *rights* refers to access to values, to reinforcers. Given this connection to physical, measurable realities, which includes behaviors and contingencies, we can focus on rights (and, later, ethics and morals) as events amenable to all the scientific consideration that we cover in natural behavior science. We define a right as unhindered access to a value, to a reinforcer. This definition comes from the contingencies, often of deprivation or coercion, that compel particular forms of verbal behavior, forms that we call statements about rights. These rights statements often take the form of claims regarding unhindered access to valued reinforcers (see Vargas, 1975; also see Krapfl & Vargas, 1977). These contingencies of deprivation or coercion also compel non–verbal behavior as specific activities that support the rights and obtain the reinforcers, activities often involved in the exercise of the rights. We generally think rather abstractly of a right while the rights claim or exercise of the right constitutes a range of more concrete behavioral events.

Let's examine rights a little more deeply. When deprivation accumulates, or when something functions as coercion by getting in the way of our access to a reinforcer, we then claim access to the reinforcer as our right. Sometimes we claim an access right individually, and at other times we claim an access right as a member of a group. Sometimes rights refer, as values, to an immediate, personal reinforcer. At other times rights refer to long–standing, traditional reinforcers. As an example of a right to an immediate, personal and individual reinforcer, after coming home tired at the end of a noisy and energy costly work shift, one might claim, as a right, access to a period of some simple peace and quiet in a home that otherwise features the more usual racket of kids and blaring stereos or radios or televisions or computer games. As an example of a right to a group–shared, long–standing traditional reinforcer, after some government agency makes illegal the defense of one's person or loved ones against immediate threats of harm—a circumstance that some describe as both deprivation and coercion—the affected group of law–abiding citizens might raise a chorus of claims for restoration of the right of self defense. The affected group in this case really includes everyone, although not everyone participates or even always comprehends the shared long–range interest in the group's endeavor. Even looking just at the twentieth century, many historical examples of tyrannical behavior begin with someone—often not a tyrannical person or government—disarming a population, sometimes for what seem like good reasons, with the tyrant or tyrannical government coming to or maintaining power more easily due to the disarmed population. However, very few if any historical examples have such tyrants or governments lasting uninterrupted for very long. Tyrannical coercion still induces the full force of countercoercion that sooner or later ends its reign.

Between and beyond those possibilities remain many of the rights that our individual and group contingency history, including our national political contingency history, has conditioned. Think about the Bill of Rights in, and Amendments to, the U.S. Constitution. We say a person is exercising a right when stimuli evoke behaviors that produce the values (i.e., the reinforcers) to which the right pertains. However, one need not exercise a right to be eligible to make rights claims about the related reinforcers. For example, all law–abiding citizens in the U.S. can make reasonable rights claims about an individual, constitutional, First Amendment right to speak and believe "freely" (i.e., without coercion in their contingencies) or a Second Amendment right to keep and bear arms. However, only a subset of all U.S. citizens may be exercising those rights during any given period or in any particular location.

Indeed, not everyone who could make a claim, for a right to unimpeded access to particular valued reinforcers, needs to exercise the claimed right as the only avenue producing the benefits of that right. For instance, FBI (Federal Bureau of Investigation) statistics for 2011 indicate ongoing reductions of violent crimes in general, and of murders in particular, both of which are down about 50 percent over the last 20 years to a more than 40–year low, a period during which the number of U.S. states with right–to–carry–concealed–arms laws increased to over 40. (You can check more recent years for comparison.) These trends typify states that implement Second Amendment supporting laws. The opposite trend typifies states with governments that enforce laws essentially requiring that citizen adults and children become victims— wounded, raped, maimed, or killed—if the local police are unable to reach the scene before harm occurs during a crime against the citizens. Technically, while some could argue that these data currently convey correlational relationships, others could argue that state legislatures have already organized and carried out repeated and reasonable equivalents of experimental research that makes these statistics convey functional relationships. Either way we can scientifically appreciate the conclusions and implications. All citizens benefit when criminal activity decreases as a result of, or at least in the presence of, laws that recognize the right of responsible armed self defense, as in the right–to–carry–concealed–arms laws of most states. In such states, *any* potential victim could be a law–abiding citizen legally carrying a concealed self–defense firearm. This status makes, through processes like generalization, *all* such potential victims—whether carrying or not—less evocative of the illegal behaviors of criminals or would–be criminals. So everyone, even those who are not carrying, benefits from this right.

Maintaining rights necessarily involves some risks. Continuing our Second–Amendment example, the group advantages of the benefits from rights activity can offset the few possible individual disadvantages, say, from tragic firearm accidents. Nevertheless the contingencies of risks induce risk–management efforts. Again, according to government statistics, these accidents have steadily reduced during decades of ongoing and broadly based citizen education programs in firearms safety. Such outcomes justify the continuation of these programs. Your own perusal of government reports at your local library, or search of government web sites, will provide you with the latest statistics, which have been along the lines of these examples for decades now.

In addition to governments, other groups implement sanctions against, or establish arrangements to protect, various rights claims. Perhaps the most powerful, and so sometimes dangerous, of these groups include religions and corporations (e.g., see Skinner, 1953, Chapters 21–26).

Another set of rights that everyone needs to exercise, and about which everyone could make access claims, pertains to *green* rights. These include rights to clean air and fresh water, to a healthy atmosphere and an intact ozone layer, to pesticide–free food and safe transportation, to enabled recycling and sustainable living, *to a population level within the planet's carrying capacity,* and so on. Given humanity's long–range requirement for such rights, let's keep them in view as we move on to consider ethics and morals.

Ethics

While the term *values* refers to reinforcers, and the term *rights* refers to access to values (i.e., to claims of clear access to reinforcers) the term *ethics* refers to the behavior of respecting those rights claims for unfettered access to valued reinforcers. Indeed, we define ethics, and ethical behavior, as behavior respectful of rights claims. Those who respect our rights claims earn the label, "ethical" or, rather, their behavior of respecting our rights claims earns the label, "ethical behavior," and we appreciate the ethics that we say they "show" by respecting our rights claims.

Verbal shortcuts for bodies mediating responses. By now the kind of subtle agential phrasing present in that last sentence likely elicits some reader wincing or annoyance responses, because readers know that no inner agents exist either to display ethics or to order the body to show respect for rights claims. So, what induces that seemingly agential phrasing? Perhaps the complexity of these topics is evoking an authorial extension of verbal–shortcut status to that kind of phrasing due to the economy it provides, as discussed in some early chapters. Or perhaps the author is under contingencies to avoid adding dozens of pages at various points, pages that would further develop a behaviorological grammar of greater economy but that otherwise go off topic. More likely, the verbal–shortcut agential phrasing, for both the author and the reader, arises from more mundane but scientifically reasonable sources. Consider that, while behavior exists during its occurrence, the body exists before, during, and after the occurrence of the behavior that the body mediates, which it mediates due to the evocative effects of the related stimuli. Thus, due to the simple respondent conditioning that inevitably happens as behavior occurs in the presence of a body, both the behavior and the body, and later perhaps just the body, evoke our energy–economical verbal shortcuts.

The success of that conditioning leads to "I" and "we" and other personal pronouns no longer referring to inner agents but merely *to bodies mediating evoked responses,* a usage to which we continue to adhere. The reinforcing value of this accurate scientific phrasing leads to the conditioning of more accurate general verbal behavior, and to claims to rights regarding more accurate verbal behavior. Your and my using pronouns this way, with reference to bodies mediating evoked responses, not only helps reduce distracting occurrences of passive voice phrasings but also gives us one type of agent–less active voice as part of conditioning a new, more scientifically friendly grammar. Let's apply this to clarify the basic point about ethics. While stimuli evoke one's (i.e., a body's) labelling of those who respect (i.e., of bodies that mediate respecting) our rights claims as *ethical,* only the behavior of respecting the rights claims actually earns the label *ethical behavior.* Reducing the inner agents to mere verbal shortcuts enables appropriate use of the economical pronouns.

Ethical communities. Since some further discussions will involve the term *ethical community,* let's define it. We define an ethical community as a group of people who share respect for one or more rights that each one holds in common with the others. While the label only occurs now, we have encountered ethical communities already. Recall that broad group of all law–abiding citizens in the u.s. who share respect for the constitutional Second Amendment right of individuals to keep and bear arms. They constitute an ethical community. We can easily recognize other large and small ethical communities, sometimes mutually supportive or overlapping, at other times neutral with respect to each other, and occasionally at odds with each other or in other ways in competition. At one end of the range of such groups, you can find, for example, a group that respects a single apartment–complex owner's right to recycle old, cleaned toilets by using them as outdoor flower pots. Meanwhile another group in the same locale respects the rights of others (e.g., tourists) to see sights lacking such possibly offensive flower pots.

You can find many other groups elsewhere on the general ethical–community continuum. You can find a group respecting the rights of other animals to the preservation of their natural habitats. You can find a group respecting the rights of children to an effective education (i.e., an education based on practices grounded on scientific principles and data, such as the best practices from Project Follow Through; see Watkins, 1997; also see Ledoux, 2002.) You can find a group respecting the rights of medical patients and behaviorally disturbed clients to effective treatments. You can find a group respecting the rights of people to earn and enjoy the fruits of a living wage. You can even find a small group that respects the rights of group members to take whatever they want from others. The larger group, however, from whom they take whatever they want, describes the members of this smaller group as criminals. Among many more groups respecting various rights, you will find a large group respecting the rights of humanity to a planetary home free of overpopulation and pollution and so on. All of these and so many more constitute ethical communities.

In another impact on ethics from respondent conditioning, the occurrence of coercion, perhaps in the form of punitive enforcement practices, respondently conditions negative emotional reactions, particularly of group members, to the stimuli that accompany responses *of members,* or outsiders, that disrespect the community's ethics. As a result even slight deviations from the conditioned accepted practices of the ethical community automatically elicit these aversive emotional reactions from which one escapes only by returning to and maintaining the group's ethical practices. Cults, which overlap little with other groups, provide an extreme example that few consider ethical. As a less extreme example, after the extensive and sometimes opposed operant and respondent conditioning during life and medical school, a doctor may experience sympathy for a terminally ill patient who requests help in arranging an earlier and more dignified end, rather than waiting for the otherwise guaranteed extremely anguished end. Even before considering such alternatives, the question itself, which contradicts ethical and legal aspects of medical school conditioning, elicits strong negative emotions. Given the doctor's conditioning history, these circumstances evoke medically acceptable steps, such as drugging the patient into a stupor that persists while other processes then lead to a less painful demise. The patient dies with less dignity but the doctor escapes not only the aversive emotional reactions but also the accusations of unethical behavior that could lead to jail time. That others would argue strenuously against such jail time would be of limited consolation to an incarcerated physician. If you find the themes of this example interesting, Lawrence Fraley addresses them extensively in his 2012 book, *Dignified Dying—A Behaviorological Thanatology.*

Also, recognize the automatic conditioning of positive emotional reactions, particularly of group members, to the stimuli that accompany responses that respect the community's ethics. Again, cults provide an extreme example. For groups whose membership overlaps other groups, more common examples would include the emotions experienced as feelings of success and belonging that stimuli elicit when these stimuli indicate other's respect for your values.

The same conditioning makes similar stimuli elicit the emotions experienced as feelings of in–group camaraderie, solidarity, and mutual respect and support.

We have now seen some aspects of how the natural science of behavior addresses ethics (and values and rights, and soon, morals). These include connections of ethics to respondent processes via the inevitable pairing of body and behavior. They also include the conditioning of positive and negative emotional reactions to stimuli associated with ethical and unethical behavior respectively. These also include some connections between operant conditioning and ethics, via rights, values, added reinforcers, and the subtracted reinforcers that occur as negative emotions reduce after stimuli evoke escape behavior. These interconnections show that, as with all other behavior, ethical behavior is a function of the variables operating in past and present, operant and respondent contingencies. This is part of why we can call behaviorology also *a natural science of philosophy,* the rubric (i.e., category name) under which these and other ancient–question topics usually appear, and an area comprised of verbal behavior with which behaviorology deals. No mystical accounts achieve status as relevant explanations of values, rights, or ethics.

The same applies to morals. However, before we move on to that topic, let's consider an additional and common aspect of ethics. Most of our discussion so far pertains to ethics among people with the relatively equal status of peers. But what about ethics when some of those involved hold power of some sort over the others? Remember, ethics concerns respecting the rights of others, respecting the others' claims to unhindered access to their values, unhindered access to the things they value, their reinforcers. But those holding power could easily be in a position to disrespect the rights of those under them. They can exert disrespect simply by arranging or allowing interference with the rights of others, without the circumstances affecting their own rights. Or they can exert disrespect by arranging or allowing interference with the rights of others in ways that enhance their own access to their own reinforcers, which comes under the label, "conflict of interest."

What can prevent the occurrence of that kind of power play? Prevention stems from the ethics inhering in appropriate but competing contingencies (e.g., requirements to respect rights to privacy). These enable society's access to the variables that could prevent unethical power plays. We may not yet have enough appropriate but competing contingencies in place. News media regularly report on bosses taking advantage of subordinates. But in the resulting general social contingencies reside the variables that induce the overriding, nearly abstract, but insufficiently powerful ethics against such power plays. These variables include virtually everyone's experience of someone who holds power over them having behaved in ways that violate their rights, usually but not always in small ways. This can range from a bigger, older teenage sibling having fun teasing you by hiding your glasses, to a manager applying subtle pressure to get a subordinate to pick up the tab for lunch, to a boss setting sexual favors as the price for promotion.

The contingencies of nearly everyone experiencing those sorts of power–play circumstances induce people to reject the behavior comprising such power plays as generally (and bordering on abstractly) "unethical." Most governments enact laws against the most severe power–play forms. Some governments even enact laws against the less severe forms. While these laws add an additional layer of consequences to violations, society's usual ethical training avoids most violations. This conditioning leaves stimuli indicative of a beginning violation eliciting negative emotional reactions. Escape from these reactions hinges on the occurrence, instead, of behavior consistent with society's general ethics. Thus the contingencies induce some resistance to taking advantage of power relations to enhance one's own reinforcers at the expense of others by violating their rights. Most professions have ethical standards as well as requirements for ethics courses, or continuing education about ethics, to assure the maintenance of ethical behavior and the conditioning effects that produce it.

However, moving into, and beyond, the area of "society's general ethics" actually moves us along into the topic of morals. When the legal kinds of ethical countercontrols prove inadequate to ensure compliance, society begins to call upon morals and morality to take up the slack. Unfortunately, shifting a behavior from "unethical" to "immoral" not only enables quite an increase in enforcement power, but also enables some dangerous opportunities for unethical coercion and abuse.

Morals

With complex topics, repetition can bring benefits. So, the term *values* refers to reinforcers. The term *rights* refers to access to reinforcers that are values. And the term *ethics* refers to the behaviors of respecting rights claims for unfettered access to valued reinforcers. Next in this reinforcers–values–rights–ethics sequence is the concept of *morals,* a term that refers to ethics that have become abstractions, with "abstraction" meaning something that cannot stand alone, as we will gradually see. As abstractions, morals may lose their connection with the contingency realities that ground values, rights, and ethics.

Ethical behaviors not only respect others' rights claims but also, as contingency processes generalize their scope, some aspects of them take on the status of *characteristics* of stimuli, especially characteristics that cannot stand alone. Our verbal conditioning then evokes our speaking of this new status as *abstraction.* This phenomenon exceeds the conditioned reach of our "ethics" term, and so evokes a different term. The conditioned term for ethics at an abstract level is *morals,* akin to the "redness" of our next, simpler, example.

Increasingly complex contingencies regularly make functional stimuli out of stimulus characteristics that cannot exist alone. The conditioning that produces abstraction, however, operates long before ethical conditioning reaches abstract levels. For example, early conditioning leaves various behaviors of children under the control of colors. Colors are characteristics that *together with other characteristics* can comprise a stimulus, but a color itself cannot stand alone.

Just ask someone to hand you a red, and only a red, but not a red this or that. A red "this or that" shares the characteristic of "redness" with other characteristics that together make a stimulus that they can hand to you (e.g., a red hat or a red bowl). But the redness cannot exist alone. Even when speaking about this, we respond differentially. When conditions evoke our verbal behavior about this characteristic appearing *along with* other characteristics, we say "red." But when conditions evoke our verbal behavior about this characteristic hypothetically standing *alone*, we say "redness." When a stimulus characteristic cannot exist alone, when it can have no real existence apart from other stimulus characteristics, we then apply the term *abstraction*. In this example, *redness* is an abstraction. You can extend this pattern to a vast list of stimulus characteristics, any of which, when they cannot stand alone, we call abstractions.

Here is an example of ethical conditioning that reaches the point at which ethics become abstract and so evoke speaking of morals. When you were young, your parents might have started with the admonition not to tease your brother; treating your brother poorly constituted unethical behavior. Additional ethical training produced the ethical behavior of treating all family members well, which later extended to people, and pets, in the neighborhood, school, city, and so on. At this point treating any of these poorly constitutes unethical behavior. If this pattern of conditioned ethical extensions grows, it could reach the point where it becomes the more abstract admonitions that treating well every living organism everywhere is *moral*, and that harming any living organism anywhere is *immoral*. Such training (i.e., conditioning) extension converts concrete ethics into abstract morals. The problem is that morals, being abstract—trying to stand alone, like redness—are thus overextended and thereby disconnected from their origin contingencies, leaving little or no room for exceptions.

The abstraction that turns such ethics into morals often instantly raises virtually unresolvable problems. For example the ethical conditioning of some people leaves them thinking that their use of antibiotics, to save the life to which they have a right, is ethical, because it respects their rights claim to the valued reinforcer of a cure for an infection threatening their life. But extending their ethical conditioning, to an abstract, moral level, leaves them thinking that their use of antibiotics is immoral, because it requires the destruction of the micro organisms responsible for the infection. While such conundrums leave us wary of morals, even better reasons exist for wariness.

Some problems of the jump from ethics to morals partially derive from the conditioning of stimulus connections between morals and other abstract terms. This conditioning makes the word "moral" evoke responses similar to those that the word "goodness" evokes. This stands in contrast to the responses that the word "badness" evokes, which in turn are similar to the responses that the word "immoral" evokes. All these are abstractions. They cannot exist apart from more specific stimulus characteristics. Further extensions carry on to related word dichotomies, such as acceptable and unacceptable, allowable and disallowable, tolerable and intolerable, and not punishable and punishable.

Our conditioning further leads us to respond to the stimuli controlling those morals–related dichotomies as intrinsic qualities. This distinguishes them from ethics, because we respond to behavior as ethical or unethical on the basis of extrinsic criteria regarding specific rights claims. We can measure ethical behavior as actually supporting the claims, and we can measure unethical behavior as actually opposing the claims. In either case, the determination of ethical or unethical depends on specific, external criteria. However, conditioning induces us to respond differentially to moral and immoral behavior on the basis of (i.e., under the control of evocative stimuli regarding) whether the behavior comports with some general, intrinsic goodness or badness characteristics respectively, characteristics that conditioning has made functional but that cannot stand alone. This abstract status of morals, as verbal stimuli, somewhat divorces them from the contingencies that generate them. This can lead to problems just as rules that no longer reflect the contingencies that they describe—because the contingencies changed—can lead to problems.

Another, related result stands out when we compare ethics and morals. *A change in circumstances,* which we can measure, can lead to a change in our assessment of a particular behavior either from ethical to unethical or from unethical to ethical. However, once conditioning processes compel classifying a behavior as moral, we continue to respond to the behavior as inherently good *regardless of changes in circumstances.* Similarly, once contingencies establish classifying a behavior as immoral, we continue to respond to the behavior as inherently bad regardless of changes in circumstances. Morals can extend to the conditioning of large numbers of people, all of whom can then become involved in punishing "immoral" behavior. However, due to the abstract level, changes that happen in the concrete, ethical contingencies behind some morals often fail to induce respective changes in the related morals. As a result many people end up punishing, as immoral, behaviors that actually are ethical. For example many people currently punish, as immoral, behaviors related to ethically, humanely, working to decrease the human population to more sustainable levels. But circumstances have changed (see the last chapter). The behavior of working to decrease the human population humanely, to more sustainable levels, needs currently to be the moral behavior. Working against this task, which speeds humanity toward species extinction, is currently immoral behavior. As you can see, our scientific principles and practices show that morals are not, cannot be, and never were absolute or written in stone.

In addition a moral behavior always occurs along with the body that mediates it, so the inherent goodness gets extended to the body through the usual conditioning process of pairing. By the further extension of our culturally conditioned predilection for inner agent accounts, the now inherent goodness of the body gets even further extended to the "person" whom we then consider as inherently good. This evokes even better treatment for his or her moral behavior. The same applies to immoral behavior, extending inherent badness first to the body and then to the "person" whom we then consider as inherently bad. This evokes even worse treatment for his or her immoral behavior. This shows us the increase in enforcement power to which we alluded earlier. When reinforcers become dependent on enforcing the status of a behavior that a powerful group considers unethical, that circumstance evokes the groups' verbal behavior of overextending claims that shift a behavior description from "unethical" to "immoral" for the rest of the culture. Since the immoral behavior remains abstractly bad independently of circumstances, that shift allows, even encourages, more extreme forms of enforcement of the current morality. This opens the door to easy and all–too–often permanent enforcement methods for possibly misconstrued morals violations. We call them possibly misconstrued, because morals are products of behavioral contingency processes and so are not written in stone. They can become harmful when the contingencies change. They can become harmful in ways similar to the ways in which rules become less helpful, even harmful, once the contingencies, which the rules state or describe, have changed.

These developments should be raising all sorts of red danger flags for you. Pursuing these flags here, however, would take us to levels of detail inappropriate for this book. Still, such details help clarify the extent of behaviorology's natural–science analysis of values, rights, ethics, and morals, so I encourage you to pursue them via the references, particularly Chapter 25 of Fraley's 2008 book, *General Behaviorology* (also, see Fraley, 2012).

As a final point, presuming some inherent goodness or badness of "persons" also misconstrues "person" as a mystical inner agent (or as a representative of some sort of inner agent). While a later chapter presents details, scientifically we instead construe "person" as the potential and actual repertoire of behavior that a body is capable of mediating due to both its genetics and its conditioning history. Here, however, construing the person as an inner agent shows us a source for some of the common objections to this natural science, behaviorology. With inner agents being discredited, these objections often get miscast first at the level of ethical concerns, and then at the level of moral concerns. The moral–level objection to behaviorology tends to evoke a culture–wide condemnation. However, to the extent that humanity's survival requires behaviorology, a culture–wide moral condemnation backfires.

By definition natural science bars mystical accounts from its explanations, and in behaviorology this means barring inner agents from its explanations of behavior. So anyone whose conditioning has induced accepting inner–agent accounts objects strongly to behaviorology not on some technical or intellectual or scientific grounds but on moral and related emotional grounds. The claim is, "How dare those behaviorologists set up their natural science against the accepted moral reality of not only our mystical, theological maxi–god that moves mountains but also against our mystical, secular mini–gods that move arms and legs!" These mini gods, of course, refer to inner agents of every sort. Such people see the inner agent as good; after all, it started out as the theological soul before its label changed to the more secular psyche or mind or self or person or personality, and so on. So they see the behaviorologists' scientific exorcism of inner agents as automatically and inherently bad, even evil. And the traditional morality of good and evil further conditions people to remain good by fighting evil. As was the case with some other scientific perspectives (e.g., Darwin) the available data suggest that society has still insufficiently conditioned resistance to carrying out that admonition about fighting "evil." This admonition gets carried to unethical, even immoral, extremes when the object of the admonition is not something evil, but is necessary science! Witness, for example, the attacks on natural science in general and on evolutionary biology and behaviorology in particular. We need some survival enhancing change in "morality."

References (with some annotations)

Fraley, L. E. (2008). *General Behaviorology: The Natural Science of Human Behavior.* Canton, NY: ABCs. This is a 1,600–page, graduate level text.

Fraley, L. E. (2012). *Dignified Dying—A Behaviorological Thanatology.* Canton, NY: ABCs.

Krapfl, J. E. & Vargas, E. A. (Eds.). (1977). *Behaviorism and Ethics.* Kalamazoo, MI: Behaviordelia.

Ledoux, S. F. (2002). A parable of past scribes and present possibilities. *Behaviorology Today, 5* (1), 60–64. This is a parable on the 20–year, billion–dollar American education research effort called *Project Follow Through,* the outcomes of which the American education establishment tends to ignore, to the detriment of students, teachers, schools, and communities across the country and even around the world.

Skinner, B. F. (1938). *The Behavior of Organisms.* New York: Appleton–Century–Crofts. Seventh printing, 1966, with special preface: Englewood Cliffs, NJ: Prentice–Hall. The B. F. Skinner Foundation (www.bfskinner.org) in Cambridge, MA, republished this book in 1991.

Skinner, B. F. (1948). *Walden Two.* New York: Macmillan. In 1976 Macmillan issued a new paperback edition with Skinner's introductory essay, *"Walden Two* Revisited." See the caution in the Bibliography about a fix for some confusing phrasing in this book.

Skinner, B. F. (1953). *Science and Human Behavior.* New York: Macmillan. The Free Press, New York, published a paperback edition in 1965.

Skinner, B. F. (1971). *Beyond Freedom and Dignity.* New York: Knopf.

Vargas, E. A. (1975). Rights: A behavioristic analysis. *Behaviorism, 3* (2), 120–128.

Watkins, C. L. (1997). *Project Follow Through: A Case Study of Contingencies Influencing Instructional Practices of the Educational Establishment.* Cambridge, MA: Cambridge Center for Behavioral Studies.

Wyatt, W. J. (1997). *The Millennium Man.* Hurricane, WV: Third Millennium Press. See the caution in the Bibliography about a fix for some confusing phrasing in this book.

Chapter 23
Some Scientific Answers to Ancient Human Questions—II:

What comprises knowledge and knowing?

Natural science begins accounting for consciousness. ...

$\mathcal{Q}$uestions about consciousness evoke substantial curiosity. Given the long, pre–scientific history of the topic, many questions occur, some from a mystical perspective. Other questions, however, provide a starting point for initial scientific answers, the kind we offer here. What is it? How does it work? What part does it play in life and survival? We begin with a preview. It sets the stage for the rest of the chapter by pulling together some of the points about consciousness that we mentioned in previous chapters.

Stage–Setting Preview of Consciousness

In his 1963 article, "Behaviorism at fifty," B. F. Skinner pointed out that the question of consciousness held a central challenge to the natural–science analysis of behavior (also, see Ledoux, 2012a, 2012b). Using the vision modality for convenience, Skinner had described consciousness as *"seeing* that we are seeing*"* (which we call "conscious *seeing*"). But he excised any mysterious implied inner agent who "does" the seeing by pointing out two general kinds of contingencies. Our physical environment supplies the kinds of contingencies that condition seeing in the first place (called "unconscious seeing") while our verbal community supplies the kinds that condition both our conscious seeing and our reporting of what is seen. The thing seen evokes our initial unconscious seeing responses. These in turn evoke the seeing/reporting conscious responses. Actually, the thing seen need not be present, because other real variables, often as part of processes that we can call imaging or picturing, can evoke the unconscious seeing response, which can then evoke conscious seeing/ reporting responses. Equally pertinent, when current independent variables are insufficient to compel the conscious part to occur, it does not happen.

The verbal community conditions such seeing and reporting, because benefits accrue to it when those events occur. In common terms (i.e., using the linguistic economy that "agential" verbal shortcuts sometimes provide) more effective social organization and discourse arise when the verbal responses (reports) of what we did, are doing, and are about to do, provide stimuli that share in evoking the responses of verbal community members. As members of that same verbal community, we also benefit when our own seeing and reporting share in evoking our own subsequent responding, for such reporting also evokes our own hearing responses that then naturally supplement the controls on subsequent responses. Often those reporting and hearing responses occur covertly as one type of the conscious neural behavior called thinking. Consider, for example, the thinking that happens during assembly of a bicycle after a box arrives containing all its parts. The thinking that all the laid–out parts evoke, during assembly, aids the process immensely.

Such thinking represents a common and vital addition to the controls on subsequent behavior (since single stimuli seldom control responses). As with all neural behavior, this thinking behavior can be difficult to separate from the neural physiology that mediates it. Still, as with all behavior, independent variables must evoke the occurrence of all neural behavior, including thinking. While all these events may be complex and occur so rapidly that they strain our measurement technology, they are nevertheless occurring entirely naturally. No inner agent of any sort is "doing" the seeing, observing, thinking, measuring, evaluating, reporting, or mediating.

While we sometimes benefit from its occasional economy (e.g., the "what we did, are doing, and are about to do" in the previous example) common language usually dampens or curtails scientific sensitivity to the natural status of human behavior. Having developed under primitive conditions that seemed to support the superstitions of personal agency, the common language *per se* unsurprisingly contains explicit and implicit references to inner agents (e.g., personal pronouns). Thus, even though it seems comfortably familiar to most audiences, avoiding common language in scientific discourse is often best. However, the technical language of natural behavior science, which works to exclude agential implications, can still sound overly complicated to new audiences even as these audiences experience an improving scientific sensitivity to the natural status of behavioral phenomena, including consciousness.

Some examples relating to the central concern about consciousness may help. While these use the seeing sense modality, other examples could use other sense modalities (e.g., hearing, taste, smell, touch). As an example of unconscious seeing, a hiker, engaged in a focused conversation with a companion will, under direct stimulus control, step over an unconsciously seen football–size rock on the trail. Later, however, the hiker cannot describe the rock as it was not consciously seen. If one is to *re–see* it later, due to a contingency involving a request for details, one must have first seen it consciously.

Conscious seeing examples are necessarily more complicated as they usually begin with unconscious seeing. For instance, under some current, relatively simple contingencies involving functional chains of external and internal (neural) stimuli and responses, one sees a favorite kind of car (i.e., physical stimulation in the form of light energy reflected onto the retina from a favorite car, perhaps on a dealer's lot, evokes an initially unconscious neural visual car response). It is a "favorite" kind of car, because it occurred (i.e., paired) with certain past variables. Later, unconsciously and consciously seeing the favorite car occurs again under other contingencies, often with the car absent, as when, unable to get to work, seeing our old, broken down, rusty wreck in the front yard evokes seeing the favorite car replacing the wreck. Still other variables can evoke such conscious seeing. When, at a grocery store, we see an acquaintance who sells cars, that person not only evokes consciously seeing both our wreck and our favorite kind of car (neither of which is present) but also evokes the responses of describing the favorite car, asking where to buy one, how much it will cost, and so on. Still other contingencies may evoke consciously seeing the favored car type, as when the clerk at an airport car–rental counter asks us, "What type of car would you like to rent?" The question may evoke only the verbal name of the favorite car, but often the question evokes consciously seeing the favored car which then evokes observing and reporting its type to the clerk. Indeed, saying the type of car may not occur until after consciously seeing the car, a kind of visual thinking reported as a public of one.

    Conditioning and response chains. Those responses—unconscious seeing then conscious seeing, and thinking, and sometimes reporting—are typical examples of the natural phenomena of responses chaining into response sequences (i.e., responses, as real events, evoking other responses; see Hayes & Brownstein, 1986). These chains usually include neural responses, all in the present, all new, and not requiring the thing seen to be the current source of evocative stimulation. The same holds for other sense modalities. A physically present object transferring energy to neural receptors of any sense modality can be an evocative stimulus for either muscular responses or neural responses or both. If the necessary conditioning (i.e., neural restructuring) has occurred, then once some stimulation evokes a response, that response—as a real event— can evoke a further response, which can evoke yet another response, and so on, chaining according to the current contingencies.

    Those cascading chains of sequential relations, though at times obscure, are not mystical. They merely involve two kinds of behaviors, covert and overt, each serving evocative stimulus functions for subsequent responses of either type. And, especially when contemplating interventions, behaviorologists quickly trace back the links in any causal chain to search in the *accessible* environment for the functional public antecedents of the covert events. By tracing the functional relations back to events in an accessible part of the environment, they locate potentially changeable independent variables. This affords control over the subsequent internal and otherwise less accessible neural parts of the sequence as well as control over the external parts.

    That interface between physiology and behaviorology enhances our understanding of consciousness, and research in both of these natural sciences continually expands and further clarifies their accounts. Physiology addresses *how* consciousness happens while behaviorology addresses *why* consciousness happens. Recall our earlier example of physiology addressing *how* striated (i.e., skeletal) muscle contractions occur (e.g., as functions of neural processes that constitute the mediation of overt behavior) while behaviorology addresses *why* a body mediates behavior, that is, *why* those innervated muscle contractions occur (e.g., as functions either of eliciting stimuli or of evocative and consequential stimuli). Behaviorology accounts for the functional relations between independent variables, such as a table blocking the path through a dining room, and the dependent variables of body–mediated *behavior*, such as the muscle contractions that this obstacle evokes thereby taking the body around the table. Such entirely natural events, at both levels of analysis, extend throughout the range of overt and covert behaviors during every human body's whole lifespan, so we will see this kind of pattern when accounting, even interpretively, for the neural behaviors of consciousness.

Of course, behaviorology accounts for consciousness without conjuring or invoking any kind of mystical inner agent. We set aside the ancient, cultural traditions that have consciousness coming from, or being a characteristic of, a mind, or any other mystical inner agent. Consciousness is *not* a vehicle through which an agent's supposed psychic powers initiatively consider the effects of real stimuli and then magically and spontaneously drive the body to produce real responses of the agent's knowing and self–observant choosing. Instead, we introduce the term *consciousness* as our term for the natural chaining of cascades of neural impulses that comprise the neural behavior–behavior relations that play functional supplemental roles in the contingencies controlling more accessible overt behaviors. A common chain of consciousness behaviors includes some speedy but identifiable parts that we call raw sensation, awareness, recognition, comprehension, observation, and covert—then overt— reporting, with the complexity of such chains dependent on the presence of verbal repertoires from prior overt conditioning. With our now more informed perspective, we can begin to address consciousness with a little more detail.

Pre–Scientific Views of Consciousness

In the last couple of chapters, we began providing current, scientifically consistent, answers for some of humanity's ancient questions. We started with verbal behavior, the medium in which such questions and answers occur, and continued with the interrelated sequence of reinforcers, values, rights, ethics, and morals. Now we begin to answer ancient questions about consciousness. Before considering several aspects of consciousness that involve reviewing realities "inside the skin," we first review a little about the pre–scientific views of consciousness. While these arose in the past, many continue in the present.

Past Pre–Scientific Views of Consciousness
Pre–scientific contingencies usually produced traditional views that regard consciousness as an untestable inner agent, or as a characteristic of one. At various times this left people describing it with general terms such as soul or psyche or mind or self, or with more specific terms such as "the mind's eye."

Whatever the mystical agent, people often considered consciousness as part of the causal armaments with which the agent directed a body's behavior, and with thought and language either indicating consciousness or comprising some of the main means through which consciousness operated.

The contingencies extant long ago produced those mystical notions about consciousness before more recent contingencies of life produced the kinds of scientific verbal–behavior supplements available today. These now enable a scientific accounting for consciousness. The mystical notions have persisted, however, because their occurrence also included the conditioning of a prevailing fallacious presumption that they were the best account, as well as the only possible account, of human nature. Gradually the contingencies of traditional cultural lore, affecting ever larger numbers of individuals, built organized forces of theological mysticism that survived across generations for many thousands of years. This survival occurred through means such as the political powers of various religious states as each backed its own version of mysticism through a history of coercive contingencies. Whenever contingencies of cultural conditioning or persuasion failed to produce behaviors consistent with their particular mysticism version, religious states responded with coercive contingencies including war, arrest, torture, or death. Nevertheless, as the increasingly complex contingencies of life gradually gave rise to natural science, mystical perspectives increasingly became describable as superstitious, which brings us to some *continuing* pre–scientific views of consciousness.

Continuing Pre–Scientific Views of Consciousness

More recently those organized forces of theological mysticism and superstition have expanded to include forces of secular superstition and mysticism. Such forces arise through disciplines like psychology, which makes claims to science status merely because it employs some scientific methods. This occurs even though at the same time psychology adamantly disregards, and even actively opposes, the assumptions of naturalism that both fundamentally inform natural science and give scientific methods most of their value. This lack of connection with natural science is a major part of *both* why behaviorology separated from psychology when contingencies showed the doomed–to–fail status of attempts to change psychology, *and* why this natural science of behavior is not, and never really was, any kind of psychology. The contingencies on all these forces of superstition, theological and secular, induce much opposition to natural science in spite of the thorough reliance of these forces on the vast array of science–based products that support living quality lives while rendering their opposition. The superstitions that these forces support include continuing the traditional views of consciousness as somehow indicative of, or characteristic of, inner agents.

In turn natural science today inherently opposes superstition. Indeed, the histories of psychology and behaviorology (see Fraley & Ledoux, [1992] 2015) show that shortly after the traditional views began expanding from theological to secular superstitions, some general cultural contingencies supporting natural science began inducing development of the natural science of behavior. As a result of such developments, including the rise of the philosophy of science that we call radical behaviorism, Thompson (2008) provided an article summarizing some solvable difficulties in moving "self awareness" into natural science.

Consciousness in other animals. An extension of the ongoing, pre–scientific views regarding consciousness involves denying any possibility of consciousness in other animals, simply because "they are not human and only humans have consciousness," or so the reasoning goes. Yet we find consciousness in humans, because humans have the brain parts that are capable of mediating the complexities of consciousness behaviors. So the first question to ask regarding consciousness in other animals concerns whether or not they also have the requisite brain parts. If their physiology contains these parts, then scientifically we should proceed by presuming that they are capable of at least some consciousness responses. Thereafter, however, the questions and answers go well beyond our coverage here. In any case elaborating the natural science analysis of consciousness helps counter all the mystical, superstitious trends that misconstrue consciousness, and this necessarily leads us inside the skin.

Scientific Views of Consciousness—Inside the Skin

Behaviorology covers a multitude of independent and dependent, stimulus and response, variables in contingencies. While some occur outside the skin as overt events, and others occur inside the skin as covert events, *all* are natural events.

The covert events make up parts of functional behavior chains. These chains begin with stimulus–energy changes outside the skin that, by affecting sensory neurons, continue inside the skin. There they affect nervous system structure through a series of events that lead back outside the skin through the mediation of behaviors that occur via innervated muscle contractions. Outside the skin these muscle contractions affect further events, the occurrence of which induces other stimuli, other energy changes at receptor cells. These energy changes (e.g., reinforcing stimuli functioning also as evocative stimuli) then provide more links in the chains of functional behavior events.

In general behaviorologists focus on the events constituting the overt parts of a functional behavior chain while usually deferring to physiology colleagues regarding the internal events. However, with topics like consciousness, we must focus more on the functional chains of behavioral events occurring inside the skin. Our discussion pertains to dealing with these covert events, but in terms of the laws of behavior rather than the laws of physiology.

That discussion of events outside and inside the skin should help us recall a related though basic classification of behavior in terms of the manner of behavior mediation. The more familiar behavior type, which we call *neuro–muscular* behavior (i.e., innervated muscle behavior) occurs through the mediation of evoked or elicited neural impulse chains that ultimately include motor–neuron impulses inducing muscle contractions. The other type, the type on which this chapter focuses, we simply call *neural* behavior as it occurs through the mediation of evoked or elicited firings of neurons, or bundles or cascades of neurons, *without* inducing muscle contractions. Indeed the neurons firing *is* the mediation and we cannot readily separate the firing and the mediation from each other. Each of these behavior types, neuro–muscular and neural, overt and covert, can occur as respondent or operant behaviors.

With our focus turned to inside the skin, we begin our account of consciousness with an over–simplified definition of consciousness as neural behavior. The behaviors of consciousness manifest as *pure neural processes* generally lacking muscle contraction components. Our discussions will expand this definition.

Reduced Accessibility

To begin, recall our discussion about our reduced access to real but covert events. This reduced access makes research more difficult but not impossible. The skin remains a scientifically unimportant boundary. The laws of the universe operate the same way on either side of the skin. Besides, covert status only reduces—not eliminates—access to these events. Overt events, occurring outside the skin, can affect more than one human organism. For *any* affected humans, many of these events evoke observing and reporting responses. On the other hand, the covert events affect only the one human organism in which they occur. But each such human functions as a *public of one,* so many of these events evoke the observing and reporting responses of the one human in which they occur. Many have access to overt events, and the publics of one have access to many of the covert events inside them.

Those reporting responses, which stimuli evoke, involve verbal behavior. But talking about them involves referring to sense modes. Understanding our most common sense–mode receptors, and the types of stimuli that affect them, helps elucidate the sources of the overt and covert stimulation that evoke observing and reporting responses, for either the public or the public of one.

The usual five sense modes. We are already familiar with some stimulus–and–receptor pairs as our usual five sense modes. We traditionally cast sense modes in agential response terms (e.g., see, hear, taste, smell, touch). While the response status accurately portrays these modes, the agential casting—although occasionally satisfactory as ordinary verbal shortcuts—remains scientifically inappropriate. So here we cast sense modes as stimulus–and–receptor pairs from which we can move into non–agential response terms.

Here are the traditional five sense modes, with stimulus–energy type, related receptors, and response type: (a) Light (or visual) stimuli affect photoreceptors, producing vision responses (or, more agentially, visualizing, imaging, and seeing responses). (b) Sound (or auditory) stimuli affect phonoreceptors, producing aural responses (or, more agentially, hearing responses). (c) Chemical stimuli on the tongue affect chemoreceptors on the tongue, producing gustatory responses (or, more agentially, tasting responses). (d) Chemical stimuli in the nasal passage affect chemoreceptors in the nasal passage, producing olfactory responses (or, more agentially, smelling responses). And (e) surface pressure (or tactile) stimuli affect mechanoreceptors at the body surface, producing contact responses (or, more agentially, touch, or touching responses).

Five more sense modes. Other stimulus–and–receptor pairs are less familiar than the traditional five sense modes. While our list here remains incomplete, these other types of receptors, and the type of stimulus energy that affects them, include these five, the first of which is an in–body counterpart to the surface sense of touch: (a) Deep pressure stimuli affect mechanoreceptors in the body. (b) Stimuli from movements in muscles, tendons, and joints affect kinesthetic receptors, and relate to coordination. (c) Stimuli from movements of the body in space affect vestibular receptors, and relate to balance. (d) Heat and cold (i.e., less heat) stimuli affect thermoreceptors. And (e) aversive stimuli, both at the body surface and in the body, affect free nerve endings. The firing of these receptors, also at the body surface and in the body, leads to pain responses (or, more agentially, "hurt" responses as in "I'm hurt").

Note that our discussion uses general, common terms to highlight valuable physiological facts. You can find the related details and technical terms in a good physiology textbook. Such a text not only covers laws and facts from the natural science of physiology, but also rejects and excludes superstitious material from either theological cultural lore or fundamentally mystical secular disciplines that can taint content with agentialism.

Sense–mode covert behaviors. In all the sense modes, before a stimulus energy trace can affect a receptor, the energy trace must reach the necessary physiological threshold. When the energy reaches this threshold, the receptor transduces the energy to neural impulses. Sometimes the energy trace, through these impulses, ultimately produces an overt response with few if any covert chain components. This can happen through a process that we call *direct stimulus control,* which is the process that, for example, can produce some relatively safe driving while other contingencies are producing covert daydreaming responses. At other times a stimulus affecting a receptor induces nerve impulses that, upon reaching the central nervous system, produce the first behavior, a covert behavior, in a chain that includes covert behaviors then overt behaviors. We call these first covert behaviors *raw–sensation behaviors,* which involve a crude kind of awareness behavior. We can conceptualize these raw–sensation behaviors as respondent behaviors that the stimulus elicits by affecting a receptor.

In a later chapter, we discuss the implications of those raw–sensation behaviors for the nature of reality. Here, however, our interest concerns what happens next, what happens when, and after, a stimulus affects a receptor. The answer pertains to a phenomenon that, thanks to Hayes and Brownstein (1986) we call "behavior–behavior relations."

Behavior–Behavior Relations

What happens when and after a stimulus affects a receptor? Upon reaching threshold the receptor transduces the energy into nerve impulses. A transduced energy–trace transfer to neural structures of any sense mode can function as an eliciting or evocative stimulus—for simplicity, here we emphasize evocative stimuli—for either neuro–muscular responses or neural responses or both, depending on the history of such energy changes (e.g., past conditioning) and the state of the nervous system at the time. As real events, any such evoked responses, neuro–muscular or neural, can function evocatively for other neuro–muscular responses or neural responses, and so on.

That happens regardless of which kinds of behavior–mediating neural structure are involved. As a real energy change, any response can evoke other responses either when *a directly genetically produced* neural structure mediates it or when a neural structure *that various continuously operating conditioning processes have changed* mediates it. If the necessary gene–produced structure or the necessary conditioning–produced structure (i.e., neural *re*structuring) has occurred, then once some stimulation evokes a response, that response—as a real event—can evoke a further response, which can evoke yet another response, and so on, chaining according to the current set of operating functional relations in the contingencies. In this way a sequence, or chain, of behaviors occurs. We call these relations *behavior–behavior* relations, because behaviors, as real events, are functioning as stimuli for further real events. This connects directly with consciousness, but we need several more background points before the connections become fully apparent.

Physiologists continue to provide evidence and descriptions of the kinds of structural, nervous–system changes that stimuli produce, and the function of these structures and structural changes in mediating behaviors. Explanations for behavior, however, do not reside in the neural structures themselves, which do not cause behavior. Behavior explanations reside in the functional effects of stimuli in the changing of neural structures, and in the subsequent mediating function of the neural structural changes that occur between the energy traces at receptor cells and the responses that the energy traces evoke or elicit.

For example, on the behaviorological level, we acknowledge the necessity of practice that produces reinforcers if practice is to improve or refine a behavioral skill. At the physiological level, the reinforcers that follow practice change the neural structure in ways that support skill improvements. Thompson (2008) describes the outcome of a series of physiological studies as demonstrating "that dendritic spine growth and synaptogenesis occur in [the] motor cortex as a consequence of reinforced practice, and that when reinforcement ceases, the number of such newly formed synapses regresses" (p. 142). Even ignoring the technical terms, the evidence supporting our interpretative analysis remains clear. The synaptic growth not only stems from the occurrence of reinforcement but also provides mediation of a performance that we describe as more skillful.

The explanation for why the more skillful performance occurs, however, resides not with the improved structure but with both the stimuli that evoke the performance and the reinforcers that the performance produces, reinforcers that improve the structure. No inner agent drinks in the reinforcing attention from practice and, wanting more, decides to direct the body in the production of a more skillful performance. The how and why of that skillful performance inhere in the physiological and behaviorological accounts at their respective levels of analysis. Even our still–developing accounts of these relations require no mystical events, no inner agents.

Note that our discussion emphasized covert behaviors as operant, evoked behaviors. However, covert behaviors, and the behavior–behavior relations that construct covert behavior chains, can occur as either respondent or operant behaviors. Let's look at each of these in turn.

Covert Respondent Relations

Some covert responses occur respondently. When a stimulus, an overt or covert stimulus (which might be a previous covert response) elicits a covert response that a genetically determined neural structure mediates, we describe both the stimulus and the response as *unconditioned*. On the other hand, some overt or covert stimuli produce no effect on the genetically determined neural structure related to the behavior of concern, so we call these stimuli *neutral*. The occurrence of some pairing (i.e., conditioning) of a neutral stimulus with a stimulus that already elicits the covert response, however, alters neural structure such that when this previously neutral stimulus occurs, the now changed structure mediates the covert response. We again say that the stimulus elicits the response although, in this case, we describe both the stimulus and the response as *conditioned*.

For example let's say that some unspecified but overt or covert stimulus induced a covert leg–muscle cramp just when the blaring horn and screeching brakes from a rapidly decelerating truck—in front of which you had just stepped—elicited the covert chemical dump of a strong emotional reaction, a reaction far in excess of the usual reaction that a bothersome leg cramp elicits. Later you are reading quietly in a library when another unspecified stimulus induces another covert leg cramp. We would not then be surprised if this new covert leg–cramp response now served the conditioned stimulus function of also eliciting another covert chemical dump as a conditioned response, a dump also exceeding the usual leg–cramp reaction.

How does that example work? That conditioned respondent involves the pairing (i.e., the overlapping or successive occurrence) of an eliciting stimulus (i.e., the blaring horn and screech of brakes) and a neutral stimulus (i.e., the leg–muscle cramp that lacks an eliciting function with respect to the kind of strong emotional reaction that a blaring horn and screech of brakes can elicit). The energy traces from the pairing of these stimuli transfer energy to nervous–system structures, energy that alters the structure such that the previously emotionally neutral stimulation—the leg cramp—comes to function as an eliciting stimulus of the strong emotional response (because the body has changed, not the stimulus). When this happens we describe the previously neutral stimulus as a conditioned stimulus and the response that it elicits as a conditioned response.

Respondent seeing. When we consider the covert neural behaviors of consciousness, we find that some, perhaps many, start as covert respondent behaviors. While this applies to any sense mode, the vision mode seems the easiest to conceptualize. The unconditioned stimulus energies at our photoreceptors elicit the initial "unconscious" seeing responses as raw sensations. This is unconditioned respondent neural vision behavior. How contingencies improve these, from ambiguous blotches of shifting shape and color to focused and detailed objects and events, goes beyond our purview (for some details, see Fraley, 2008, Chapter 27). When other stimuli occur together with the stimuli that elicit unconditioned seeing responses, in the pairing process, then those other stimuli come to elicit the vision responses as conditioned seeing. For example take a bird watcher who regularly sees a member of an ordinary, uninteresting species fluttering among the leaves of a particular kind of tree. Seeing this bird shows little evidence of functioning as a reinforcer, while the pairing of this bird type and this tree type has been extensive. We should not then be surprised if this bird watcher reports seeing such a bird in a tree of this kind when a video camera running at the time shows no such presence. The conditioned tree stimulus elicited the bird–seeing response. Seeing that bird in its absence constitutes covert conditioned respondent neural vision behavior.

Covert Operant Relations

Many covert responses (i.e., neural behaviors) occur operantly. We have covert operant behavior when an overt or covert antecedent stimulus (which might be a previous covert response) evokes a covert response. Such responses often evoke further covert responses as consequences. At other times, such responses chain to further responses that include overt responses the mediation of which affects—as in "operates on"—the environment in ways that produce stimulus energy changes as a consequence of the response. These changes are stimuli that transfer energy back to the nervous system thereby producing (i.e., through the operant–conditioning process) neural structural changes that further establish the functioning of similar future evocative antecedent stimuli. With all the stimuli in our internal and external environments, and the continuous overt and covert interactions among them, both respondent and operant conditioning processes continuously affect each of us on a moment by moment basis from before we are born until we die. This builds the extensive, unique and still natural history—and thus the unique neural structures— that each of us accumulates as we live. These ongoing processes also build the extensive overt behavior repertoire as well as the extensive covert behavior repertoire, parts of which we call consciousness.

Operant seeing. When Skinner (1963) described consciousness as *"seeing that we are seeing,"* which we call *"conscious* seeing," he was alluding to the operant contingencies of the verbal community that condition both our conscious seeing and our reporting of what we see. The thing seen elicits our initial "unconscious" seeing responses which in turn evoke the seeing/reporting conscious responses. Skinner could have used any sense mode but, again, the vision mode seems the easiest to conceptualize at present.

Once stimuli have elicited raw–sensation vision responses, these responses can chain to vision–awareness responses (i.e., Skinner's "seeing that we are seeing" responses). All of our operant laws of behavior apply to these responses, and they produce many kinds of covert operant behavior.

However, let's clear up a potential confusion: The neural behavior does not produce the vision; instead it *is* the vision. Extend this with respect to consciousness. For ease of discussion, we say that the functioning of neural structure mediates behavior. But this invokes a standard shortcut for the more complicated reality that this functioning of neural structure actually *is* the mediation of behavior, for we cannot separate the two. Furthermore, just as the functioning of neural structure does not mediate behavior but rather *is* the mediation of behavior, so also neural behavior does not produce consciousness. Instead some contingency–induced natural functioning of neural structure *is* consciousness behavior. Again we cannot separate the two. Overt behavior shares this concern. Contingency–induced muscle contractions do not produce behavior; they *are* the behavior.

Before adding more details, here is one example of covert operant neural vision behavior. You wait at the busy airport arrivals gate to pick up a friend who has been away for a year abroad. In the crowd of arriving strangers, some people may share some facial features with your friend. As a result, through generalization, several of those strangers may evoke your seeing your friend and calling out her name. As a public of one, you can see that you *saw your friend* and called her name. But others can only hear you call her name in the presence of strangers. You can analyze *seeing her* as the stimulus that evoked your calling her name. But others analyze the similar facial features as evoking, through generalization, your calling her name. In either case such seeing and calling produces reinforcement only the last time when one of the faces turns out actually to be your friend's. Such seeing exemplifies operant seeing.

Again, you also *saw* that you were seeing your friend. You could report that you *knew* you saw your friend each time you called out her name. We could also say that you behaved the *reality* of your friend's presence and so called out her name, which we discuss in a later chapter. The consequences of your calling response evoked your knowing, *after* each calling response, that it was incorrect, except for the last calling response. We can take such *seeing that you were seeing* your friend as an example of a neural response of consciousness. You were conscious of seeing your friend. We can also say that you were aware, or self aware, of seeing your friend (although no "self" saw…).

"Self awareness"

The basic problem is that "self–awareness" implies "awareness of self." But no self exists, as a legitimate scientific concept, to be aware of. While we can define the term *self awareness* scientifically, the very use of "self" in the term carries this inherent agential implication. The implied problems get compounded when people use self awareness interchangeably with consciousness. To reduce confusion, we generally avoid the self–awareness term. However, some uses of it seem legitimate so let's explore them a little.

The term *self awareness* has different usages, some of them problematic. In one case when stimuli in a mirrored space evoke appropriate responses, such as young children or pigeons responding appropriately to their mirror images (which we discussed in a previous chapter) people say that the child is self aware. The question of the pigeon being self aware only emphasizes some problems with this term for humans. In another case when stimuli evoke a person's overt behavior that produces overt stimuli, and these stimuli then evoke, from the person, overt verbal–behavior reports about the behavior and stimuli, people say that the person is self aware. For example after stimuli evoke musical instrument playing, and the quality of the produced sounds evokes the verbal report, "I played that piece really well that time," people say that the player is self aware. In such cases "self aware" may simply, and poorly, substitute for "a public of one." Acknowledging the player's public–of–one status, we may also accept that the quality sounds first evoked the neural behavior of observing, as in hearing, the sounds, which then evoked the overt verbal report.

We also previously mentioned a constraint on the relevance of past or future events on any behavior, including self–awareness behavior. While past or future stimulus events seem to affect our behavior, they actually *cannot directly* elicit, evoke, or consequate responses, because both responses and stimuli occur only in the present, which makes all behavior *new* behavior (with responses grouping into response classes for experimental analysis). Regarding the future—be wary of teleology—neither future stimuli nor future responses have yet happened, although *presently* operating contingency rules can seem to bring future events into play. And, regarding the past, neither stimuli nor responses exist in any kind of storage (e.g., file cabinets or flash drives of the mind). Some people prefer the metaphor of neural structure being the storage facility, but this still too easily misleads us. *Current physiology*—neural structure—mediates every behavior, overt or covert. Every behavior occurs under the functional control of *current* overt or covert stimuli regardless of the complexity, multiplicity, or interactivity of these stimuli or responses. Recall our Law of Cumulative Complexity. Even memories, traditionally meaning past stimuli and responses, are not stored. They are *new* responses that *current* stimuli evoke and that *current* neural structures mediate, *neural structures that have their current structure because conditioning processes changed them both at, and since, the time of the original instance.* For this reason they often fail as accurate re–behavings of past events, a point pregnant with implication for eye–witness testimony.

Knowing. Where the terms self awareness and consciousness really begin to overlap, and where we begin really to prefer to speak of consciousness, concerns what we mean by the term *knowing*. The movement of objects (e.g., rain on a window) or the behavior of another organism (e.g., some bees hovering around some flowers) can evoke your body mediating some covert behavior such as seeing the rain splattering or the bees hovering. With a history of appropriate conditioning (e.g., conditioning that has made stimulus indications of strong weather, or the dangers of insect stings, effective evocative stimuli) this covert seeing behavior, as part of a chain, can evoke your body's mediating a covert observing response of the rain or the bees. Even just reading words like these usually evokes such covert responses. The covert observing response then evokes a covert verbal report of some related covert observation behavior, such as, "An umbrella would be handy" or "Keep away from those flowers." If audience variables are present, this covert verbal report may merge into an overt verbal report of the same or similar content. Sometimes the audience variables will be potent enough that the covert observing response directly evokes the overt verbal report. In any case, we would say both that you *know* what evoked the overt verbal report (i.e., the rain or the bees) and that the covert verbal report constituted a kind of consciousness responding that we call thinking.

That same kind of thing happens with a body's "own" behavior. When a stimulus evokes the body mediation of some overt behavior, and this behavior then evokes the body mediation of some covert behavior, such as seeing or hearing or feeling the overt behavior, and this covert behavior then evokes the body mediation of a verbal report (possibly covert then overt) of the covert behavior, we can describe this sequence in two main ways. Traditionally we would agentially say that *you* are *self* aware, that *you* know or are aware of what *you* were *doing*. But no inner *self* exists to be aware, no inner *you* exists to *do* the behavior or to know about it. Instead we can say scientifically that the body (i.e., "you," as a verbal shortcut for the mediating body) *knows* or is conscious of the behavior that stimuli make the body mediate, not in the sense of giving the body inner–agent status, but merely in the sense that a stimulus evoked a sequence of completely natural (i.e., non–mystical) behavior–behavior functional relations, a sequence (i.e., a chain) that included purely neural consciousness responses that we call *knowing*.

Even though you know that no inner agent "you" exists to do the knowing, rephrasing these points accordingly remains difficult. Even a successful attempt works mostly by detailed elaborating, turning each sentence into three, and making book editors upset. So our legitimate verbal shortcuts continue.

Brain, Chaining, and Consciousness

Covert verbal behavior constitutes a common kind of purely neural consciousness behavior. We often use the term *thinking* to describe this kind of neural behavior, although thinking need not be verbal, and happens in other sense modes as well. While overt verbal behavior, along with other overt behavior, involves the behavior–capable body parts that we call muscles, covert behavior, including covert verbal behavior, involves the behavior–capable body part that we call the brain, which also has parts.

The role of the brain. When we say that consciousness manifests as purely neural behavior, lacking muscle–contraction components, we allude first to the brain as a behavior–capable organ. However, the brain is far more complex than a muscle, and the brains of some species are far more complex than the brains of other species. So the behaviors of which brains are capable cover a vast range. Even so, these behaviors share one characteristic; they are all covert, which necessarily means they are less accessible than overt behaviors.

Physiological equipment and procedures provide a kind of access to what is happening in the brain when behaviors, both overt and covert, occur. Still, we must remain respectful of disciplinary expertise. I am a behaviorologist, not a physiologist. So I defer to my physiology colleagues for more accurate elaboration of points pertaining to the brain per se. And they defer to behaviorologists for elaboration of the functional effects of stimuli on the operation of brain parts, particularly as this operation pertains to the mediation of behavior, covert and overt, because behavior mediation comprises a major brain function.

The brain, as the major component of a nervous system, mediates—not originates or initiates—behavior that occurs as a function of other real variables. This puts the brain, as a physiological organ, at the interface of external and internal environmental energy exchanges. Furthermore, as a behavior–capable body part, the brain mediates not only overt, neural–muscular behavior, but also covert, neural behaviors—inseparable from the mediation—including those we consider as consciousness behaviors such as thinking. In this particular disciplinary interface, behaviorology accounts for specific functional relations between real, independent variables and real, dependent variables on both sides of the skin, including in the brain. Meanwhile brain physiology accounts for the structural changes that are occurring as those behaviorological–level independent and dependent variables interact. Both disciplines, of course, also concern so much more, in their separate subject–matter domains, than behavior and mediation.

Neural behavior chaining. Contingencies in the external and internal environments of peoples' daily existence feature the occurrence of energy exchanges that already affect people's behavior, or that condition people's behavior. As a result some behaviors, including neural behaviors, function as real, independent variables evoking further behaviors, including more neural behaviors, as real, dependent variables, which evoke further behaviors, including more neural behaviors, and so on. This process can build strings of behavior. These behavior strings include some of the types of consciousness behaviors that we have already mentioned, such as thinking—verbally or non–verbally and in any sense mode—and feeling, and observing, and knowing, and seeing that we are seeing, and hearing that we are hearing, and so forth.

Such behavior strings provide typical examples of the natural phenomena of responses linking together in a chain of responses (i.e., responses, as real events, evoking other responses) usually including neural responses, all in the present, all new, and none requiring that the thing seen or heard, or felt (and so on) be the current source of evocative stimulation.

Those neural chains interweave with reinforcement feedback loops featuring functional relations outside the skin between stimuli and neuro–muscular behaviors. The neuro–muscular behaviors change the environment, producing the reinforcement feedback into the nervous system such that neural structures change. These changes leave the neural structures, which mediate the covert and overt responses, more susceptible to the energy traces from the covert and overt evocative stimuli, which a behavior could provide. And some of these reinforcing effects may affect responses supportive of survival.

Behaviors and Sequences of Consciousness

Having dealt with many pieces of our consciousness puzzle separately, the time has come to consider them together. Our account of consciousness in terms of neural–behavior chains usually involves several types of consciousness behaviors. In the sequence in which they usually occur, after respondent raw sensation and basic operant awareness, the common terms for two types of consciousness behaviors that usually occur first are recognition and comprehension. Other operant types of consciousness behaviors typically occur after these two, but their order depends on the prevailing contingencies. The common terms for some of these other types are knowing, observing, explaining, and thinking, which also covers covert verbal reporting. Any of these—sensation, awareness, recognition, comprehension, knowing, observing, explaining, thinking, and so on—can each evoke covert or overt non–verbal behavior, and most can evoke further covert and overt verbal behavior. After raw sensation, the sequence of awareness, recognition, and comprehension often chains further to verbal, and still natural, knowing, observing, thinking, or explaining. These usually lead to overt behaviors including reporting, which supplements other controls and thereby induces more effective responding to the world around us.

Even in these interpretative analyses, given the speed at which neural impulses travel, chains of these consciousness responses can occur in the blink of an eye, although they often take a little longer. Thinking, of the covert *verbal* behavior kind, can take nearly as long as equivalent overt verbal behavior.

Let's focus on the common sequence of sensation, awareness, recognition, and comprehension for an example. Continuing with the convenience of our visual sense mode, consider a living adult human body with a behavior repertoire from the usual life–long and thus extensive conditioning history (i.e., an adult *person*, using "person," as a verbal shortcut...). If an object passes by, a foot or two from the person's face, the raw sensations—occurring when light–energy traces bounce off the object and strike the retina—evoke *awareness* responses that manifest as shifting spots of shape and color. These awareness responses then evoke (i.e., as real events these responses function as stimuli and evoke) the *recognition* responses of basically knowing—from past conditioning—that the properties (i.e., the shifting spots of shape and color of the awareness responses) comprise a butterfly. These recognition responses may even include the covert or overt verbal behavior of saying "butterfly." These recognition responses then evoke *comprehension* responses that could involve any or all of the wider covert aspects of this body's behavior repertoire—if previously conditioned—with respect to butterflies. Thus the recognition responses could evoke comprehension responses such as (covertly, and not necessarily verbally) "That is a Monarch butterfly, a really big and beautiful one, the first one in the garden this season, and seeing one this early makes for wonderful feelings about the potential success of pollination this season." Other comprehension responses could include further poetic and scientific implications of the presence of Monarch butterflies in that location at that time of year, and so forth. Much of this we might further call parts of thinking and explaining. Also, other covert or overt stimuli could evoke overt verbal forms of these responses at any point during their covert occurrence.

We can *conceptualize* the neural awareness–recognition–comprehension behavior chain as seeing (or hearing, or other sense mode) some object or event (i.e., awareness), knowing what it is (i.e., recognition), and understanding how it relates to other events (i.e., comprehension), all in the sense of the covert differential responding to those aspects of it. We *summarize* such covert responding with the term *consciousness*.

A related concern, however, pertains to the ongoing conditioning involved in these sequences, which continuously increases the speed at which each chain component—awareness, recognition, and comprehension—evokes the next. The time duration of these components was never extensive given the speed of nerve impulses, even masses or cascades of them. Hence the conditioning quickly succeeds in shrinking the duration of each component to the point that sometimes they all *appear* to occur simultaneously, even instantaneously. People come to seem to instantly behave awareness, recognition, and comprehension not separately, sequentially, but all at once. This raises the need for us to repeat that we all know better by now not to see ghosts—inner agents—in these events, or in our phrasing of them, for all these events are still, and inevitably, occurring quite naturally. No other reasonable option exists.

We should also take note of some other characteristics of consciousness responding. These concern consciousness appearing as a single–channel covert process often separate from a concurrent occurrence of other overt behaviors happening under direct stimulus control. While we have encountered each of these terms (i.e., single–channel process, and direct stimulus control) before, here we relate them to consciousness in the context of our favorite daydreaming–while–driving example, along with repeating that this combination is dangerous. Hopefully contingencies have conditioned drivers to resist it (along with texting and other cell phone use while driving).

To begin, recall that consciousness responses provide a supplementary focusing of the evocative effects of stimulus control. Without the consciousness responses, some stimuli fail to evoke behavior, or evoke only part of a response pattern, possibly the wrong part in the sense of a part that fails to produce reinforcement. On the other hand, the unchanging or familiar stimuli of a regular or consistent environment seldom require the supplements that consciousness responses provide. In this case the controlling stimuli may not even evoke consciousness responses, and instead control responding directly. Directly operating stimuli control vast areas of human behavior, although people tend to consider these areas less important than the equally vast human behavior areas that benefit from consciousness responding. You may have already surmised this point from the content of the examples that we used in earlier parts of this book.

Consciousness responses, however, involve easily evoked neural responses that other stimulus–response chains replace just as easily. To keep them going, they must quickly, even continuously, produce reinforcers. In the most common scenarios, each next evoked neural response in a chain serves not only the function of evoking another response but also serves the function of reinforcing the previous response. We called this *dual function of a stimulus* when we saw it among overt behaviors as we described the intervention technique of backward chaining. Here, however, the covert consciousness responses, as real events, serve these stimulus functions. Nevertheless, the relatively constant stimulus barrage that we are under from the internal and external environments can easily induce fleeting changes in stimulus control from one neural response chain to another to yet another, and so on. If a particular chain is continuing, looking for some compensating reinforcing effect is appropriate.

A contributing factor to the brief duration of many consciousness response chains is the apparent single–channel nature of these chains. Some evidence suggests that the brain part that mediates consciousness is capable of only one chain happening at a time. The contingencies, or our physiology, or both, seem unable to manage even two, let alone several, consciousness chains at the same time. On those occasions when two seem to be occurring simultaneously, closer inspection reveals that the chains are rapidly alternating and so are still occurring only one at a time.

For example consider driving on the 100–mile–long trip from Sacramento to Lake Tahoe on u.s. Highway 50. This rather scenic two–lane highway, over the Sierra Nevada mountains in California, occasionally grips the edge of steep, high cliffs. As the car approaches some curves, those curves, as stimuli, evoke steering responses for which staying safely on the roadway provides appropriate reinforcement. This consequence of staying on the road is contingent upon the necessary steering responses, which in turn are contingent upon the evocative stimuli that changes in road direction—the curves—supply. If the curves are the sharp ones along the steep cliffs near the pass that leads down to the lake, then adequate steering responses may require the supplementation that consciousness response chains provide. Your eyes and all your consciousness resources remain glued to the road with the intensity of a pilot on final approach to landing. The children arguing in the back seat, and even the beautiful, high sierra scenery, fail to evoke other consciousness chains, neither ones alternating with each other nor ones alternating with the chain focused on the curves. The single consciousness channel remains completely preoccupied for the duration of those particular cliff–hugging sharp curves. On the other hand, once you reach the valley floor, and the curves become simple, gentle direction changes. Then the scenery and the children's activity again begin to control not only some of your overt behavior chains but also your single–channel covert-behavior chains, either in slow or rapid alteration, but not simultaneously.

Actually the available data (e.g., see Fraley, 2008, Part III) are not yet entirely adequate for full confidence in the current single–channel conclusion. So far this is the conclusion to which the available data lead. However, we may discover, from accumulating additional data, some circumstances under which more than one simultaneous channel is possible for consciousness response chains. Or we may discover particular processes or procedures the potency of which allows successfully conditioning the occurrence of simultaneous, rather than alternating, consciousness response chains (i.e., dual, or multiple, "channels"). At the time of this writing, however, consciousness response chains occurring only as a single–channel capability remains the most parsimonious option. Perhaps a general limitation is one "channel" *for each behavior–capable body part, including brain parts;* this still allows a wide range of simultaneous behaviors that involve various, separate body parts and brain parts.

Back to our driving example. Regarding the gentle curves on the valley floor, another consideration is that *you need not be aware* of those curves for them to exert their evocative stimulus function on your driving responses. Perhaps the children are napping and your single–channel neural consciousness responses are engaged, say, daydreaming about a week of fishing in the beautiful lake waters, and the great trophy fish you are sure to catch. The daydreaming consists of a sequence of thematically related, often mostly visual responses, each one evoking the next while each one also reinforces the previous one. Even while the neural daydreaming response chain completely occupies your consciousness response–chain channel, preventing awareness responses with respect to the curves, the curves still function as effective evocative stimuli for your overt steering and other driving responses (and continuing to stay on the road still functions as the effective reinforcer). Again, our term for the curves successfully evoking your steering responses is *direct stimulus control.*

Another example of single–channel consciousness chains, with direct stimulus control over concurrent overt responses, involves some types of needlework. This activity can occur while the television has control of your single–channel consciousness response chains. The stimulus aspects of the needlework exert direct stimulus control over your needlework–related responses. On the other hand, if you were working crossword puzzles while listening to the news, then each of these would be controlling its own consciousness response chain. But likely these would alternate with each other, rather than occur simultaneously, because *both* chains seem to involve the same body parts, the parts mediating verbal behaviors.

Again, back to our driving example. At some point a change in energy–trace types refocuses the consciousness channel. As you stop at a red light, you wonder (another neural behavior of consciousness) how you got there, or where the last mile or three went. You neither crashed nor heard sirens wailing nor got pulled over, so your driving responses must have been adequate. However, the driving responses occurred under the direct stimulus control that the curves exerted. Those overt driving responses occurred right along with—that is, at the same time as—your covert daydreaming responses. Neuro–muscular (i.e., overt) behaviors, and neural (i.e., covert) behaviors seem to occupy different neural channels, involving different behaving body parts. While neural behaviors like consciousness seem limited to a single channel, neuro–muscular behaviors seem to have many channels available (i.e., many different overt responses, *involving **many** behaving body parts,* often occur simultaneously). In any case these daydreaming responses prevented your being aware of the steering responses, because the needed channel, so to speak, was already occupied with these daydreaming responses. You may have experienced something like this many times. It is not magic (although it seems magical). It is just the behaviorological laws of nature at work, laws which operate without our needing to be aware of them or the variables involved.

Since we have talked about the necessity and value of reinforced practice, you can experience the benefit of practice by analyzing how parts of our previous examples line up with various parts of the general consciousness behavior chain of raw sensation, awareness, recognition, and comprehension, possibly mixed in with knowing, observing, thinking, explaining, covert reporting, and overt behaviors. For example, analyze the music–practice chain we covered earlier in this chapter under "self awareness." Or analyze the rain chain, or the bee chain, both of which we covered earlier under "knowing." The more thoroughly you comprehend the scientific account of consciousness, the more comfortable you will be with the interpretive accuracy of your analyses.

Conclusion

Conditioning involves energy transfers between the environment (internal and external) and the body in ways that, as our physiology colleagues can show, trigger cascades of neural impulses that variously induce several processes. That is, stimuli *always* produce any neural impulses. Some of these impulses induce—and constitute—chains of consciousness responses, while other impulses induce the greater energy expenditures involved in bodily movements (i.e., neuro–muscular responses). Still other impulses induce the altered neural structures that constitute a different body that mediates (not initiates) both covert and overt behavior differently on future occasions. The greater energy expenditures come from nutritionally derived bodily reserves. When happening, all these intertwined environmental/neural/behavioral processes move along at such a rapid pace that they may seem undetermined, particularly when we try to encompass behavior in general across a time frame beyond a few moments, because events can quickly outpace our limited measurement technology. This, however, is a not a problem with nature. This is a problem stemming from the limitations of our measurement technology, limitations that contingencies constantly drive us to reduce. Meanwhile we manage our resulting residual ignorance (see Fraley, 1994) with various methods (e.g., probability theory) as necessary. Nevertheless, all these processes remain *lawful.* They all remain *entirely natural.*

As all those events unfold, including chained consciousness behaviors, no capricious inner agent makes them happen. Any behaviors that happen had to happen (and if a behavior failed to happen, that was because it could not have happened) under the present contingencies. As we pointed out in earlier chapters, including self agents in accounts for behavior is not only redundant and misleading but also dangerous and irresponsible because the resulting reduced effectiveness in problem solving can cause harm.

On the other hand, as natural scientists we respect the natural functional history even of extremely complex and multiply–controlled response chains such as those involved in consciousness responding. These cascading chains of sequential relations, though seemingly obscure, are not mystical. They simply involve two kinds of behaviors, covert and overt, each serving evocative, and reinforcing, stimulus functions for ongoing responses in chains of either type. And, when contemplating interventions, behaviorologists quickly track back along the links in any causal chain to search in the *accessible* environment for the functional public antecedents and consequences of the events. By tracing the functional relations back to events in a more accessible part of the environment, they locate potentially changeable independent variables. This affords control over the subsequent internal and otherwise less accessible parts of the sequence as well as control over the external parts.

All this complexity in behavioral and physiological events can seem amazing, wondrous, awesome, even overwhelming. But it is all natural, and exemplifies our *Law of Cumulative Complexity:* The natural physical/chemical interactions of matter and energy sometimes result in more complex structures and functions that endure and naturally interact further, resulting in an accumulating complexity. Our account of consciousness as covert, neural–behavior chains—variously with raw sensation, awareness, recognition, comprehension, knowing, observing, thinking, explaining, and reporting components—is *cumulatively complex* and *entirely natural.*

References (with some annotations)

Fraley, L. E. (1994). Uncertainty about determinism: A critical review of challenges to the determinism of modern science. *Behavior and Philosophy, 22* (2), 71–83.

Fraley, L. E. (2008). *General Behaviorology: The Natural Science of Human Behavior.* Canton, NY: ABCs. For more on consciousness, see Chapter 27 of this 30–chapter, three–course, 1,600–page, graduate textbook.

Fraley, L. E. & Ledoux, S. F. ([1992] 2015). Origins, status, and mission of behaviorology. In S. F. Ledoux. *Origins and Components of Behaviorology— Third Edition* (pp. 33–169). Ottawa, Canada: BehaveTech Publishing.

Hayes, S. C. & Brownstein, A. J. (1986). Mentalism, behavior–behavior relations, and a behavior analytic view of the purposes of science. *The Behavior Analyst, 9,* 175–190.

Ledoux, S. F. (2012a). Behaviorism at 100. *American Scientist, 100* (1), 60–65. This article extends B. F. Skinner's 1963 article "Behaviorism at fifty." The Editor introduced this article with excerpts, on pages 54–59, which he listed as an "*American Scientist* Centennial Classic 1957" that came from Skinner's 1957 *American Scientist* article "The experimental analysis of behavior." (Skinner's complete 1957 paper was also available online.)

Ledoux, S. F. (2012b). Behaviorism at 100 unabridged. *Behaviorology Today, 15* (1), 3–22. With *Behaviorology Today* becoming fully peer–reviewed with this issue, this fully peer–reviewed version of the paper that originally appeared in *American Scientist* included the material set aside at the last moment to make more room for the Skinner article excerpts that accompanied the original article. *American Scientist* posted this paper online along with the original version at www.americanscientist.org (also, you can find the unabridged version at www.behaviorology.org).

Skinner, B. F. (1963). Behaviorism at fifty. *Science, 140,* 951–958.

Thompson, T. (2008). Self awareness: Behavior analysis and neuroscience. *The Behavior Analyst, 31* (2), 137–144.℧

Chapter 24
Some Scientific Answers to
Ancient Human Questions—III:

What is the nature of living?

Consider life, including personhood. …

*T*he last chapter extended our coverage of current scientifically consistent answers, for some of humanity's ancient questions, into the area of consciousness. We described *consciousness* basically as our term for the natural chaining of cascades of neural impulses that comprise the neural behavior–behavior relations (i.e., behaviors, as real events, evoking further behaviors) that play functional, supplemental roles in the contingencies on more accessible overt behaviors. A common covert chain of consciousness behaviors includes raw sensation, awareness, recognition, comprehension, observation, and reporting (and then overt reporting). The complexity of such chains depends on the presence of verbal repertoires from prior conditioning.

In this chapter, ancient questions that invite and demand scientifically consistent answers cover the topics of life and personhood. Our coverage introduces scientific accounts for each of these topics, even though these topics invariably and extensively overlap each other and the next chapter's topic, death. Let's look at these topics, beginning with the living body and what we mean by *living*, what we mean by *life*.

Life

Life inevitably intertwines with personhood, but agentialism can taint each of these. As usual we begin by setting aside the ancient, cultural traditions that imbue these terms with agentialism. Early and long–enduring, pre–scientific contingencies compelled ultimately unhelpful descriptions of each of these topics: *Life* referred to animation from the breath of a god, an "outer" agent. *Personhood* referred to a mystical behavior–directing, and so responsible, inner agent, religious or secular, that inhabited the animated (i.e., alive) body. And *death,* as we will see, referred to the departure of the entity from the body, leaving it no longer alive. Today, instead, *life* scientifically involves a range of levels of chemical complexity that feature processes involving energy exchanges among simple and complex chemical units that change these units and their surroundings, processes such as evolution and conditioning and even consciousness and culture. Similarly, *personhood* scientifically involves the whole repertoire of behavior that contingencies induce and that a physiological body, by its nervous system—all products of natural life processes—mediates when environmental energy changes produce behavior. And *death* scientifically describes the results when, across several steps, natural processes permanently interrupt the function of behavior mediation and reduce complex chemical units to simpler levels of chemical structures and functions incapable of mediating behavior.

Even in current non–scientific usage, the term *life* describes the status of living that an untestable inner agent animates. Supposedly the agent animates a lump of matter making it alive. Scientifically we must reject this notion not only as agential but also because it confounds our interrelated topics of the alive body, its mediated behavior, life, and personhood. To enable a scientific discussion that clarifies the interrelationships, we treat the topics separately and without the agentialism. For starters, while we define personhood behaviorally, we define life, the alive body, biologically.

A Scientific Meaning for the Terms "Life" and "Living"

In general scientific terms, *living* refers to the activity, including body–mediated behavioral activity, that results from environmental energies affecting a lump of matter. At some point—actually, not a "point"—a lump of matter shares characteristics of enough complexity that these complex characteristics participate in contingencies that induce us to call the lump of matter alive, to call it an alive body, to call it a life form or, more pertinently, to call it a form of life. Yet what are these characteristics? What is this *life* to which we refer by saying "form of life?" This determination of life status, and our recognition of it, stem from natural, not magical, functions.

We incorporate and extend the understanding of life and its characteristics as our biology colleagues see them. So for us *life* refers to forms that reside on part of a continuum of chemical complexities that *seamlessly* extends from non–life forms to life forms. In a boundaryless area between these, various realities continue to evoke debate about whether or not some forms are alive, or nearly so, or not at all. At what point would they evoke the descriptions *living* or *life*? No such *point* exists, but only a *gray* area. This chemical continuum derives from the circumstances specified in our *Law of Cumulative Complexity*, which states, as originally reported in Ledoux, 2012 (p. 10) "the natural physical/chemical interactions of matter and energy sometimes result in more complex structures and functions that endure and naturally interact further, resulting in an accumulating complexity." Examine this basic continuum:

←——Non–Life——→?—Nearly–Life—?←———Life————————→

On that continuum, as lumps of matter accumulate complexity, non–life forms *grade gradually* into life forms. At some "point" forms of chemical combinations still show too few chemical characteristics to evoke the response "life" from even the most thoroughly conditioned biologist. After this "point" come some forms of chemical combinations that show increasingly complex chemical characteristics that evoke responses such as "not quite non–life" to "almost life" to "proto life" to "maybe life" and to "life, except..." The constant shifting on how to define "life" characterizes the stimuli in this "gray" part of the continuum, stimuli that thus evoke these kinds of ambiguous responses. Biologists who study the origins of life find a range of increasingly complex chemical structures and functions (e.g., lipid barriers, RNA, DNA, prokaryotic cells, eukaryotic cells) the accumulation of which leads to the definition for an unambiguous notion of "life." Around this "point" some lumps of matter finally evoke the response "it's alive," or "it's life," while some just slightly simpler forms next to them evoke the ambiguous responses. And even simpler forms evoke the "it's not alive" or "it's not life" responses.

Indeed, *no* "point," *no* thin line exists on one side of which we find non–life and on the other side of which we find life, as if some mystical maxi–deity drew such a line. While some forms of chemical combinations show complex chemical characteristics that unambiguously evoke the term "life," no *qualitative* difference, with forms that fail to evoke this term, has occurred. Only a continuous gradation in chemical complexity has occurred, all entirely naturally. Thus the term *life,* and the chemical realities that evoke this term, merely help control our responding to complex phenomena in ways that evoke other responses that deal successfully with the world around us, ways that we call scientific, ways that may support our survival.

Some Implications of the Term "Life"

This entirely natural conception of life, and by extension personhood, if popularized throughout the culture in a timely manner, may prove to be one of the most culturally valuable developments of science. This is because it can countercontrol that most pernicious cumulative error in human history, the error that we call theological and secular agentialism, and the anti–science superstitions that these engender. The term *agentialism* refers to any and all kinds of mystical (i.e., unnatural, untestable, unmeasurable) inner agents or entities that pre–scientific, inaccurate contingencies conditioned humans to use as causes of human behavior. This agential error threatens to prevent adequate and timely human solution responses to the stimuli of global problems. In contrast the entirely natural, scientific conception of life, personhood, and so forth, supports the sustainable survival of life as we know it. This is a conception which behaviorological analysis presently provides.

The continuum from non–life to life has, of course, "another end," the end where physiological structures, and the functions they support, begin a disintegration on the level of an individual organism. This disintegration occurs relatively rapidly compared to the organism's usual lifespan. The term for such disintegration is *death*. With respect to some practical matters for individual humans, we return to the topic of death in the next chapter. Meanwhile this "other end" of the continuum from non–life to life also applies at the species level. At this level biological *extinction* becomes the appropriate term.

Our emphasis on behaviorologically answering ancient questions includes the status of all human endeavors as behaviors under the control of natural laws. These endeavors include—perhaps particularly—all human scientific endeavors. Here that emphasis extends to considering life and personhood mostly in relation to individual human organisms. Throughout this book, we have considered the laws of environment–behavior functional relations in the same manner. Such relations include antecedent and postcedent stimuli and their effects (e.g., on non–verbal and verbal behavior including values, rights, ethics, morals, and consciousness). Having touched on life (and before considering death, which we will cover in terms of dignified dying) let's elaborate on life in terms of personhood.

Personhood

Again, we set aside the cultural traditions that misconstrue the term *personhood* as representing some mystical inner agent or the behavior–initiating activities of one. We define personhood scientifically as the full potential and actual *repertoire of behavior* that a body is capable of mediating due to the structures, neural and non–neural, resulting from its genetics and conditioning history. The neural structures mediate the behavior repertoire while the non–neural structures put limits on the behavior repertoire that contingencies can condition. For example, because humans lack the necessary non–neural structures, contingencies *cannot* condition flyng behaviors merely involving flapping arms. In any case this full potential and actual *repertoire of behavior* constitute the *person,* and thus the person only continues to exist while the body remains capable of reacting to the environmental energy changes that elicit or evoke the body's mediating of the behaviors in the repertoire.

Some Personhood Considerations

Those behaviors include at least the general run of neuro–muscular non–verbal and even verbal behaviors that our laws of behavior well explain in accessible and applicable detail. However they also include the more complex neural behavior which those laws account for interpretively under labels such as thinking, consciousness, and "sentience." Neural behaviors like these, rather than representing actions of untestable inner agents, instead refer to the most complex manifestations of personhood, even if only observable by a public of one. Indeed such neural behaviors, along with all overt behaviors, rather than implying a person, *are* the person. Otherwise the term *person* too easily begins slipping back into some non–scientific agential usage.

Sentience. However, while we covered thinking and consciousness already, especially in our last chapter, the term *sentience* deserves some separate attention. Pre–scientific contingencies have compelled people to reserve this term nearly exclusively for fellow humans, using sentience to indicate something supposedly special about people. In this case, though, what people see as special about people remains outside science. Sentience implies a supposed and unmeasurable *qualitative* difference between humans and other species, an implication rather in contradiction with the finding, ever since Darwin, of only *quantitative* differences among species, including the human species. Sentience says that something about humans resides above or outside of nature, better than everything else, an unscientific status generally characteristic of putative inner agents. While the kinds of stimuli that evoke verbal–response terms like consciousness and thinking can also evoke the term sentience, such stimuli seldom actually evoke this term. Instead, the stimuli that evoke the term sentience are generally the same stimuli that evoke such terms as mind, psyche, or self. While some might see "sentience" as another term involving neural behavior, the term sentience retains agential implications that we may better handle by simply setting it aside as a non–scientific category.

Goodness and badness. Furthermore the scientific construal of personhood separates behavior repertoires from the notion of inherent goodness or badness of either "persons" or bodies. Bodies only mediate behavior that environmental variables evoke, and that behavior *is* the person. Characteristics of goodness and badness stem from the kind of contingency relations that we described as values, rights, ethics, and morals. Many cultures often consider an as yet unconditioned body, such as a newborn infant, as essentially innocent of such characteristics, and possibly not yet even "human," not until some observable *behavior* establishes personhood. Some cultures may even specify various indicators that they accept as establishing personhood. For example, in the Navajo culture, many acknowledge a baby's first laugh as representing the first expression or indication of personhood (e.g., Simpson, 2003, p. 54).

Some Personhood Implications

Our traditional, agent–laden cultural conditioning generally compels us to respond as though a body's personhood, as a characteristic of an inner agent, retains a continuous and steady–state status. Science, however, having compelled setting inner agents aside, also compels recognition responses that we actually see personhood occurring occasionally, with ups and downs. With personhood inseparable from the neural mediation of environmentally evoked behavior, and *with behavior occurring occasionally*—more at some times and less at other times—the variable status of personhood fails to surprise us. This variable status derives from a range of sources. While environmental contingencies remain responsible for much of the occasional status of behavior occurrences, and hence for some part of the variable status of personhood, another part of this variable status happens in accordance with various factors that change the body. Good nutrition, bad nutrition, disease occurrence, disease recovery, more stress, less stress, and many other factors (e.g., more education) drive changes in a body's capacity to mediate behavior. As a result, we see personhood manifesting not only intermittently but also with sometimes substantial changes in magnitude.

With personhood thus occurring as behavioral process, we see those factors bringing about repeated waxings and wanings of the person, and we can track these on various time scales throughout the time a body evokes the response "living." As the bodily effects of conditioning begin to accumulate early in life, building a more and more complex neural and neuro–muscular behavior repertoire, we see the person generally waxing, coming into existence, not as any sort of inner agent but as generally an increasingly sophisticated and interrelated collection (i.e., repertoire) of behaviors. At the point when environmental events fail completely to evoke any behavior, particularly neural behavior, due to bodily degradation leaving the body incapable of ever again mediating such behavior, the person wanes completely, going out of existence permanently. We call this phenomenon "person death" and it may occur before or along with body death. However, person death differs from body death, as discussed in the next chapter. Sometimes a body seems temporarily to lose the physiological capacity to mediate evoked neural and muscular behavior. But should the body eventually regain that capacity and so mediate some of that behavior, the person may then manifest again. All this helps us to comprehend the nature of a person, the nature of personhood, scientifically.

Many additional culture–related implications stem from the contradiction between the mystical and natural–science views of personhood. A friend and colleague, Lawrence Fraley, considers his own 2008 treatment of the topic— to which he devoted over 80 pages, ten times the length of this chapter—as still incomplete (personal communication). Some of these implications include the abstract person, the relative worth of persons, and the cultural costs of superstitiously misconstruing personhood. These cultural costs include the financial burdens (i.e., in tuition and taxes) of developing and maintaining undergraduate and graduate degree programs in mystically grounded subject matters, both theological and secular, in so many colleges and universities. Other implications involve the costs of the behaviorological engineering that the culture needs that would help reduce some of the costs of misconstruing personhood. These engineering costs include establishing undergraduate and graduate degree programs in behaviorology subject matters in colleges and universities, programs that can produce demonstrable products and outcomes which reimburse society some of the costs of the programs. And we still must mention the implication of our natural–science account of personhood fostering humanity's survival along with the rest of life on this planet. It also fosters our own further evolutionary development biologically, emotionally, intellectually and culturally.

Does Life Have Any Meaning?

Lastly, let's extend that last implication of our personhood analysis, about supporting our survival and fostering our development, to a common ancient question: *Does life have any meaning?* Many answers exist for such a question. They all may even evoke a response of at least seeming partially right, although only some will evoke a response of also seeming scientifically reasonable. Certainly, some people feel that life has meaning, and others feel it lacks meaning. Indeed the same person may feel one way at times and the other way at other times. Having addressed feelings as part of the natural science of behavior, we can empathize and recognize not only the validity of all of these feelings but also the circumstances that produce them.

Traditional (i.e., pre–scientific) cultural contingencies, however, in competition with scientific contingencies, sometimes evoke a superstitious reaction saying that a purely natural–science analysis removes any meaning from life. Natural–science analysis, though, only reasonably removes the already questionable *unnatural or supernatural meaning* from life, which remains a purely natural phenomenon. What that superstitious reaction leaves unstated, though, proves most meaningful. That is, by leaving no mystical meaning to life, the purely natural–science analysis leaves only the meaning to life that remains ever present. This meaning puts our inherent nature, as *a part of nature,* under contingencies that induce interactions with the rest of nature that affect at least the viability of our continued status as a *living* part of nature, as opposed to the status, in nature, of fossils of yet another extinct species. Will we relatively soon become another extinct species (and likely take lots of other species into extinction with us)? Or will the complexities of behavior, with which nature endows us, enable us to survive, in part by enabling behaviorology to join the science and engineering departments in universities, so as to better join the science and engineering team efforts to solve global problems? Only time will tell. Will the contingencies, of which we are a part, evoke enough listening and other appropriate activities supportive of solutions to our problems in the required timely manner? Personally participating in these contingencies, in ways contributing to solutions, seems at least to me as more than adequately meaningful. Such meaningfulness always remains present for everyone, to a degree, again, that the contingencies determine. In this sense *one basic meaning of life inheres in the contingencies on helping our species remain unfossilized, remain a non–extinct, viable species.*

Your successfully plowing through this book to this point implies that such meaning already affects you as well as me. On the other hand, if the cultural contingencies in effect for your living fail to induce such meaning for you, then so much the worse for our culture, so much the worse for our species, so much the worse for our planet. If the lack of this meaning of life extends too far among the population, then extinction may follow too soon.

Conclusion

Our treatment of this chapter's topics can evoke grave concerns about certain historical trends in our culture. In general these concerns arise from observing some long–standing contingencies driving our cultural investment so thoroughly, and for so long, in superstitious alternatives to natural science in essentially all areas, especially in the area of human nature and human behavior.

Those observations leave us concerned about whether or not more recent scientific contingencies can reverse the current trends that threaten our cultural and species survival. Far beyond this chapter's concern with life and personhood, such contingencies extend to our present scientific efforts to solve all aspects of global problems, including behavioral aspects. While we cannot know the final outcome beforehand, we can make predictions from certain past patterns of contingencies that included compelling *continued trying* (e.g., the contingencies that compelled the extensive behavioral efforts that gave rise to natural science in the first place). We can predict that these recent patterns of contingencies compelling further extensive scientific, including behaviorological, efforts to solve global problems, will succeed, because these contingencies also include components compelling *continued trying*. While only "giving up" guarantees failure and extinction, ask "yourself" what efforts your contingencies are promoting regarding supporting the natural sciences, including behaviorology, in solving global problems.❧

References (with some annotations)

Fraley, L. E. (2008). *General Behaviorology: The Natural Science of Human Behavior.* Canton, NY: ABCs. See Chapter 28 for more about personhood. Contact the author (through "contacts" at www.behaviorology.org) for autographed copies.

Ledoux, S. F. (2012). Behaviorism at 100 unabridged. *Behaviorology Today, 15* (1), 3–22. With *Behaviorology Today* becoming fully peer–reviewed with this issue, this longer and fully peer–reviewed version of the paper that originally appeared in *American Scientist* (Ledoux, S. F. [2012]. Behaviorism at 100. *American Scientist, 100* [1], 60–65.) included the material that the *American Scientist* editor set aside at the last moment to make more room for the Skinner article excerpts that accompanied the original article; *American Scientist* posted this paper online along with the original version. Chapter 1 of Ledoux, 2014 *(Running Out of Time—Introducing Behaviorology to Help Solve Global Problems.* Ottawa, CANADA: BehaveTech Publishing.) includes and extends this paper.

Simpson, G. K. (2003). *Navajo Ceremonial Baskets: Sacred Symbols, Sacred Space.* Summertown, TN: Native Voices.☙

Chapter 25
Some Scientific Answers to Ancient Human Questions—IV:

What is death and how can we better help the terminally ill? Start by understanding death and enhancing dignified dying. ...

*H*aving covered some scientifically valid meaning of personhood and life, you can begin to appreciate the greater care we may need to take as a culture regarding *quality* of life in comparison to *quantity* of lives. This is a theme to which science fiction writers have long alerted us. Until rather recent times (e.g., since 100 years ago or so) the contingencies on species survival induced and rationalized reproduction. This favored large quantities of living human bodies trumping the quality of these human lives. However, now that we have exceeded the planet's carrying capacity with overpopulation numbers that threaten our continued existence, the contingencies on species survival actually favor fewer humans. If the contingencies on the culture induce letting its citizens begin to help everyone humanely resolve the overpopulation problem, then we might possibly avoid the kind of decimation of population numbers that otherwise remains the likely outcome of current circumstances (e.g., through extreme weather and any accompanying disasters, pestilence, war, and other outcomes).

Producing more behaviorologists to work on such problems would also be of particular help to the culture, because the overpopulation problem is, among all our global problems, the most pernicious and perhaps the most thoroughly imbued with behavior components (all of which provides a topic for a later chapter). If the quantity of humans reduces humanely and gradually rather than insensitively and drastically, the necessary shift in contingencies can also begin to favor efforts at improving the quality of life.

Death

That shift in contingencies, to favor concerns about the many aspects of life quality, may have already begun with the increasing concern, in many communities around the globe, about end–of–life issues for individuals. As a result of natural–science developments leading to improvements in medicine, an ever increasing number of humans are living long enough to face death through one or another terminal illness. Let's review some of the implications from natural behavior science regarding not only death but also the processes that ever more people are calling *dignified dying.* For depth and details well beyond this book's scope, consult the book *Dignified Dying—A Behaviorological Thanatology* (Fraley, 2012).

Dignified Dying

Widespread cultural contingencies have traditionally emphasized the suffering of those left behind when someone dies, rather than the suffering of those who are dying. Throughout the historical and pre–historical tenure of our species—up until a few hundred years ago—many deaths occurred violently and most deaths occurred relatively swiftly. Also, few lived long enough to develop the kinds of conditions that produce the slow, protracted dying of a terminal illness. The recent change in this pattern, wherein more of us are living long enough to die a protracted death from a terminal illness, has altered the contingencies. This has given rise to the movement for dignified dying, a movement that emphasizes dignified reduction in suffering for the dying individuals at least as much as, if not more than, the suffering of their survivors whose life continues and for whom many helpful cultural resources already exist. Scientifically our point is to provide some additional intellectual foundations to add to the usual emotional foundations for better treating— that is, treating with more dignity—those who must undergo a protracted dying process.

A natural–science thanatology (i.e., the study of death) can clarify some of the ethical and other issues in support of dignified dying. As a start to carrying these concerns toward practical and effective levels, we will introduce five areas of consideration. (a) We consider three basically different kinds of death concerning the body, the person, and social relations. (b) We consider the extra strain on both the dying and the survivors that stems from the currently poorly recognized, and widely misunderstood, gradual degradation in the social contingencies that accompanies the gradual degradation of both the dying person and the dying body. (c) We consider the confounding of the person and the body. (d) We consider some improvements that behaviorology can provide to reduce the aversiveness of these burdens on both the dying and the survivors, including a new cultural practice that Fraley (2012) introduced, and which he calls a "Foreniscon." And (e) we consider some questions regarding medical ethics.

Three kinds of death. Bodily mediation of the behavior repertoire that constitutes personhood leads us to differentiate between two of the three kinds of death. We call these two *person death* and *body death.* The usual deterioration in social contingencies during the slow dying of person and body leads to a third kind of death that we call *social death.* All three kinds of death can occur simultaneously, or they can occur at different times across the extended time span of a terminal illness. When the three types of death occur slowly, social death usually occurs first, followed by person death, and then by body death. However, our discussion follows the opposite order, moving from the conceptually simpler body death, through person death, to the conceptually most complicted, social death.

Body death refers to the irreversible cessation of all body–sustaining physiological activity, the kind of activity that evokes the responses "alive" or "living." This term scientifically describes the results when natural processes reduce complex chemical units (i.e., an alive body) to simpler levels of chemical structures and functions, levels that never again support the function of behavior mediation, levels that no longer evoke the description "alive" or "living." Instead these low levels evoke the description "dead." At the cellular level, body death can take from minutes to hours to complete. With substantial medical equipment and efforts, the early parts of this process can stretch from days to years, because modern medical life–support machines and procedures can keep enough respondent and physiological processes operating to prevent complete cellular death. After complete cellular death, the body continues various gradual processes of disintegration that we call decomposition.

Prior to complete body death, a period of physiological functioning and basic respondent reactions may proceed *only* due to medical life supports. During such a period, operant neural behavioral function has forever ceased. It not only is completely missing but it also is no longer possible. Deterioration of the necessary brain parts causes mediation of operant neural behavioral function to cease permanently. We call this cessation "person death," and it may actually precede body death by a substantial time period.

Person death refers to the irreversible termination of the physiological functions that mediate both the evoked operant neuro–muscular behaviors and the evoked operant neural behaviors. This includes the complex neural behaviors of consciousness. At this point neither internal nor external stimuli can ever again evoke the behaviors that they previously evoked, the behaviors that constituted the person. The person is gone. Meanwhile other physiological functions may continue at a level that leaves the body "alive." Person death occurs when the person stops happening because the body parts that mediate the person (i.e., that mediate the behavior repertoire) degrade to the point where they are permanently unable ever to mediate the person again. Often body death and person death coincide. This happens, for instance, under extreme trauma such as upon contacting the ground after falling from a height, or after suffering an irreversible heart function failure. However, person death from the degradation of the body parts that mediate the person can occur slowly, particularly during protracted dying from terminal illness, and lead to complete person death before, and sometimes long before, body death occurs. During gradual person death, changes occur in some of the social contingencies between the dying person and the survivors. These contingency changes build up to what we call social death.

Social death refers to the interpersonal results that sometimes occur due to the gradual loss of contingency components between the dying person and his or her survivors prior to person death. The interpersonal relations, between the terminally ill person and her or his loved ones and friends or associates, gradually fail as the degradation of body and person brings about a reciprocal degradation of the social contingencies responsible for those interpersonal relations. All parties find this degradation confusing and distressful. Understanding how social death happens may reduce the confusion and distress from this experience.

Social contingency disruptions. Every person (i.e., the evoked behavior that every body mediates) participates in interconnected webs of mutually controlling, and generally reinforcing, contingencies with a fewer or greater number of other persons. These social contingencies comprise the bulk of interpersonal relations. At times in these contingencies, one's behavior evokes the behavior of others that reinforces one's behavior. At other times in these contingencies, another's behavior evokes one's behavior that reinforces the other's behavior. That is, under appropriate contingences, at some times your behavior produces reinforcers from others, while at other times your behavior provides the reinforcers for another's behavior. Interpersonal relations not only involve such social contingencies but also *require that all involved parties have a physiology healthy enough to mediate the related behaviors.*

When someone becomes terminally ill, the resulting progressive physiological deterioration reduces the body's capacity to mediate the usual social behaviors with respect to social–circle members. The physiology of the dying person increasingly becomes incapable of mediating behavior that provides the relevant reinforcing consequences that it provided to others in the past. The dying person's physiology also becomes increasingly incapable of mediating behavior that evokes the reinforcing behavior of others. As each of these occurs, they together progressively eat away at previously well established social–contingency relations. Business and professional associates, friends, and loved ones become increasingly disconcerted as their previously occurring behaviors, which reinforced their dying loved one, fail to occur (i.e., because the evocative stimuli can no longer occur, given the reduced capacity of the dying person's physiology to behaviorally produce those stimuli). They also become disconcerted as reinforcers for their current behaviors also fail to occur (i.e., because the dying person's physiology is also unable to mediate the behaviors that provide these reinforcers). Sometimes the coercive quality of these changed contingencies, on some of these others, produces their withdrawal from further social contact with the dying individual, and not always at an appropriate or helpful time. Also, such developments sometimes cause later painful regrets, for those who withdraw, that better understanding of these events could have avoided. We describe the end results of this process, the resulting social isolation even in the presence of some surviving friends and loved ones, as the social death of the terminally ill person.

The social–death process, the gradual degradation in social contingencies that accompany the gradual degradation of the dying person and body, exacerbates the strain on both the dying and the survivors. While nothing can completely eliminate the stress of a friend's or loved one's—or your own—protracted dying, some steps might reduce some unnecessary components of this stress. The behaviorological analysis of the contingencies surrounding protracted dying prompt new practices related not only to ethics and person/body death, but also to easing the bereavement of survivors, and especially to managing the social death experience in ways that reduce its negative impact. While we will leave a thorough discussion of such practices to other sources (e.g., Fraley, 2012) let's touch on them only after considering the confounding of person and body that natural processes automatically condition.

Confounding the person and the body. A complicating factor in *all* contingency relations, especially those surrounding protracted dying, involves the early occurring, general and long–term confounding of the person and the body. In this confounding both the person and the body function separately as stimuli. Either of these stimuli can seem to evoke behavior from other persons, *due to the ever–present pairing of the person and the body that mediates it.* That is, the person—the behavior repertoire that the body can mediate—is always present with the living body, and so gets constantly paired with it. Thus the "person" that a now dead body, or at least a now person–dead body, previously mediated, still seems able to evoke our behavior. But it is the body that is actually evoking our behavior in this circumstance, due to the past pairing of body and person. Furthermore, this body–evoked behavior is not functional. For example, asking, even demanding, that the dead person/body come back to us produces no change in the dead person/body status. Meanwhile the body continues to serve evocative stimulus functions for the behavior of others. These functions remain separate from the status, living or dead, of the body (e.g., both a living and a dead body can evoke the response "That body measures two meters from head to toe").

However, no body ever initiates the behavior that it exhibits. A body only mediates behaviors produced by contingencies, with all the behavior that a body mediates constituting the total person. At various times various parts of this behavior, this person, evoke behavior—relevant to the living status of the person—from other persons (e.g., the person–behavior of calling out "Anyone home?" at an open door evokes another person's response of saying "Come on in"). Thus, only the *person* that a body mediates (i.e., only the behavior, which occurs one or another response at a time) evokes the person–relevant behavior of other persons.

Let's go into a little more detail. The behavior *is* the person. The person is the collective set of behaviors that operant processes have conditioned a particular body to mediate under the range of external and internal environmental controls. These behaviors evoke the behavior of other persons. Inevitably, however, this means that if the person occurs (i.e., if behavior occurs) then the body is also occurring, for we cannot separate the behavioral person, a process, from the physiological, mediating body. Meanwhile, the behavior that constitutes the person continually evokes behaviors from others. This continual occurring–together status of person and body, *this constant pairing of the person and the body*, can leave the *body* also evoking responses from others, responses that otherwise only the person could evoke, as if that *body* is the person it mediates, which it is not. The body only constitutes the necessary physiological, mediating interface between environmental energies and behavioral responses, particularly person–related behaviors. However, a problem arises in that the permanent pairing, during life, of person and body leaves *even a dead body* evoking responses from others as if that dead body were the person it had mediated before death. This confounds body and person. For example the dead body can evoke responses from others such as, "Wake up. Speak to me. Tell me how to help you." Since such responses produce no reinforcers—they cannot produce any reinforcers—from the dead body, such responses undergo fairly rapid extinction.

Furthermore, a body that remains biologically alive (e.g., via medical life support) but person dead (i.e., it permanently lacks the operant, and much of the respondent, capacity for behavioral interaction with the environment) can also still evoke the previously person–relevant behaviors of others, due to the constant pairing during life of the person and the body. Whether in this state or also biologically dead, a body provides a most unreinforcing stand–in for the previously mediated person. It provides a not entirely necessary increment in the general aversiveness of the situation for survivors. But, reasserting our interest in reducing aversiveness not only for survivors but even more for the terminally ill, let's now consider a couple of new and scientifically grounded cultural practices that help us better handle end–of–life issues.

***Aversiveness reductions including the* Foreniscon.** Current cultural practices concerning death and dying sometimes originated for practical reasons (e.g., wakes) but often they originated in traditional folklore, much of which comes from a range of long–developing, pre–scientific superstitions. Since these practices arose when few efforts could stop pre–death suffering, these traditional practices also typically accept the horrors that nature visits on the dying, including the terminally ill, and their survivors. In addition these superstitious practices even push participants, both the dying and survivors, to endure every measure of pain and suffering that death, including protracted dying, has to offer, even to the extent of making a putative virtue of such endurance. When no alternatives were possible, such virtue conditioning may have helped. But alternatives, ethical scientific alternatives, are now possible. We must begin to replace the previous practices with newer, more humane practices that afford substantive relief from these horrors. Some modern practices already move in this direction (e.g., hospice care, and the sensitive counseling that even some scientifically uninformed professionals provide, which nevertheless coincides with scientific realities). Other practices to reduce the aversiveness of these burdens, on both the dying and the survivors, emerge as implications of our behaviorological analysis.

A good place to start involves increasing, through the inclusion of behaviorology in standard educational programs, everyone's comprehension of the natural processes that are operating during life, person death and, especially, social death, as well as body death. This comprehension can reduce the confounding of person and body, and mitigate some of the ensuing confusion and distress, especially during periods of slow dying.

Furthermore, such comprehension can also show us the previously unrecognized scientific rationale for the general, intuitive practice of saying good–byes before the factors leading to social death deprive the dying person of all capacity for such interactions. Fraley (2012) introduced one new cultural practice directed precisely at enhancing, formalizing, and culturally institutionalizing this general, early good–byes practice. He calls it the *Foreniscon* (short for "**For**mal **En**ding of **I**ntimate **S**ocial **Con**tacts"). The Foreniscon revolves around improving the social–death experience by arranging parts of it for both the dying and their survivors while following the pattern of ceremonial recognition of life's major events (e.g., birth, marriage). The Foreniscon involves one or more arranged, ceremonial occasions (somewhat like wakes that you attend while alive) at which you say your good–byes to various and increasingly intimate groups *while you still can say goodbye* and, importantly, after which you avoid seeing those in each group again.

An endeavor like a series of Foreniscons takes substantial planning, an effort likely beyond the capacity of a terminally ill person. Yet a person in this position likely also needs help with several other end–of–life considerations, probably including business, legal, and counseling as well as medical and other concerns. A coordinated, team approach would seem to be in order to assist in these matters with the kind of ethical and sensitive professionalism that promotes the fair and dignified conclusion of these affairs.

Along with the business, legal, medical, and other experts on such a team, at a technical level a behaviorologist team member can provide the scientific thread that ties the team together. A behaviorologist team member can also provide the necessary and appropriate professional therapeutic tools to help the dying person—and the survivors—comprehend all that is happening and deal, emotionally and intellectually, with the different parts effectively. Furthermore the behaviorologist applies her or his knowledge and skills to help reduce the problems that any survivors face during the bereavement process, particularly after the complete body death of their intimate, which can even reduce the length and negative effects of the period of the bereavement–process.

Questions of medical ethics. While all those possibilities receive further development in the resource listed in the references, we should mention one other important area for which behaviorology is relevant. All the protracted–dying events that happen under current cultural practices generally occur under medical supervision. Some of these practices, which I cannot recommend (Can you?) can include treating terminally ill patients in ways that compel them, often against their preference statements, to experience all the suffering, horror, and indignity of body degradation and person death and even social death. Sometimes this even extends to shipping such patients off to often privately owned, for–profit "hospitals" specializing in procedures specifically designed mainly to keep patients bodily "alive" for as long as possible, regardless of their suffering, while legally but unethically milking every possible dollar from the "health" care system. Such treatment also compels survivors to experience additional kinds of suffering and horror during the social death and person death, and even *beyond* the body death, of their terminally ill intimates.

Those circumstances appropriately evoke reviews of all applicable ethics, including medical ethics. The ethics of many living individuals, as well as many dying individuals, call for recognition and acceptance of their right to die with dignity, and even with medical assistance. But traditional medical ethics put many capable doctors in desperate conflict with the contingencies driving their own preferences for reasonable, compassionate, and dignified patient care. Sometimes ethically acceptable medical action, or inaction, under strained medical ethics that need updating, alleviates the *very end* of the protracted dying experience for the terminally ill. But this leaves what some would call undignified patient abuse necessarily occurring during the periods of social death and person death. Improvements in these medical ethics would not only generally improve end–of–life experiences, for the terminally ill and their survivors, but such improvements also bear substantial implications for resolving questions about medically assisted death. Some countries (e.g., Belgium, Luxemburg, and the Netherlands) along with several states and Provinces in the u.s.a. and Canada, have already made notable progress in resolving questions of medically assisted death. While you can apply the values, rights, ethics, and morals parts of an earlier chapter to these concerns, the *Fraley* resource listed in the references contains more thorough material worthy of your attention.♣

References (with some annotations)

Fraley, L. E. (2012). *Dignified Dying—A Behaviorological Thanatology.* Canton, NY: ABCs. Chapter 3 is particularly relevant to medical ethics. As stated earlier, go to www.behaviorology.org for information about how to obtain books like this one and Dr. Fraley's *Rehabilitation* book (or contact the author at ledoux@canton.edu).ᐁ

Chapter 26
Some Scientific Answers to Ancient Human Questions—V:

Are our surroundings really what they seem, and how about us? Some surprises regarding the nature of reality and robotics ...

*R*ecent chapters addressed some scientifically consistent answers to ancient human questions about life, personhood, and death. *Life* refers to a range of levels of chemical complexity that feature processes involving energy exchanges, among simple and complex chemical units (i.e., from less than single cells to greater than humans and whales) that change these units and their surroundings. Such processes include evolution, conditioning, consciousness, and culture. *Personhood* refers to the whole repertoire of behavior that conditioning has made a physiological body, through its nervous system, mediate when environmental energy changes produce behavior. *Death* refers to the results when natural processes, which sometimes involve stimulus–caused human behavior, permanently interrupt life functions and begin reducing complex chemical units (i.e., bodies) to simpler levels of chemical structures and functions, levels that can never again mediate behavior. Beyond such body death, we also considered person death, and social death, in the context of scientifically supporting death with dignity for the terminally ill. All these areas comprise natural products of natural processes.

All those topics also relate to the major topic of this chapter, *reality*, because they are part of some sort of reality. But what is reality? Of the two topics in this chapter, reality has long prompted ancient questions while robotics pertains to more recent developments. Our coverage introduces scientific accounts for each of these topics in turn.

Reality

Ahhh, reality. "Not only is reality stranger than we imagine, but it also is stranger than we can imagine." With words like these, several science commentators over the last century have characterized the results of closely examining the nature of reality. Our commentary on reality may show them as more *literally* correct than they ever imagined. Yet a century from now, science commentators will likely report both how this chapter fits right in with their newer characterization, and how it too falls short of the realities regarding reality. Ahhh, the value of self–correcting science.

Reality Knowing Versus Reality Existence

Even at this early point, however, a caution may be in order. This chapter's conclusions about reality, to which our behaviorological analysis leads, *only* concern what we can *know* about reality and how we can know it. Our conclusions remain largely irrelevant to *existence* questions, such as questions concerning whether or not reality "exists" or what "really" exists in, or makes up, reality. Here we emphasize what we can know about reality and how we can know it.

About reality. Prior to stating our behaviorological conclusion about reality, we should recognize that it develops as a direct extension from the principles, concepts, methods, practices, research, and philosophy of science that make up the behaviorology discipline. This reality conclusion develops from taking all of the laws of behavior, in everything that we have covered in this book, and seeing where the combination leads. Even if our past conditioning makes the present conclusion elicit emotional discomfort, good science says that we can still only ultimately benefit from improving our knowing of reality.

To say it baldly, our current behaviorological conclusion about reality states this: Due to the constraint of sensory neurons being our *only* source of contact with stimulation, the firings of sensory neurons evoke neural behaviors that *are* (i.e., that establish) our reality. In other words, due to the firings of sensory neurons, we individually *neurally behave* reality. That is, "Reality *establishes.*" Reality establishes, due to our neurally behaving it due to the firings of sensory neurons. This sounds confusing only because English seems to lack a better term than "establishes" for these events. The usual, typical substitutes (e.g., "presents," or "occurs," or "happens") provide little help. Reality establishes due to the firings of sensory neurons evoking our neurally behaving reality as covert, neural responses that then lead to interactions with the stimulus events indicated (i.e., established) by the sensory–neuron firings. For example seeing, hearing, tasting, and so forth, are all neural behaviors that begin occurring due to the firing of sensory neurons, and these neural behaviors establish what a body "sees," "hears," "tastes," and so on. That is, the firing of some sensory neurons begins the process of neural mediation of what one "sees," "hears," "tastes," and so on. This does not mean that the firing of neurons initiates behavior, which would be magic that natural science has already set aside. But for simplicity of discussion at this point, we take stimuli evoking the firings of sensory neurons, which starts the mediation of behavior, for granted.

The counter–intuitive part is that, because we (i.e., our bodies) have no sensors going beyond the sensory–neuron parts of our nervous system, this *evoked neural behaving of reality* is our only source of knowledge about reality, about those stimuli evoking the firings of sensory neurons that "we take for granted." This means that this evoked neural behaving of reality is *also* the only source of knowledge about the reality of our sensory neurons. Thus it is *also* the only source of knowledge about the reality of any stimuli or stimulus energy traces that produce the firings of sensory neurons. And so it is *also* the only source of knowledge about the reality of any behavior—neural or neuro–muscular—that the firings of sensory neurons subsequently induces. Indeed, as a result of our still only contacting, *through our sensory neurons,* all the sensing equipment that we consider as enhancing our sensors (e.g., from microscopes to telescopes and spectroscopes, as part of the visual sense mode in science) we still only know about the reality of this equipment, and about whatever the equipment produces, through the firings of sensory neurons.

Now if that last paragraph made sense to you rather than confusing you at least a little, or if it did not make sense but gave you a headache—which reading further may not cure—*then just skip the rest of this chapter and turn to the last chapter,* which is more important in the present–day era. You can always come back to this point later. For some—perhaps most—readers, the content of all the past chapters can provide an inadequate preparation, which is my fault as author, for this chapter's topics, in competition with so many traditional contingencies in each person's history. The rest of this chapter only contains an extension, a mostly theoretical rather than a practical extension, into an area which the science itself opened up. This extension, in the remainder of this chapter, mostly attempts to disentangle some of the confusion about reality, and it adds some related stuff about robots and robotics.

Reviewing "knowing." A little review, however, about *knowing* can help. The word *knowing* refers to parts of the cascade of purely neural responses of consciousness. These overlap in the rolling sequential chains that external or internal stimuli evoke when energy traces bring about raw sensation, awareness, recognition, comprehension, thinking, observing, reporting, and so on.

A point that I brought to Lawrence Fraley's attention, while editing his 2008 book, is pertinent here. Behaviorological analysis, based on summarizing the natural laws governing behavior, leads to the same conclusion about reality that Hawking and Mlodinow reached, through the logic of naturalism in physics, in their book *The Grand Design* (2010). Based on their conclusion, let's clarify our conclusion about reality: *Our neurally behaving reality is the sole source of knowing about reality, because we can get no closer to reality than the neural behaviors that the firings of sensory neurons evoke.*

Three different perspectives. That conclusion has the firing of sensory neurons as its beginning, according to its own, new perspective, which Fraley (2008) calls the *robotic* perspective. Yet, from a more common, current and also scientific perspective—one which we call the *environmental* perspective—that conclusion seems to start with an end point. The *environmental* perspective starts with energy traces causing the firing of sensory neurons that induces neurally behaving reality. To deal with why the robotic conclusion starts with "the firing of sensory neurons" we must differentiate *three* related perspectives.

The first perspective is the age–old *agential* perspective. This perspective constitutes a *pre–scientific* environmental perspective in which various forces in an environment, including magical forces, exert mere influences on a mystical inner agent while this agent spontaneously initiates directives that tell the body what behaviors it wants done, that tell the body what behaviors to emit. (Due to this residual agential connection, the word, *emit,* rarely appears in this book.) Since the main variables in this perspective, the inner agents, reside untouchably (i.e., neither testable nor even measurable) in a mystical realm, the perspective remains divorced from effective application in practical matters.

Having set aside the pre–scientific agential perspective, the next perspective is the current, agent–free, scientific *environmental* perspective. This perspective informs science in general, including the other chapters in this book. In this perspective only energy traces from natural, internal or external environmental stimulus events produce the bodily changes that *are* the mediation of behaviors. This perspective occurs throughout the natural sciences, and currently enables the practical, including behaviorological, benefits through which our culture justifies the expenses of scientific and engineering activities. Furthermore, in this perspective, environmental energy traces induce sensory–neuron firings that evoke our neurally behaving reality, which includes the environment. *Nothing precludes the existence of the environment, or reality, beyond our capacity to neurally behave both it and knowledge of it,* an important point to which we will return. This perspective nevertheless acknowledges that organisms lack the capacity for more direct contact with the environment, or reality, beyond what sensory–neuron firings furnish.

The third perspective is the newer, scientific perspective that we call the *robotic* perspective. This perspective *expands from* the point of recognizing that we behave reality, because we can get no closer to reality than the behaviors that the firings of sensory neurons evoke. From this perspective these sensory–neuron firings constitute the *start* of behavior, including the start of knowing and any other consciousness, covert, or overt behaviors. Thus only behavior itself is available to establish reality (i.e., to establish the sights, sounds, tastes, smells, touches, and feels, including movements and dreams, to which we otherwise seem to be reacting). Furthermore, because we are still investigating its implications, the robotic perspective currently seems somewhat removed from practicalities, although not inherently so as is the case with the agential perspective. We currently remain unable to specify how to apply it in interventions. While it also seems to be the most consistently naturalistic perspective, further research will clarify its potential practical benefits.

Fraley calls this perspective the *robotic* perspective, because it describes the limitations on perspective for a completely natural lump of organized matter, a category into which all known life forms classify, including people. I suspect that you can already sense the difficulties that some language problems are going to cause us. Our language arose under contingencies that induced the agential perspective. The relatively recent scientific contingencies of the environmental perspective have only just begun to induce linguistic "catching–up." That is, we are just beginning to develop a grammar supportive of science that we can use in place of our current, still traditional agential–perspective supporting grammar. Yet here we are, while comfortable with the familiar environmental perspective, trying to talk about the thoroughly naturalistic robotic perspective that seems to turn things all around. For this reason, and due to the level of this book, we will try—admittedly without complete success—to stick with the environmental perspective even as we introduce the robotic perspective.

The *robotic* perspective describes the limitations on perspective that a completely natural lump of organized (e.g., biological) matter can have. In spite of the difficulties, that description applies broadly, equally well, to everything. With adjustments for quantitative differences, it applies to me and to you and to all other evolving biological lumps of "living" and behaving matter. It also applies to any lumps of matter, such as electro–mechanical or digital lumps (e.g., computers and other robotic devices or robots) that contingencies induce biological matter lumps (e.g., people) to build. Thus it applies to robots behaving, and meeting scientific criteria justifying the responses "living" and "it's alive." We have not as yet reached such a point, at least not in the way that science fiction novels have described (e.g., those by Issac Asimov). Nevertheless, the glimmerings have begun. A misbehaving computer often evokes responses like "it's alive" and "it has a life of its own." *From the robotic perspective,* such comments constitute entirely legitimate responses. In this case the computer (an example of a relatively primitive life form naturally constructed through contingencies on human behavior) evokes these responses from another life form (an example of another, less primitive life form also naturally constructed but through biological evolution).

Recomputing the Behaving of Reality

Now, this may be a good time to recall that these conclusions about reality only consider what we can know about reality and how we can know it. Here we keep these conclusions, and their connections with knowing behaviors, entirely separate from questions regarding whether or not reality really exists, or what really exists in any reality that exists.

Still, the point of elucidating those three perspectives concerned why our reality conclusion starts with what our scientific environmental perspective leaves us responding to as an end point rather than a starting point. Turning that conclusion around might more easily show the sequence that reaches that conclusion, although turning it around switches from the robotic perspective to the environmental perspective. Consider together some of our behaviorological discoveries that we have otherwise covered across a variety of chapters: (a) In the process that we called *direct stimulus control,* some behaviors, especially neuro–muscular behaviors, occur under direct control of particular stimuli, as in our driving–while–daydreaming example, without supplementation from neural–only and single–channel consciousness–behavior chains. (b) Energy traces from internal or external stimuli evoke neural chains of consciousness behaviors, which cannot happen any other way (e.g., they cannot happen agentially). And (c) these behaviorological processes ultimately address the accessible independent variables that are responsible for *why* behavior happens (and we can apply them in helpful interventions). Meanwhile physiological processes address the accessible independent variables that are responsible for *how* behavior happens (i.e., how the physiology mediates behavior).

When we put those three puzzle pieces together, we get an environmental–perspective view of reality. *This perspective assumes* (which is just another kind of behavior) *reality in the first place* without necessarily characterizing it further. In this perspective we still behave reality, with our reliance on the firings of sensory neurons—as the sole source of knowledge about reality—constraining our behaving reality. However, in this perspective, reality "establishes" *further* into the event chain. Following a standard environmental–perspective time line, the environmental–perspective chain begins with energy traces, from whatever exists that constitutes reality, inducing the firings of sensory neurons. These firings evoke neural behaviors that establish reality and that chain to other neural and neuro–muscular behaviors.

A difference in perspectives. The environmental perspective *begins* with the energy traces. The robotic perspective *begins* with sensory neurons firing that induce neural behaviors that establish reality. This reality includes the reality of the sensory–neuron firings, the reality of the energy traces that induced the firings, the reality of the things/events that produced the energy traces, and even the reality of the sensory neurons. That is, in the robotic perspective, the *sources* of the energy traces, the energy traces, the sensory neurons, and the sensory–neuron firings due to these energy traces, all remain parts of the reality that the neural behavior establishes. However, from the environmental perspective, the neural behavior establishes the reality of these parts only *after* the energy traces induce sensory–neuron firings that evoke the reality–establishing neural behavior. Confusing?

To remove one source of confusion, let's stick with the familiar environmental perspective. Even though our comments often relate to both perspectives, we cannot easily talk about both of them, or from both of them, at the same time. For instance in both perspectives—but stated from the environmental perspective—improvements in our neurally behaving reality result from the ongoing conditioning that affects the neural structures. Yet the functioning of the neural structures comprises the neural behaviors that establish reality. This is probably still confusing. We soon get to a series of examples that can help reduce the confusion.

Some reality implications. The current evolved state of our physiology provides no other access for inputs (i.e., energy traces) of any kind from further afield than our sensory neurons, as extensive or as limited as they might be. All inputs must come through our sensory neurons (e.g., photoreceptors, phonoreceptors, and so on). This carries some important implications. For instance, any inputs that induce sensory–neuron firings that fail to evoke the *necessary* neural behaviors of knowing with respect to them (i.e., to the inputs) remain inputs about which we know nothing. They affect other behavior, neural and neuro–muscular (e.g., through direct stimulus control) but their reality remains *un*established and so unknown. As another and related implication, inputs that induce sensory–neuron firings that fail to evoke *accurate* neural behaviors of knowing with respect to them (i.e., to the inputs) remain inputs for which our neural knowing behaviors reflect an inaccurate establishing of reality. This leaves their reality *mis*established even while they then affect other behavior, neural and neuro–muscular, also inaccurately. We will see examples of both **unestablished** reality and **misestablished** reality.

Reality summary. Energy traces from environmental events affect sensory neurons the firing of which evokes neural behaviors, of which some establish reality. Other neural behaviors of consciousness constitute what we previously described as knowing. However, the only contact points available to physiological organisms (e.g., the famous carbon units so common in science fiction, like us) are the sensory neurons at the interface of the physiology and whatever else *presumably* exists. But our only knowledge of what exists comes from sensory neurons firing, which produce the behavior that establishes the reality of what exists. So we can only know of what exists by behaving it. We can get no closer to reality than the behaviors establishing reality that the firings of sensory neurons evoke.

Here is how that would come out if we stated it from the robotic perspective. Humans (and likely—to the extent possible—other animals) neurally behave reality. That is, sensory neurons fire, inducing the neural behavior that establishes the reality of whatever elicits or evokes (or consequates) their further neural or neuro–muscular behaviors. This includes the sights, sounds, tastes, smells, touches, feels, movements, and dreams, as well as the sensory–neuron firings that evoke their neural behavior and the energy traces that induce the sensory–neuron firings, and even the sensory neurons and the rest of the nervous system and physiology, along with whatever they "know" about reality.

Some reality examples. Regardless of which perspective we use, natural sciences have on several occasions clarified that realities differ, and change, according to the repertoire of the body that is behaving the reality. For example, consider a dining table from the 1930s in your family home then and today. Non–physicists behave the reality of an inch–thick tabletop through which they cannot push a coffee cup regardless of how much force they can personally exert in the effort. They behave the reality of a table top as "solid." Yet professional physicists from the 1930s, making an instrumented examination of a small sample of the same tabletop, behave the reality of a complex atomic structure consisting mostly of empty space. And professional physicists of today, making a more advanced, instrumented examination of that same small sample of the same tabletop, behave a reality of even greater complexity, greater precision, and greater confusion or at least obscurity to the non–physicist. Realities differ.

Returning to **un**established reality, a lack of behaviors of knowing, regarding particular inputs, leaves the reality of these inputs unestablished. This means that, while these inputs can still affect us, they remain unknown to us.

As an example, once again consider daydreaming and driving. Energy–trace inputs from the road ahead induce sensory–neuron firings that evoke only driving responses, under direct stimulus control. Meanwhile other stimuli are evoking knowing responses on the single–consciousness channel (i.e., we "know," and can often later describe, the theme, content, and direction of our daydream). As a result, we remain unaware of (i.e., we lack knowledge of) the road stimuli that are controlling the driving responses. They remain unestablished, and we cannot report them later even though they controlled our driving responses with at least minimal adequacy.

Meanwhile **mis**established reality involves behaviors of knowing some stimuli as the causes of some sensory–neuron firings that are actually not the stimuli that induced the sensory–neuron firings. This kind of inaccurate behaviors of knowing, regarding particular inputs, leaves the reality of these inputs misestablished. This means that these inputs likely affect subsequent behavior with similar inaccuracy.

As one example, once again consider our seeing–ghosts account (from an earlier chapter). The energy–trace inputs from vague and partial stimulus sources, which in this case were from both the external and the internal (due to conditioning history) environment, induced sensory–neuron firings that evoked neural behaviors of consciousness and knowing, along with neuro–muscular behaviors. All of these behaviors were quite unrelated to (i.e., were inaccurate with respect to) the actual external–environment inputs. As a result, the behavior–established reality about the inputs (i.e., what became known) was not the actual inputs. Reality was misestablished as seeing ghosts.

Here is another example, one like so many mundane examples that each of us has experienced. Indeed, these events happened to me and, under contingencies compelling the recording of simpler examples than those previously entered in my notes, I recorded this observed example. However, like all *"I"* personal–pronouns, the *I*s in this example never imply inner agents. They only imply a physiology, "my" physiology, mediating evoked consciousness behaviors. Here is the written record of this example, which follows the actual time line. Note that, as is the case with all the other examples, this is not an example of what the term "introspection" means. This public–of–one report includes observed covert neural behaviors as well as overt neuro–muscular behaviors:

One of the foot paths to my campus office crosses a small, forested island. Walking that path one day shortly after the college had cleared the debris left over from the now completely melted snow, I observed that I was seeing, perhaps 50 feet ahead at the end of a curving part of the path, a stone the size of a large watermelon at the edge, and in the shadow, of a bush that was about five feet tall and about eight feet away from the path edge. That is, I neurally behaved the stone, presumably from light–energy traces inducing some sensory–neuron firings in my eyes, which established the reality of the stone, along with, of course, the path and the bush and so on. If the path had not taken me close past the stone, I likely would never have behaved it again, and thus left it a stone forever so far as I could *know*. However, the path did take me closer to the stone. As I started to pass the stone, I observed that I was seeing letters on it. That is, I was behaving letters on it, establishing the reality of the letters, and also observing further neural responses such as "What? Letters? On a stone?" Then, as the letters were evoking these neural questions, I also observed that I was behaving the "stone" changing shape—Whoa! Spooky! (in the respondent, elicited–emotion sense)—apparently in a sudden gust of wind the reality of which I was also behaving. Nearly simultaneously, I began behaving (i.e., establishing the reality of) the "stone" as a discarded gray grocery bag with the name of a store printed on it. Behaving it as a stone misestablished its reality. Was it a grocery bag all along? Careful; tricky question. So far as I *know,* no, it was not, although so far as I or anyone might *presume,* yes, it was. However, presuming involves a further behavioral step away from the events, and it is *not knowing*. It is an additional behavior that comes from less direct but combined lines of suggestive evidence. Having behaved the object as a stone, one can only know it to be a stone. Only after behaving it later as a bag can one know it to be a bag, and presume that it had always been a bag.

We all behave "realities" such as the variety in those examples, even those of us from whom such examples tend to evoke snide giggles. Ahhh, reality. It is not as sure a thing as past conditioning compels us to think, or so present conditioning compels us to think. And so the question arises, what if any supports are available for the *behaved confidence* that reality evokes in its apparent existence? Does reality actually exist? We have no way to know, given the constraint that only sensory–neuron firings establish reality. To answer this question, as with those examples, recall that we are still limited to addressing, not *whether* reality exists, but only how we can *know* about any reality. Nevertheless, some people say that, since we cannot know that reality exists, independently of our neurally behaving it, then it does not exist. Others would argue, though, that since we cannot know that reality exists, then we also cannot know that reality does not exist. So what can we say?

Reality supports? The behaviorological characterization of reality as minimally introduced in this chapter—that we neurally behave reality under the limitation of sensory–neuron firings—allows, in that it fails to preclude, the non–directly knowable existence of a reality, of an environment, of a realm beyond our neural behaving. Inevitably a discussion, of how we can know any reality, evokes (Provokes?) questions of the *existence* of reality, of the environment, of something, some realm, outside us. This includes a realm, of course, that everyone else must inhabit since otherwise they are not real (Right?). They are only inside our own heads (i.e., your head and my head, etc.). So, let's paraphrase what a famous film character once stated. *Of course this is all happening inside our heads. Why should that mean that it is not real?*

To prevent this chapter's content from inducing some perhaps unnecessary stresses, let's look at some starting points, for further discussion elsewhere, about these questions. And questions they must remain, because the kind of answers that would fully satisfy us require contact points forever unavailable to us. They remain unavailable because, again, the only source of *knowledge* that *anything* exists—including any particular thing as well as any reality, environment, or realm outside us—is our evoked behaving of them due to the firings of sensory neurons. We are stuck inside that constraint.

However, in the same way that we cannot *know* that reality, and so forth, exist, we also cannot *know* that they do not exist. *We lack contact points* with the relevant places—if those places exist—beyond our sensory neurons, the reality of which we must also behave. Can we say nothing else about any reality of reality? Technically, the sensory–neuron contact–point constraint leaves little else to say in terms of what we might otherwise consider the most direct evidence of reality. *We can, though, consider some possible, suggestive, indirect lines of evidence,* even though these too are under the same constraint: To know them we must behave them. Still, let's try them.

Some seemingly sensible supports for the reality of reality seem to exist, although these more indirect lines of evidence still ultimately merely involve more behavior that also remains subject to the same sensory–neuron firings constraint. Nevertheless, in support of the self–correction routines of natural science, we will list the labels of some of these indirect supports, and follow this list with a variety of practical examples. We will not, however, further analyze the examples to connect them with particular types of indirect support.

Such indirect supports strengthen our confidence that, despite lacking more direct proof, something (e.g., reality, environment, realm) exists outside us. Of course, confidence is but another one of the variety of neural behaviors of consciousness. These indirect supports include the contiguity, consistency, reliability, potency, and even patterns, including illogical patterns, extant in our ultimately behaved data, along with the process and outcomes of direct stimulus control and our very continuation and survival.

Here in an example supportive of our confidence with respect to the reality of reality. It takes the form of this question: "Why would ... whatever happen, *if no reality exists* that somehow supplies the energy traces that induce the sensory–neuron firings that evoke neurally behaving the reality of whatever happened that provided those energy traces?" Here is the example. Why would someone behave a noisy car (or a screeching cat, or an arguing neighbor, or a crashing burglar, or the sweet sounds of a capable street musician, or an alarm clock) disturbing their sleep, if... (i.e., if... *no reality exists* that somehow supplies the energy traces that induce the sensory–neuron firings that evoke neurally behaving the reality of the noisy car [or any of the other listed possibilities] that provided those energy traces)?

Here are several more examples supportive of our confidence with respect to the reality of reality, each including, but not repeating, that whole question.

Why would a letter carrier behave a dog appearing from behind a bush and biting her or his behaved leg, if...? Why would a college graduate behave bird droppings landing on his or her behaved, long–haired head and then dripping onto her or his behaved new suit as he or she behaves walking into a behaved building where behaved job interviews occur, if...? Why would you start down your basement stairs (Yes, I know, "behaved basement stairs.") on a hot August afternoon only to stop short as you behave a four–foot layer of backed–up dirty sewer water covering the floor, if...? Why not just behave clean water, and take a swim? No? Why would that not work?

Such examples can go on virtually endlessly. Here is a last set: Why would a body, when driving a car over the crest of a hill on a curving road, behave a flock of turkeys, or a deer, or a family of skunks, or a batch of baby lambs on the road around the bend at the top, which may or may not also require behaving a swerve, or a hard stop, or getting out to pet the lambs, or hitting the deer and wrecking the car and maybe terminating the reality behaving of (i.e., killing) a passenger, if...?

With respect to reality and knowing it and behaving it and questions of its existence, what are we to make of the very last possibility in that last example? When we behave someone dying, the reality behaving of *that* body ends. She or he behaves no more reality. (Or, saying the same thing without the confusing pronouns, the now dead physiology no longer mediates any reality.) Yet *we* continue to behave reality. Was not that body behaving the reality of us while still alive? If the now dead physiology no longer mediates any reality, what about us? Yet *we* continue to behave reality. Is this another indirect line of evidence that reality exists? We wonder because, while our reality behaving ends at death, we presume—a behavior that our death stops—that our death leaves everyone else still behaving their reality, even in their reality, which presumably then exists beyond us, in some form not knowable except by behaving it, and so on.

We might never behave satisfactory answers because, again, all those example questions continue, "..., *if no reality exists* that somehow supplies the energy traces that induce the sensory–neuron firings that evoke neurally behaving the reality of whatever provided those energy traces." Yet in each of those and any other examples, we can get no closer to reality than neurally behaving it, along with the neural consciousness behaviors of knowing, all of which the firings of sensory neurons evoke. However, altogether, as natural–science self–correction routines operate, such supports may strengthen our behavior of confidence that something really exists as reality. Such reality, of course, remains subject to the sensory–neuron firings constraint—as Stephen Hawking, Leonard Mlodinow, and Larry Fraley pointed out—that invariably limits what we can *know* about reality. We still can only neurally behave reality. So let's move on from this topic and take a little look at robotics.

Robots and Robotics

So far we talked about reality, including some about the robotic perspective. Now we talk a little about robots and robotics. The robotic perspective and robotics are related but not identical topics. And unlike the nature of reality, the nature of robotics is not so "surprising."

The term *robotics* refers to the study of "life forms" that we call *robots* that arise naturally through the conditioned behavior of other life forms (e.g., humans) rather than through biological evolution. However, in a vital scientific sense, the term *robot* applies to *both* of these life–form types, a finding that results from comparisons of these two types of life forms. The most meaningful comparison concerns the origins of both of these life forms occurring completely naturally under the Law of Cumulative Complexity. Thus both life forms react to energy changes in a way appropriately described as robotically— with no differential or unequal judgment implied—even though this term originated around one of these life forms. This "robotic" connection certainly helps the other life form (i.e., us), pre–scientifically conditioned to see its status as somehow unique and above the laws of nature, to follow more closely now that ancient dictum, mentioned many chapters back, to "know thyself." Understanding how the naturalism of robotics applies to humans certainly helps us know our humanity, our human nature, better. We also entertain some implications that these comparisons hold for the building of robots.

While we could discuss robotics from any of our available perspectives— agential, environmental, or robotic—we derive our discussion of robotics from our standard, scientific environmental perspective. We thereby avoid any terminological confusion that could stem from trying to deal with both robotics and the robotic perspective at the same time. More importantly, trying to address robot construction along with the prediction and control of robot behavior from the agential perspective is as doomed to failure as is trying to predict and control human behavior, or the behavior of any other biological robot (i.e., organism) from the agential perspective, as we have regularly noted.

So the term *robot* applies both to life forms arising through biological evolution and to life forms arising through engineered construction. Both life– form types are entirely natural, complex clumps of organized chemical matter. Both *operate* through entirely natural processes. Both *arise* through entirely natural processes, namely biological evolution and engineered construction. Both of these construction types also constitute entirely natural processes, although engineered construction depends on the extent of the conditioning processes that biological evolution made possible (e.g., so far only humans, but not rabbits, construct robots). And both life–form types rate the label *natural organism,* regardless of whether they are biologically evolved or factory engineered. We can even see Mark Twain on this same track when (in his anonymously published last book *What Is Man?*) he supported the view that circumstances entirely determine an individual's life, that humans are biological machines. We simply add the clarification that humans lack inner agents. On the same wavelength, science fiction writers regularly treat machines as human, and posit evolving machine cultures lacking any further presence of the engineers—biological or machine—that originally built their earliest cultural members. Nothing in nature says that such notions must be wrong. Indeed many laws of nature provide the basis upon which to predict that such notions may prove right. Current contingencies, however, (even beyond those supporting organized superstition) may or may not prove adequate to the task of leaving citizens emotionally comfortable with these possibilities.

So, how is a robot, particularly a non–biological robot, made? Leaving structural considerations to others (simply because our science of function has less to say about these considerations) professionals in the field of robotics, whom we can call *designers,* have been at this task for decades, and continue to make progress. However, they also regularly run into conundrums. Without taking the natural science of behavior into account, the traditional contingencies under which they labor still wreak some havoc by inducing further doomed–to–failure attempts to replicate the spontaneous workings of inner agents. Others emphasize gradually successful attempts at more and more comprehensive *pre*programming to determine a robot's functioning.

Overall our elucidation of the laws of behavior and how they work with biologically evolved robots provides the kind of input that can make much easier the task of engineering the kind of functioning life form that we would traditionally call a robot. When basically familiar with behaviorology, some robot designers have seen, and others may well see, the value of designing for operant–style, as well as respondent–style, conditioning to determine a robot's functioning, rather than for preprogramming. Designing for the energy feedback mechanisms of operant conditioning will probably prove particularly promising. From that point the discussion can considerably broaden. The value of stimulus–control processes (e.g., generalization and evocation) quickly becomes clear, as these processes necessarily interact so thoroughly with operant feedback mechanisms.

Fraley (2008, pp. 1547–1563) extends the behaviorology–for–robot–construction discussion in several areas. These include energy sources, reproduction, pre–installed experience, planned diversity, social life, quality of life considerations, and even providing robots with respondent emotions, events that, by design, would affect the intensity of their current responding in the same way that emotions affect the intensity of a human's current responding. As more researchers and robot designers, particularly more thoroughly behaviorologically informed researchers and robot designers, become involved in robotics, developments can only expand further.

Conclusion

We should be clear that examining robotics and reality, especially reality, may prove at the moment to not be much of a practical matter. These represent more a matter of acknowledging some greater extents of behaviorology, the science of the contingencies responsible for behavior, especially human behavior. Whatever the status of reality, and however we know it or cannot know it, we daily deal successfully with "reality," something which our continued existence shows. While our reality conclusions may presently bear only on some aspects of the robotics field, only time will tell whether or not either of these produce any other practical benefits. For instance, we may have to recognize that how our improved natural–science understanding of reality and robotics might benefit solutions to global problems remains an unanswered question. Perhaps behaviorology's more accurate and complete, natural–science account of behavior, including scientific behavior, will help produce even more accurate and complete accounts in the subject matters of other natural–science and engineering disciplines. Perhaps contingencies will induce some readers to take the next steps along these lines even while the rest of us implement solution aspects from the other parts of this book, other parts of behaviorology's contribution to helping the natural–science team efforts to solve global problems. This is the topic of the next, and last, chapter.✧

References (with some annotations)

Fraley, L. E. (2008). *General Behaviorology: The Natural Science of Human Behavior.* Canton, NY: ABCs. See Chapters 29 and 30 for more on reality and robotics.

Hawking, S. & Mlodinow, L. (2010). *The Grand Design.* New York: Bantam. Especially see Chapter 3.☙

Chapter 27
Adding More Scientific Answers to Current Human Concerns:

Where is humanity going, and can we get to a good place? Consider overpopulation and sustainability as well as behaviorology contributing to the natural–science solutions to global problems. …

We began the *Introduction* chapter (i.e., the Preface) with an observation to which we can finally return. During the last decades of the twentieth century, traditional natural scientists (e.g., physicists, chemists, and biologists) were turning their attention to solving the many major (and minor) problems around the globe. In this period they have increasingly realized that these problems and their solutions extensively involve human behavior. Thus both the problems and especially the solutions must also involve changes in human behavior. Yet most of these natural scientists were generally unaware of the scientific contingency causes of human behavior. They called for the establishing of a science of behavior, as they were unaware that a now over 100–year–old natural science of behavior, called behaviorology, was available to address its part in the solutions. The centenary year of behaviorism—2012—provided an occasion to review this discipline for them, if ever so briefly, in the form of an abridged article, "Behaviorism at 100," in the first issue (January 2012) of the centenary volume of the journal *American Scientist,* available at www.americansccientist. org. Expanding on this article, the unabridged, peer–reviewed version appeared in volume 15, number 1, of the journal *Behaviorology Today* two months later, available at www.behaviorology.org (Ledoux, 2012a, 2012b). To support the teamwork of all the natural sciences in solving local and global problems, this book has provided some basic elaboration of the contingency science and engineering behind these articles. Now, here in our last chapter, we consider more closely the relation of this science to solving global problems.

We must first recognize that, because contingencies control all behavior, either through positive controls or through coercive controls, misuse of contingencies is possible, for instance, to abuse or suppress people, or for purely personal gain. However, making knowledge of *all* the laws of behavior, both positive and negative, available to everyone constitutes perhaps the most valid resistance to the misuse of these laws. This provides a major reason for founding and expanding programs and departments of behaviorology in our colleges and universities. What does your local college or university offer? If you think it too little, you have a right, as a tax payer, to ask for more.

Such programs and departments would also enable traditional natural scientists to become more familiar with behaviorology, thereby assisting their team efforts to solve these problems. Instead, many—perhaps most—have experienced exposure mainly to popular cultural and academic views on human nature and human behavior of the kind that fundamentally mystical disciplines offer. These disciplines include some that even use scientific methods to support claims of mystical origins of events.

Some traditional natural scientists attempt to rescue the human nature and human behavior topics from theological and secular mysticisms by trying to shoehorn them into strictly evolutionary, genetic, or physiological accounts. This step deprives them of the benefit of *access* to the contingency engineering needed for their team efforts to solve global problems.

Exposure to popular mysticism can lure other natural scientists into accepting and repeating mystical accounts when they venture beyond the range of their disciplinary training. Yet other *natural* scientists have made substantial progress on behavioral fronts. When traditional natural scientists become more aware of this progress, the accuracy of their own work expands and the risk of resorting to mystical accounts shrinks. Previous chapters have provided some general coverage of this progress, some of which we will touch on again here.

Two Major Cultural Contributions of Behaviorology

Consider two related contributions to the culture that behaviorology supplies. One concerns replacing the problematic inner–agent accounts of religion and psychology with natural–science accounts. The other concerns ending the unnecessary and counterproductive tradition of polite but actually now dangerous compromises between religion and natural science. These two contributions help us apply the more parsimonious, experimentally based findings of behaviorology to its share of more focused contributions toward the solutions for global problems.

Replacing inner–agent accounts. When some philosophers began psychology, they maintained the theologically grounded inner–agency category to which they attributed many phenomena including human behavior. In essence psychologists secularized the theological *soul*, separating it from its theological history by renaming it, calling it the psyche (thus, "psychology") and then calling it the "mind." Later they added "self" and "person" (and other labels) to the list of essentially equivalent but still unmeasurable agents.

Those behavior–initiating, body–directing, inner self agents, and all their related and still mystically based sub parts and processes, continue formally in psychology today. This happens in spite of the nearly century–long accumulation of evidential material, from natural behavior science, undermining any support for their continued assumption.

That is a discipline–level problem. Some individual psychologists would just as soon see inner–agent accounts dropped in favor of natural–science accounts. But they do not openly or obviously seek, individually or collectively, this kind of change in their psychology discipline. Thus they share its non–natural status. The retention of non–theological inner agents as core explanatory concepts in psychology confirms its status as a *secular* mysticism. Its commitment to those agents makes it the leading light for a range of similar secular mysticisms in the form of other non–natural disciplines. These usually operate in social–"science" departments due solely to their interest in people, since they cannot qualify for natural–science status.

The contingency accounts in behaviorology replace these inner–agent accounts. And behaviorology operates most appropriately in college natural–science administrative units.

Reducing dangerous compromises. Behaviorology is also well positioned to clear the air about the mistaken compromises that can arise due to those two inner–agent sources. These compromises began several hundred years ago when religious authorities yielded the physical world to the emerging proto–scientists of the time. In return these proto–scientists yielded life, human nature, and human behavior to religious authorities. These proto–scientists assented to this compromise, perhaps observing the mistreatments of Giordano Bruno and Galileo. (We say "proto–scientists," because the words "science" and "scientist" arose only after another hundred years or more.) Later, in light of scientific advances, religion grudgingly yielded the subject matter of life to science. But science still assented to the continuing compromise of religion hanging on tenaciously to human nature and human behavior with claims that science cannot address these topics. We have already seen the falseness of these claims.

Today, we see those compromises as unjustified, because all real subject matters, including human nature and human behavior, are amenable to study by science. The historical jury is still out, though, regarding the outcome of the current struggle, between the cultural forces of the natural sciences and the cultural forces of superstition and mysticism, over whether or not to continue the unjustified compromises between them. We may also judge such compromises as counterproductive as they provoke debates that delay action. But we need action on global problems, and we are running out of time. These debates possibly put human survival, above subsistence minimums, itself in jeopardy due to the behavior components inhering in both global problems and their solutions. As natural behavior science, behaviorology can account for and deal with human nature and human behavior more accurately and effectively than religion can. So supporting behaviorology increases the chances that humanity will be able to survive by solving these problems within the appropriate time frame before we must suffer their worst effects. Successful solutions could prevent another impending Dark Age, or worse.

Some circumstances can easily make those kinds of compromises look appropriate, which can compel our falling for them. For example, our culturally conditioned sense of fairness makes agential accounts and natural–science accounts seem mere equals, which can make compromises seem appropriate. However, these two really are grossly unequal. The basis of both of these sets of accounts are assumptions, theological–style assumptions for psychology's inner–agent accounts, and natural–science assumptions for behaviorology's environmental–contingency accounts. However, recall that no one can prove or disprove assumptions. Theologians and psychologists induce mystical assumptions off–the–cuff or, alluding to tradition or common lore, from various kinds of beliefs. In contrast, scientists induce natural–philosophy assumptions from the consistency of the results and developments over several hundred years of accumulating experimental evidence and effective engineering practices that virtually everyone hopes would remain available. These two types of assumptions, naturalistic/scientific and mystical/superstitious, are indeed *not* equal in any meaningful way. Humanity can no longer afford the delays in global problem solving that compromises between them can induce.

The contingency accounts in behaviorology reduce the appeal of unneeded, and now dangerous, compromises. This is especially apparent in the contributions behaviorology brings to the team efforts to solve global problems.

Contributions summary. We have touched upon but two major contributions to the culture that behaviorology supplies. In the first we examined the problematic inner–agent accounts of religion and psychology so that we can replace them with natural–science accounts. In the second we recognized as unnecessary and counterproductive the tradition of polite but actually dangerous compromises between religion and natural science. We can instead move forward, no longer burdened with dangerous compromises, to apply the more parsimonious, experimentally based findings of behaviorology to its share of contributions toward solutions of global problems.

The range of behaviorology's contributions, of course, extends well beyond those two. In earlier chapters we touched on some areas in this range. Before extending the range further regarding global–problem solutions, we should clarify the confounding difference between natural science and social "science."

Natural Science and Social Science Defined

Several commentators (e.g., McIntyre, 2006) have put forth calls for the development of a natural science of human behavior (like behaviorology). The difference between natural science and social science is relevant to understanding the place of those calls. We will follow this with an example that can show the crucial cultural value of a fully supported behaviorology.

The main difference between natural science and social science pertains to the definitions of each, which need not be mutually exclusive. People define natural science and social science in various ways. Rather than review the range of definitions, here are workable definitions that avoid some all–too–common impolite comparisons, and instead highlight some positive aspects that help us sort out these two (see Ledoux, 2002). We define *natural sciences* as disciplines that insist on dealing solely with natural events (i.e., with real variables, independent and dependent, in nature) using scientific methods. On the other hand, we define *social sciences* simply as disciplines that are interested in people. Some social sciences also use scientific methods, but very few of these "deal solely with natural events." Those that do "deal solely with natural events" thereby meet the definition of natural science as well, and we should designate them as such. For example epidemiology, which is basically the scientific study of disease patterns, especially epidemics across a geographical area or society, is an offshoot of biology that deals solely with natural events using scientific methods; so it is a natural science. Yet it is clearly interested in people, and many universities teach it out of a social–science department.

We can discern a similar pattern for behaviorology itself. Behaviorology scientifically studies behavior, especially human behavior, also as an offshoot of biology. It also deals solely with natural events using scientific methods. So we are correct to call it a natural science. Yet it is clearly interested in people, and sometimes it too is taught in a university social–science department. Thus also calling it a social science might in principle be acceptable. But actually trying this has produced negative outcomes. One college administered a solid, 14–course behaviorology undergraduate major equivalent—documenting the courses as natural–science courses—from within the social–science department. When the lone behaviorologist retired, four psychologists killed this effort.

What is curious is that calls for the development of a natural science of behavior have come as much from social scientists as from natural scientists. These other researchers, who made these calls, were apparently also unfamiliar with the development, even existence, of behaviorology. Perhaps this happened because, prior to its declaration of independence in the late 1980s, behaviorology was hidden under a label (i.e., behavior analysis) that a non–natural discipline (i.e., psychology) claims. Furthermore, the general lack of readily available access to any *separate* natural science of behavior could have prevented researchers from contacting it. As one example of such a call, Lee McIntyre, a researcher at the Center for Philosophy and History of Science at Boston University, published a book–length call in 2006 for a natural science of behavior (McIntyre, 2006) entitled *Dark Ages—The Case for a Science of Human Behavior.*

With the natural science of behavior, now as behaviorology, being 100 years old, such calls are not really necessary. What is necessary is the end of theology's and psychology's continued attempt to monopolize the cultural knowledge about human nature and human behavior. Colleges and universities need to end this monopoly by embracing behaviorology departments and programs, at least on a level playing field. On that kind of field, science can, as it has before, show how its value for the culture is greater than the value of superstition and mysticism, theological or secular.

Perhaps such calls are justified by helping to level the playing field and so make behaviorology more available. Consider an example of potentially crucial cultural value. This example concerns technologies and their use, like birth control technologies that can humanely aid the kinds of population reductions our world needs if we are to improve the chances of human survival at a level above subsistence minimums. The still growing problem of the increase in greenhouse gases, regardless of whether or not these arise from human activity, leads to global warming, with its increasingly dangerous changes in climate and weather patterns that could effectively destroy our global living standards over the next 100 years. While we environmentally try solving these problems with sustainable lifestyles, the problems themselves remain among the indications that *we have exceeded the carrying capacity of the planet.* This includes the planet's capacity to provide not only enough food, clothing, shelter, and so forth, for all the people, but also enough resources for the range of other planetary animal and plant species that are so necessarily intertwined with human survival.

To return to levels of resource need and use that are compatible with the planet's carrying capacity, perhaps one of the most important steps, if not *the* most important step, is to reduce *in humane ways* our overall population level. The broad range of widely available birth–control technologies (e.g., various practices, compounds, and devices), and the various behavioral technologies that could support their use, would contribute much to our overall success. This is yet another example of the behavior components that suffuse our global problems and the apparently required part that our natural behavior science must play in the solutions.

Even though those birth–control technologies are available, who will use them? As a result of their past conditioning histories, many people throughout the world, perhaps at this point a majority, exhibit a hard–core, mysticism–based resistance not only to the birth–control technologies but also to their use and to the behavioral technologies that would enable their easier adoption. Indeed, this superstitious resistance even extends not only to our natural behavior science but also to natural sciences in general, even though the resistors cannot live without the products of science. These may thus be safe from the destruction that those resistors sometimes talk of visiting on the culture of science. But is that the only threat to science and its products, many of which make the very development of sustainable lifestyles possible?

The risks are large and looming! The successful application of the technologies of birth control, and the sciences behind them, could make a meaningful contribution toward a smaller world population, one compatible with the planet's carrying capacity, while also reducing one of the greatest single sources of greenhouse gases. However, if the resistance to these technologies and sciences succeeds in preventing their use, then this resistance may by default succeed in preventing the overall and lasting solution to global problems. From such an outcome, we and our children's grandchildren for many generations to come—if our species does not succumb to an earlier than predicted extinction—may end up experiencing another thousand–year Dark Age in which science is again lost. Then people must gradually and painfully rediscover science, as old and new varieties of mysticism and superstition continuously abound while whoever is left chases rats with bats to survive.

Is that a kind of alarmism? Perhaps it is. But what if, too late, it turns out to be correct rather than alarmist? Contingencies have left me leaning toward the side of solving problems faster than might be needed, emphasizing prevention while avoiding "too late." How about you? Are we already too late?

For instance, you may recall the importance we gave, and which many still give, to efforts to achieve "Zero Population Growth." The general global culture has made quite a bit of progress along these lines, quite a bit but not nearly enough. Indeed one can argue that *mere* zero population growth is today no longer a viable option. Instead we really must somehow practice "Humane Population Reduction" on the broadest possible scale with every humane practice and technology at our disposal (i.e., no bombing or poisoning or other form of mass murder of whole groups, societies, or cultures). This is an especially important part of solving, as in mitigating, all interrelated global problems in a manner that helps keep us within the still shrinking time frame that is available before we must experience the worst effects that these problems will throw at us. We want to minimize how much we will still have to suffer, when this time frame ends, while still adapting in whatever ways may still be available to us.

Humanity has already made some attempts at humane population reduction. For example, whatever the process, problems, or successes of China's one–child policy, we really must acknowledge not only the foresight but also the appropriateness of their attempt. And then we must coordinate even more successful policies. We need policies successful at reducing the superstitious resistance to the birth–control practices of humane population reduction. And then we need policies successful at implementing those practices, along with all other humane practices, to bring about substantial reductions in the world population relatively quickly (e.g., in two to three generations).

Remember that, essentially, overpopulation makes every other global problem worse. If humane practices to reduce overpopulation fail, other natural processes can and will *inhumanely* succeed at population reduction, through famine, pestilence, war, and an increasing number and variety of natural disasters of increasing proportions (e.g., size, destructiveness). But events need not develop that way, especially if we increasingly take into account the behavior components that suffuse our global problems—of which overpopulation is but a particularly important example—by accepting the essentially required part that our natural science of behaviorology must play in the solutions.

Dangers of Unbalanced Educational Playing Fields

Unbalanced educational playing fields present additional dangers. Consider the resources lost when educational programs remain disrespectful of science. Societies that allow such programs also automatically allow wasted research funds to investigate what is not really there, while mis–training legions of students to replace the professionals expended on the crusade. All this diverts resources that could otherwise support the development and dissemination of the findings and applications of natural science, including behaviorology.

That leaves resources in short supply to replace the fictional inner accounts for human behavior with scientific accounts. The importance of providing those resources grows exponentially as efforts to mitigate environmental destruction languish for lack of widespread access to natural behavior science. The traditional natural sciences intuitively recognize these resources as vital for the team effort necessary to solve global problems due to the extensive behavioral components in both the problems and their solutions.

Some examples of scientists recognizing the behavior connection can help. Dai Qing, a Beijing–based water policy specialist, expressed concern about the chronic water shortages for Beijing's 17 million people. She said, in a *USA Today* article (2008 June 20, p. 9A), that there "will never be enough unless the citizens of Beijing change their behavior and water usage." Also, articles in publications of the *Union of Concerned Scientists* regularly delineate the kinds of behaviors needed to achieve solutions to some problems. An article under the title "Climate Action in Your Hometown" (*Catalyst*, Spring 2008, p. 20) mentions *launching* a local shuttle service (and getting people *using* it), and *conserving* energy, and citizens *agreeing* on, and *participating* in, solution development and follow through. All of these involve behavior, and efforts to increase them benefit from a behaviorological analysis.

Perhaps most important are calls that traditional natural scientists make for a natural science of behavior, as part of ending the grip that the forces of organized superstition have on the culture, especially with respect to human nature and human behavior. Authors like Richard Dawkins, who wrote *The God Delusion,* have argued persuasively that behavior must be approached naturalistically. But again, such authors often seem unaware that what they seek is already largely available through the behaviorology discipline. The continuing series of court cases over the teaching of evolution in biology classrooms also all too regularly reminds us of the difficulty of winning the cultural battle against superstition. Entrenching behaviorology thoroughly among the natural sciences provides a potent, pertinent, and practical partner in these struggles with superstition. (For a related and enlightening exposition, see Skinner's 1971 book, *Beyond Freedom and Dignity,* especially Chapter 1.)

In the mean time, we are running out of time. Way back in 1972, scientists at MIT released the book, *The Limits to Growth* (Meadows *et al*, 1972; also see Hayes, 2012, for an update). This book examined many of today's worrisome environmental parameters while addressing how much time we had before the human population outgrew the carrying capacity of our world and its resources. They concluded about 100 years. That was over 40 years ago, and the parameters they examined have worsened faster than expected. In 2012 Megan Gambino quoted Dennis Meadows, one of the book's authors, as saying that already, "We're at 150% of the global carrying capacity." That cannot and will not last very long; the percentage is climbing. We are quickly overspending our food, water, air, and other resources, and nature's accounting will soon cost us dearly as these resources run out and world population drops drastically from losses as masses of people die of starvation, thirst, disease, war, and so on. Any chance we may still have to avoid or reduce such scenarios depends on our increasingly and comprehensively dealing *scientifically* with human behavior.

Extending Reinforcers Through Morals to Survival

We previously discussed the interconnected series of reinforcers, values, rights, ethics, and morals. Here we can review, apply, and extend this series to our concern for sustainability and human survival. Our reinforcers are or become our values. The stimuli in various contingencies evoke our claims that we have rights regarding unrestricted access to these values, these reinforcers. Next, ethics involve the respecting of our rights claims for unobstructed access to our valued reinforcers. And our conditioning evokes our applying the label *ethical* to those who respect our rights claims, and further evokes our applying the label *ethical behavior* to their behavior of respecting our rights claims. Then, we speak of morals when, through processes like generalization and other conditioning extensions, ethical conditioning reaches the point at which ethics become abstract, that is, become morals, involving stimulus characteristics that cannot exist alone. This enhances enforcement potential but at the risk of damage to our culture and survival through both insensitivity to reasonable exceptions and determined resistance to needed change.

Also, consider again the difference between unconditioned reinforcers and conditioned reinforcers, for this difference produces the dichotomy not only between absolute and relative morals but also between absolute and relative ethics, rights, and values. We consider unconditioned reinforcers as *inherently* valuable, based on their function due to genetically produced neural structures, and due to their role in species and individual survival. On the other hand, we regard conditioned reinforcers as only *relatively* valuable. We say relative because these stimuli, initially lacking reinforcer functions, gain the status of values only through the conditioning process that bestows a conditioned reinforcing function on them. *Without this process occurring, they lack value status.* So, for example, food and sex are inherent values while money and music are only relative values. Does that make food and sex *absolute* values?

Let's discuss that question. In affecting the status of values, that difference in the origin of various reinforcers flows through the whole sequential framework of these concepts, from values to rights to ethics to morals. In all of these cases, we can only consider those stimuli that are grounded in unconditioned–reinforcer status, and hence hold inherent–value status, as having any sort of claim about status as "absolute" values, and thus as absolute rights, absolute ethics, and absolute morals. This can induce some to argue that, as unconditioned and so inherent values, food and sex are indeed absolute values (and rights and ethics and maybe morals). Meanwhile other stimuli, being grounded in conditioned–reinforcer status, retain a status of relative, in the sense of arbitrary or conditional. If the conditioning happened, then the status of value—and hence of right and ethic and maybe moral—begins, but if no such conditioning happens, then the stimuli are not values, and so on.

However, are claims to absolute status absolute? Indeed not. While some stimuli that evoke *inherent* also evoke *absolute*, the term *absolute* invites inaccuracy. For example, the unconditioned reinforcer status of sex leads—and has long led—to the claim that procreation (i.e., procreative sex) is an absolute value forever necessary for species survival. In one sense this is correct, and some people argue that therefore the culture should encourage relationships that support procreative sex as the only legitimate form of relationship. But in another sense, the notion that procreative sex is an unquestionable, absolute value, necessary for survival, is dangerously wrong. Beyond debates about issues like free love versus the nuclear family, and heterosexuality versus homosexuality, we can see that procreative sex, rather than being an absolute value, right, ethic, or moral, is an absolute disaster. Consider simply that the current level of procreative sex among humans on this planet is leading to the demise of planetary life, including humanity. As we mentioned earlier, our human population is currently already at over 150 percent of the planet's carrying capacity. Such a condition cannot continue for long without portending disastrous effects of truly momentous, even obscene, proportions. Does this not make procreative sex (as opposed to other forms of sex) at least a serious if partial detriment to species survival? Humans will always produce enough babies for species survival. However, other, non–procreative forms of sex, between consenting adults of the opposite or same gender, currently contribute to our species survival by not producing babies and so humanely help reduce the population growth rate. The resulting survival benefits increase as a function of how quickly new culture–wide ethics encourage the broadest range of non–exploitative relationships with sexual practices unrelated to pregnancy. As we already pointed out, with *"zero* population growth" being a quite inadequate option, contingencies are inducing options like these.

Humanity has taken too few population–control steps too slowly. Shall we just wait until disasters destroy two or three billion of us? These running–out–of–time circumstances make stimuli that *reinforce alternatives to procreative sex*—thereby producing decreases in population reproduction rates—reasonable values, rights, and ethics to include in any expansion of green interventions. These interventions work for the survival not only of our species at this time but also of the many other species whose survival various web–of–life contingencies tie to whether our actions wreck or save the planet as a place safe for life.

Until a hundred or so years ago, procreative sex *was* actually a stable support of our species survival, but that is not now the case. Yet today, *under changed contingencies,* our most common and age–old cultural institutions still stridently strive to induce problematic feelings of guilt, shame, or sin over any other forms of sexual activity. But procreative sex now works against our survival by leading to ever higher population levels. Can contingencies change us to increase support for the long and respected history of family planning, and even improve on it by allowing, even encouraging—and without religious persecution or secular prosecution—widespread use of birth–control technologies and treatments and, again, various forms of sex including those between consenting adults, such as forms between same gender or opposite gender couples that do not produce babies? If we fail to reduce population levels *humanely*, then disasters, like those induced by global warming, *will* reduce our population *in*humanely!

Note that the dangers of *moral* pronouncements, which we already discussed, prevented my including such a pronouncement in those comments on the current survival value—and rights and ethics—of alternatives to procreative sex. Whether or not—or how much, for how long—moral status accrues to those non–procreative sex alternatives remains a question of naturally occurring ongoing cultural conditioning.

Beyond the interplay of sex and survival, think about how developing the behaviorology discipline enhances our human potential to deal effectively with the values, rights, ethics and morals of clean air, fresh water, a healthy atmosphere, an intact ozone layer, pesticide–free food, safe transportation, enabled recycling, sustainable living, a population level within the planet's carrying capacity, and so on. For example, contingencies that induce analysis and adjustment of our routine cultural arrangements will condition the effectiveness of green values like the reinforcers of lifestyle sustainability at a lower population level. These then routine contingencies can (a) better condition effective rights claims regarding all components of a healthier planet, and (b) better condition the ethics of respecting these rights claims. However, we should remain aware of the possible dangers of our routine cultural contingencies conditioning too much moral status for those ethics. Such moral status would tend to render the ethics too resistant to change when circumstances again change. For instance something better than, but incompatible with, current recycling efforts might develop, but excess moral servitude to present recycling methods might delay or prevent implementation. Thus, overly moral conditioning could contribute to contingencies that work against humanity, in ways similar to the ways in which rules become less helpful once the contingencies, which they state or describe, have changed.

Interdisciplinary Developments

We can share and apply more than morals about what we have discovered regarding human nature and human behavior. We may finally be arriving at some scientifically, as well as emotionally, satisfactory and difference–making answers across science disciplines working to solve human problems.

Contributions to green behavior. Based on its informing philosophy of radical behaviorism, and beyond experimental and practical contributions in general, behaviorology makes other important contributions. Some of these pertain to the capabilities of traditional natural scientists. One major current area involves working together on *behaviorological green engineering* projects, an area we can also call *green contingency engineering* (e.g., working on overpopulation concerns as a foundation for achieving sustainable lifestyles). Behaviorological scientists and practitioners already work in this area. Again, so many of the seemingly intractable problems facing humanity today involve problems of human behavior as much as problems of physics or chemistry or biology. Examples include out–of–control population levels, increasing climate extremes from global warming, water and air pollution, potable water depletion and the rising risks of water wars, resources depletion, and loss of species through higher extinction rates due to habitat destruction and so on.

The *solutions* to those intractable problems also involve human behavior. A special section in the fall 2010 issue of *The Behavior Analyst* begins to address this consideration with ten articles on "The Human Response to Climate Change." The wide range of topics in this special section is evident from the article titles, which the Supplemental References provide for you on the last page of this chapter. Later, Grant (2011) extended these topics with good data on the negative effects of overpopulation and consumerism. While he relies unnecessarily on agential terms, something that you can take as the verbal shortcuts that they must be, he includes a range of positive and broadly scaled solution activities that go beyond individualistic interventions.

After the introductory remarks in that special section, paleo–climatologist Lonnie Thompson (2010) sets the stage for the other papers with his article entitled "Climate change: The evidence and our options." After reviewing the evidence and discussing the relative merits of mitigation, adaptation, and suffering, Thompson stresses the connection between human behavior and global problems, and their solutions. He concludes that "There are currently no technological quick fixes for global warming. *Our only hope is to change our behavior* in ways that significantly slow the rate of global warming, thereby giving the engineers time to devise, develop, and deploy technological solutions *where possible*" (p. 168, emphases added). Similarly, Douglas Larson makes the behavior connection when discussing the problems and solutions regarding Devil's Lake in North Dakota (see Larson, 2012).

Others have also made the crucial behavior–connection point, in some cases even earlier than Thompson. For example, in a 2007 speech, Frederick A. O. Schwarz Jr., the 17–year leader of the Natural Resources Defense Council (NRDC), said, "Global warming is the greatest threat we face, but it is not the only threat… Too many wild places are disappearing, too many species are being snuffed out, and too many babies are being born with bodies and brains damaged by man–made chemicals and pollution… To win [these battles]… *we must change how people think—and how they act*" (Schwarz, 2008, p. 60, emphasis added). In acknowledging the importance of changing people's behavior as part of solving world problems, Schwarz was implicitly encouraging the traditional natural sciences to coordinate with an effective natural science of human behavior in green engineering efforts and the movement toward sustainable lifestyles.

Completing such tasks must be a team effort. The players are the natural sciences of energy, matter, life forms, and life functions (physics, chemistry, biology, and behaviorology) as well as all the natural science and engineering disciplines related to these, because the complex problems facing humanity, and hence the complex solutions, involve aspects of *all* these disciplines. Will we cooperate in time? In his paper Lonnie Thompson also pointed out, "… our future may not be a steady, gradual change in the world's climate, but an abrupt and devastating deterioration from which we cannot recover" (p. 165). As Thompson describes, we must mitigate the problems while that is still an option, or we will be stuck with adaptation and suffering. The message is clear. Time is running out for efforts to solve world problems, including developing programs to train more people in *all* the relevant natural sciences, including behaviorology, so that they can work more effectively on solutions.

How much time remains before we are stuck with adaptation and suffering? Research continually shows estimates to be overly optimistic. The more than 100–years that behaviorologists presumed, as they moved on formal independence in 1987, shrank to less than 100 years about a decade later, and more recently to 50 years *or less!* In a special article for Earth Day 2010—with the subtitle, "Want peace? Solve the energy crisis"—Walter Simpson (2010) makes this point: "Climatologist Jim Hansen said in 2006 that he believed we had just ten years to make substantial progress reversing current carbon dioxide emission trends or we would be unable to avoid the worst consequences of climate change" (p. G2). According to that math, we have until 2016. Hmmm…

However, the question can no longer merely be how much time is left to fix overpopulation and global warming before the worst effects overtake us. Various media reports can leave readers with the distinct impression that the worst effects are *already* beginning to overtake us. The resulting scenario could lead to some rather ultimate, previously mentioned results including a series of deep population reductions leaving survivors with another millennium–long dark age or worse. Our best tool, scientific knowledge, will likely all but disappear and need reinventing, while the many unhelpful mysticisms endure with perhaps only some changes in flavor. So asking "How long do we have?" retains little value. Instead the reality that, in any case, *"We are running out of time!"* must prompt us continually to move ahead on solutions.

Those solutions require all natural scientists to work together. "STEM" must include behaviorology. In part behaviorologists moved decisively for formal independence when they did, so that their science could contribute its share to the expertise and coordinated efforts needed to solve such problems within the necessary time frame. Under these circumstances, they considered that their *not* declaring independence, and instead spending much energy over many more, likely fruitless years in further efforts to change psychology, would be essentially an irresponsible mistake. In agreement, other natural scientists are welcoming behaviorologists to the roundtable of basic sciences for the coordinated efforts that solving major problems requires.

Further contributions to fellow natural scientists. The behaviorology discipline makes additional contributions to the capabilities of other natural scientists. As mentioned, after becoming basically familiar with behaviorology, scientists in many disciplines are more able to remain naturalistic in dealing with subject matters at the edge of, and beyond, their particular specializations. This reduces slipping into the compromising use of common, culturally conditioned, superstitious agential accounts. With behaviorology familiarity, they may also add desirable details to accounts within their specializations. For example, when natural scientists (e.g., Sam Harris or Michael Shermer) say that science accounts for morals and values, mentioning the controlling relations that behaviorology describes for these topics strengthens their point. Also, behaviorology provides the students of natural scientists with a natural–science alternative to the non–natural disciplines that most of these students must currently study when covering behavior–related subject matters.

For their part, other natural scientists can also help themselves by contributing to behaviorology in several ways. They can support the wider availability of academic behaviorology programs and departments located within their university or college natural–science units. They can also increase their own familiarity with behaviorology, which may be particularly valuable in the efforts to solve world problems, as we discussed in the earlier chapter on recombination of repertoires research. Our world needs these contributions now. They increase people's contact with behaviorology, which reduces or avoids the increased difficulty in solving problems that stems from culturally conditioned susceptibilities to behavior–related superstition and mysticism.

The need to locate behaviorology programs and departments within university or college natural–science units proves difficult to meet because, as a result of the historical circumstances of the origins of their discipline, many academic behaviorological scientists and engineers remain scattered among academic departments of non–natural disciplines. Traditional natural scientists can quickly help solve this problem by promoting the addition of behaviorology courses and programs in their own larger academic units.

For most people a meaningful amount of contact with behaviorology will occur when behaviorology is a requirement in high school science curricula along with the other foundation natural sciences of physics, chemistry, and biology. To achieve that, science teachers must have behaviorology courses available in their college training programs. To make those courses available, faculty to teach them must be trained in this discipline. And for *that* to happen, programs and departments of behaviorology need to become more widely established at colleges and universities. This would also generate increased development of basic research and behaviorological engineering applications, including those contributing to solving personal, local, and global problems on which natural scientists in general are already working together.

While Ledoux (1989/2015) reported the consensus among behaviorologists regarding some departmental curricula for various academic levels, one of the obvious places from which to grow behaviorology programs and departments is from within departments of biology, especially within strictly natural–science schools. Skinner (1963) recognized early in his "Behaviorism at fifty" article that the *natural* science of behavior was an offshoot of biology. As he elaborated the connection in *The Shaping of a Behaviorist* (1979, pp. 16–76) even though he was earning his doctorate through the psychology department at Harvard University in the 1930s, much of Skinner's work occurred with W. J. Crozier who headed the physiology section of Harvard's biology department and who had been associated with biologist Jacques Loeb. Both Crozier and Loeb not only emphasized studying the whole organism, including its movement (behavior), but they also emphasized studying the causal mechanism of selection, which Skinner subsequently adapted from biology and applied to behavior. While those activities essentially started this natural–science discipline, modern behaviorology now features its own level of analysis and, as we have seen, can stand alone on its own disciplinary merits. These disciplines complement each other, but are not logically dependent. Consequently, a biology department would be only a good temporary home for behaviorology.

Contributions to other disciplines. Other disciplines may also find value in behaviorology. Over the last several centuries, authors in many disciplines have written much that mentions causes for human behavior. Occasionally these writings invoke stars as causes. More often they invoke selves, or other inner agents, as causes. Only a rare writer has invoked contingencies, and then, while on the right track, contrary contingencies induce embarrassment, worries, and apologies. Yet human knowledge and effectiveness could grow in leaps and bounds in, for example, historical analysis or literary analysis, if new writings took behaviorological contingencies into account. This benefit likely holds for other disciplines as well. Consider paleontology, archeology, anthropology, botany, geography, zoology, linguistics, ecology, and so on. Indeed, naming a discipline that could not benefit from taking behaviorology into account seems rather difficult, since every discipline connects with human behavior if only through its practitioners being human. Understanding human behavior could help understanding in any discipline.

Intelligence, Consciousness, and Survival

Consciousness clearly can and has benefited humans, particularly as a primary process supportive of the verbal behavior through which we accumulate both individual and cultural archives of knowledge that extend our intellectuality. However, the age–old misconstrual of consciousness, as involving mystical inner agents, detracts from that intellectuality by helping superstition to go unchallenged. The age–old misconstrual of the nature of human nature, of life, of living, of behavior, and of consciousness behavior, might just represent the greatest analytical error of human history. The implications of this error compound across so many aspects of today's human culture, and threaten its very survival.

Coincidental contingencies have led a surprising array of disciplinary domains into superstitious, analytical dead–ends. During times when humanity's numbers posed little threat to environmental stability, the superstitious assumptions continued relatively benignly, and possibly with occasional short–term benefits. However, humanity's population has climbed well beyond the Earth's carrying capacity. The resulting environmental stresses currently, and predictably on into the future, present us with increasing numbers of survival tests at an increasing rate. Have we passed them? Can we continue to pass them? Do humanity and the rest of life have a future?

Current environmental data suggest clearly that humanity can no longer afford its traditional reliance on, and modern compromises with, superstition and mysticism. Culturally, our intellectual history provides the basis—the natural sciences—for understanding, solving, and preventing our problems. These sciences include behaviorology, the science that accounts for consciousness and shows that by itself consciousness cannot provide a complete solution; it is one factor, albeit a potentially helpful one, among many.

Behaviorology itself can help solve global problems. But how helpful can its existence be when superstitions are so thoroughly entrenched in the culture that a person can get advanced degrees in various superstitious disciplinary domains at most major universities, and yet rarely encounter behaviorology at these same institutions? Meanwhile even many natural scientists are not as yet under contingencies to assure that their own students contact the natural science of behavior, rather than the currently more prevalent mystical disciplines of behavior. They need instead to assure the availability of behaviorology courses and the necessary professor–producing programs. Will humanity become a victim of its own ignoring of these concerns, or a beneficiary of their resolution? Will we resolve our conundrum between science and mysticism in a timely enough fashion? Fraley (2008, p. 1098) puts the question this way. "Will we witness it [the resolution] safely from the secure perspective of an intellectual alternative that portends cultural survival or from the comfortably seductive perspective of an intractable mysticism that leaves us blissfully imperiled?"

With humanity, evolution has produced a species with extensive capacity for both consciousness and intellectuality. Are these enough to prevent the destruction to which coincidental contingencies are leading by producing collective behavioral mistakes? Without invoking inner agents, having written this book makes me think that I must think that they are enough, but I can certainly be wrong. They will be enough only if more humans, especially more natural scientists, come under contingencies to participate in expanding educational options for far greater numbers of humans in the sciences, especially, at this juncture, in behaviorology.

This science helps us understand ourselves and our shared place in the web of existence that we call life, and thus helps the realities of global problems evoke more effective solution behaviors among more people. Behaviorology experimentally studies and interprets human nature and human behavior, and provides the derivative engineering technologies for effectively addressing accessible independent variables in ways that bring about improvements in behavior. The contingencies of global problems are inducing us to implement these engineering technologies not only at home and work, in education and diplomacy, and in interpersonal relationships, but also in the global problem–solving applied–behavior fields of recycling, sustainable lifestyles, the management of dangerous asteroids, resource and biodiversity protection, dealing with overpopulation, and so on.

Beyond its immediate domain, behaviorology coordinates with other natural–science disciplines. It even overlaps with some of these, and understanding these overlaps helps build our interconnected perspective.

Beneficial Disciplinary Overlaps

Among the natural sciences, behaviorology is one of the foundation life sciences (along with biology) rather than one of the foundation physical sciences (such as physics or chemistry). The life sciences stretch across a continuum of analysis levels, from molecules to cultures. We find the sub–cellular and cellular levels of the organism at one end of this continuum. In the middle we find the level of individual organisms. And on the other end we find the level of groups or populations of organisms, especially cultures.

"Culturology." We use the name *biology* for the *sub–individual* disciplinary level of the life–science continuum. For the *individual* disciplinary level of the life–science continuum, we use the name *behaviorology.* However, no name has covered the natural–science, behavior–respecting *group or population* disciplinary level of the life–science continuum. Sociology might have worked, but attempts to turn it into a natural science (see Fraley & Ledoux, 1992/2015) remain unsuccessful. Another discipline, anthropology, contains a possible contender. This area features a natural–science philosophy of science, namely the cultural materialism of Marvin Harris (Harris, 1979; also see Vargas, 1985). However, no separate *disciplinary* name for a natural–science anthropology has arisen. So, since 1986 I have been using the term *culturology* as the label to fill this gap (Fraley & Ledoux, 1992/2015, p. 147). This label provides a conveniently short replacement for "anthropology informed by cultural materialism." Presumably, in due time, natural–science anthropologists will provide their preferred name for their discipline. Stay tuned.

Each of those three life–science disciplines studies functional relations at its own level of analysis. Biology studies the functional relations both in the history of species and in the physical and chemical processes of individuals from the sub–cellular parts to the whole organism. Behaviorology studies the functional relations between environments (both internal and external) and the behavior (both overt and covert) of individual organisms during their lifetimes. And culturology studies the functional relations in the behavior of social and cultural groups, particularly involving group–produced effects that can outlast the lifetimes of the individuals that make up the group (e.g., education).

The overlaps. Each of those disciplines, however, also overlaps somewhat with the others. Biologists and behaviorologists share interests in the physiological mechanisms through which the body mediates behavior, particularly purely neural behavior. Behaviorologists and culturologists, meanwhile, share interests in the operation of the laws of behavior because, while the same laws apply at both levels, outcomes can differ due to the complexity increment that comes from dealing with groups of interacting individuals rather than with single individuals. Furthermore, some applied fields (i.e., an area where one applies a foundation science discipline) of interest to behaviorologists, such as solving global problems, reside as well, if not more so, in the province of culturologists. Figure 27–1 illustrates the positions of these three disciplines along a life–science continuum.

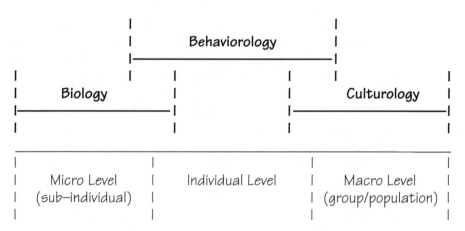

Figure 27–1. Disciplinary coverage for the three main levels of analysis in the life sciences.

The study of ecosystems, species evolution, and the behavior of animals in groups by some biologists points to a disciplinary overlap also between biology and culturology. So you might redraw Figure 27–1 as three intersecting circles. Try it. Each circle would represent one of these disciplinary domains, while the areas where the circles overlapped could then represent the shared–interest area of the intersecting disciplines. (See Chapter 6 of Fraley & Ledoux, 1992/2015, for additional details, including yet another diagrammatic option for indicating the overlap between biology and culturology.) These disciplinary overlaps provide further areas for applications. Let's revisit an important one.

Cultural—and Green—Engineering

Developing the behaviorology–culturology overlap helps apply it to the cultural–practice engineering that supports solving global problems. We call this area *green contingency engineering,* an area particularly relevant to many pressing issues including the humane reduction of population levels and the building of sustainable lifestyles. Perhaps the culture could currently derive the most benefits by first expanding behaviorology into this educational area, from which graduates could then extend it into the needed practical areas. Consider that a degree in *Green Behavior Engineering* (or *Behaviorology and Green Engineering)* includes basic coverage of the full roundtable of foundation natural science and engineering disciplines (e.g., physics, chemistry, biology, behaviorology) so that graduates can contribute to any and every area of solutions for global problems. We are sitting on the brink of a breakthrough to substantive successes in slowing global warming and solving global problems by building a more complete science and engineering team—a team with members from *all* the natural sciences—to address these concerns. How long will we merely sit on this brink?

We have perhaps been sitting on that brink ever since Rachel Carson's 1962 book, *Silent Spring,* appeared over 50 years ago. William Souder's 2012 biography, *On a Farther Shore: The Life and Legacy of Rachel Carson* (from Crown Publishers) takes us back to that time, refocusing our attention on *Silent Spring* as the origin of the movement to save and preserve our environment. (You can find Julianne Lutz Warren's review of Souder's biography on pages 146–147 of the March–April 2013 issue of *American Scientist* under the title "Crafting a narrative of care.") This biography also reiterates the legacy of controversy surrounding the movement from its very beginnings. The controversy is rooted in the notion that humans are supposedly the masters of nature, a notion that stems from the traditional but erroneous cultural view that we are somehow—usually agentially—above, or outside of, nature. This view continues to pressure us even as we finally set it aside and face current realities with the kind of humility that must be a part of long–term solutions.

Sitting on that brink, we have spent less time devising and implementing long–term solutions while spending more time arguing about short–term interests, an activity akin to fiddling while Rome burns. Can we now move beyond this brink, and work seriously—with all the relevant natural sciences including behaviorology—on the long–term solutions instead?

Education may comprise a helpful arena, at least one with less controversy. Educational campaigns, about steps that ordinary citizens can take to help solve global problems, will be vital components in the efforts to clean up and protect our planetary home. Currently these campaigns stress some of the crucial *behaviors* that contribute to solutions for these problems. For example the folks at the Environmental Defense Fund (EDF) broadly disseminate a range of materials. Among these are two lists of such behaviors, one containing ten "precycling" tips and the other presenting ten steps that fight global warming (with both lists still likely available at www.edf.org). Other organizations, such as the Union of Concerned Scientists (UCS) and the Natural Resources Defense Council (NRDC) provide similar and related materials.

To emphasize the importance of recognizing that *behavior* components comprise a major portion of the problems and their solution that demand our attention, components for which behaviorology provides the relevant natural science, here is a quick version of the ten EDF steps that fight global warming: (1) Recycle used materials… (2) Wash clothes in colder water… (3) Install low–flow shower heads… (4) Run the dishwasher when full, and without heat… (5) Replace standard light bulbs with CFL or LED bulbs… (6) Plug window and door air leaks… (7) Replace appliances with energy efficient models… (8) Walk, bike, carpool, or use public transport… (9) Adjust the thermostat seasonally… (10) Share these simple steps with others…

Such intuitively composed lists can only get better as contingencies increase the exposure of authors to more behaviorological knowledge and skills. For example each of those ten steps involves an explicit *behavior* (e.g., recycle, wash, install, and so on). Providing lists of behaviors, like that list, addresses the middle term in our fundamental three–term contingency regarding basic environment–behavior functional relations. Additional steps concern also addressing, orchestrating, and engineering relevant aspects of the first term of the needed contingencies, which covers the stimulus changes that *evoke* these behaviors successfully. And more steps concern addressing, orchestrating, and engineering relevant aspects of the third term of the needed contingencies, which covers the stimulus changes that *reinforce* these behaviors successfully. The culture then increasingly supplies the full contingencies that generate and maintain these behaviors as increasingly standard, even institutionalized, cultural practices. With such practices we can begin to reverse the retardation of human intellectuality that the age–old cultural dalliance in superstition breeds, particularly with respect to the theological and secular purveying of agential superstitions and the activities that these breed against the productive cultural practice that we call natural science.

EPILOGUE

*W*e have now come just about full circle. People's increased understanding of behaviorology makes them move to apply it to the widest range of humanity's concerns, using the behaviorological technologies that we can derive from the principles and concepts of this natural science of behavior. How can we help this happen? Basically, the next step involves you and me and other readers supporting, perhaps even agitating for (Dare I say "campaigning for"?) establishing more university behaviorology programs and departments. We all require these to meet growing needs so that soon an increasing proportion of our world population operates with an expanded, behaviorologically informed repertoire as a component of their general science education. This would lead not only to more traditional natural science in the education of behaviorologists, but also to more behaviorology—and less superstition—in the education of the general population and traditional natural scientists.

Why is that important? Recall the value of the interdisciplinary approach to the engineering of solutions for global and local problems that we covered in earlier chapters. This approach is not only valuable but necessary, because humanity is running out of time to solve these problems before their implications overwhelm us, forcing us to experience their worst effects. To design and implement such solutions effectively requires *all* the natural sciences to coordinate their efforts. And these sciences include not only the traditional ones focused on energy, matter, and life forms (e.g., the sciences surrounding physics, chemistry, and biology) but also the sciences focused on life functions (i.e., behaviorology and culturology). Given the acknowledged, substantial behavior components of global problems and their solutions, humanity needs all of these natural sciences working together, with reasonable familiarity with each other, if the solutions are to occur in a timely fashion.

That familiarity supports and enhances the interdisciplinary field of green contingency engineering. The culture, or at least cultural survival, puts this interdisciplinary cooperation at the forefront of efforts to solve global problems. Examples of concerns on which traditional natural scientists and behaviorologists work together, or even on which just *behaviorologically informed* traditional natural scientists work, include humanely reducing overpopulation (the necessary foundation for solving *many* global problems), establishing sustainable lifestyles, keeping the air and water clean, and preserving habitats and resources and species diversity, to name but a few.

In addition, go beyond the coverage of this book. Its real value might be in its setting the stage for your repertoire in the behaviorology discipline, and its applications, to expand beyond these essentially beginner's considerations. Delve deeper, studying and applying the full range of conceptual and practical details that grace the pages of behaviorological science journals and disciplinary textbooks. For example, see Ledoux, 2014, and check out the resources in the bibliography at the end of this book, and visit www.behaviorology.org.

Even if people have only a minimal conditioned repertoire in behaviorology, they would still be more likely to produce, or consider as viable, solutions to global problems that at least intuitively, or better, through design, take behaviorological realities into account. This holds particularly for those whom we call scientists, because they can also already apply a thoroughly conditioned repertoire in one or another traditional natural science. As a result the behavior–related components of the solutions could develop as reasonable behaviorological interventions with an increased likelihood of success, thereby supporting the physical, chemical, and biological solution components.

The alternative of continuing to stumble along with relatively little success, from attempting solutions that stem from *only* behaviorological natural science, or from *only* traditional natural sciences, or from superstitious cultural lore, could be reduced or entirely avoided. This would be a major benefit even if it only meant that solutions would not involve the usual knee–jerk suggestions such as "Make pollution (or whatever) illegal and punishable" (since nature will punish plenty for failures anyway). Instead, solutions would involve designing interventions that included new or enhanced reinforcing contingencies for behaviors consistent with improving environmental health. These contingencies should involve the long–term best interests of most people, and would help generate and stabilize the behaviors required to maintain environmental health. Indeed, as many more people gain more extensive behaviorological repertoires, and begin to apply them to global problem solutions, so much more is possible and will, I think (and we all hope) happen.

And so, dear reader, we approach the end of this book and this journey together, a journey ultimately related to concern for our planetary home. Our journey showed you something about behaviorology, the natural science of the contingencies for WHY human behavior happens, a natural science to help solve local problems, and global problems in a timely manner. Other journeys await.

Perhaps you are wondering why this book only really introduced the value of behaviorology for solving global problems in this last chapter. Why does this book really *only* point out the principles, concepts, methods, and practices of this natural science? Why did the book never spell out exactly *how* to apply each principle or concept of behaviorology to solve all those problems? The reasons are several; here are two of them: (a) The topic of how, thoroughly, to apply behaviorology, to cover its share of the efforts to solve global problems, requires several *more* books based on new research by new behaviorologists from those new university departments and programs that we must all demand. As yet too few behaviorologists are available for this work. Perhaps you will become one of the new ones. And (b) since thorough treatment of this topic extends well beyond my own expertise, and likely beyond the expertise of any single professional, it might best come from teams of authors, teams from *all* the basic natural sciences—including many new, doctoral–trained behaviorologists— working to solve these problems together. Possibly you will be a member of such a team. After all, *whenever* contingencies have compelled behaviorologists to address particular past problems, successful interventions have followed. Problems, whose solutions need broader teams of behaviorologists and other natural scientists, should similarly see solution successes.

In any case our current global problems, with their behavior components, loom in our collective face. Contingencies must compel enough of us to participate in the implementation of solutions through understanding and applying behaviorology as well as other natural sciences. Then we can together prevent humanity, and life on this planet, from running out of time. After all, contingencies are what cause human behavior, not stars or selves.♣

References (with some annotations)

Carson, R. (1962). *Silent Spring.* Boston, MA: Houghton Mifflin.

Fraley, L. E. (2008). *General Behaviorology: The Natural Science of Human Behavior.* Canton, NY: ABCs.

Fraley, L. E. & Ledoux, S. F. (1992/2015). Origins, status, and mission of behaviorology. In S. F. Ledoux. *Origins and Components of Behaviorology—Third Edition* (pp. 33–169). Ottawa, CANADA: BehaveTech Publishing. The original publication date was 1992.

Harris, M. (1979). *Cultural Materialism: The Struggle for a Science of Culture.* New York: Random House.

Gambino, M. (2012 March 16). Is it too late for sustainable development? *Smithsonian Magazine* (at http://www.smithsonianmag.com/science-nature/Is-It-Too-Late-For-Sustainable-Development.html).

Grant, L. K. (2011). Can we consume our way out of climate change? A call for analysis. *The Behavior Analyst, 34* (2), 245–266.

Hayes, B. (2012). Computation and the human predicament. *American Scientist, 100* (3), 186–191. This article considers the impact of later developments in computer modeling on the conclusions in *The Limits to Growth* (see Meadows *et al*, 1972).

Larson, D. W. (2012). Runaway Devil's Lake. *American Scientist, 100* (1), 46–53.

Ledoux, S. F. (1989/2015). Behaviorology curricula in higher education. In S. F. Ledoux. *Origins and Components of Behaviorology—Third Edition* (pp. 173–186). Ottawa, CANADA: BehaveTech Publishing. An extension of this paper, on curricular courses and resources after 25 years (1990–2015) of practical curriculum–development experience, appears as the "Addendum to Appendix 3" later in this same book (pp. 314–326).

Ledoux, S. F. (2002). Defining natural science. *Behaviorology Today, 5* (1), 34–36.

Ledoux, S. F. (2012a). Behaviorism at 100. *American Scientist, 100* (1), 60–65. *American Scientist* published this paper with introductory excerpts, on pages 54–59, that the editor listed as an *"American Scientist* Centennial Classic 1957." These excerpts came from: Skinner, B. F. (1957). The experimental analysis of behavior. *American Scientist, 45* (4), 343–371. (Skinner's complete 1957 paper was also available online with this paper.)

Ledoux, S. F. (2012b). Behaviorism at 100 unabridged. *Behaviorology Today, 15* (1), 3–22. This peer–reviewed version of the paper included the material that the *American Scientist* editor had set aside at the last moment to make more room for the Skinner article excerpts that accompanied the original article in *American Scientist.* Chapter 1 of Ledoux, 2014, expands this paper.

Ledoux, S. F. (2014). ***Running Out of Time—Introducing Behaviorology to Help Solve Global Problem.*** Ottawa, CANADA: BehaveTech Publishing. See *"...and Related Resources"* (back on p. xiv) for how to obtain this book.

McIntyre, L. (2006). *Dark Ages—The Case for a Science of Human Behavior.* Cambridge, MA: MIT Press.

Meadows, D. H., Meadows, D. L., Randers, J., & Behrens III, W. W. (1972). *The Limits to Growth—A Report for the Club of Rome's Project on the Predicament of Mankind.* New York: Universe Books.

Schwarz Jr., F. A. O. (2008). *Onearth,* Spring. The article was the author's speech on the occasion of his retirement after leading the NRDC for 17 years.

Simpson, W. (2010 April 18). Earth Day 2010. *The Buffalo News,* G1–G2.

Skinner, B. F. (1963). Behaviorism at fifty. *Science, 140,* 951–958.

Skinner, B. F. (1971). *Beyond Freedom and Dignity.* New York: Knopf.

Skinner, B. F. (1979). *The Shaping of a Behaviorist.* New York: Knopf.

Vargas, E. A. (1985). Cultural contingencies: A review of Marvin Harris's *Cannibals and Kings. Journal of the Experimental Analysis of Behavior, 43,* 419–428.

Thompson, L. (2010). Climate change: The evidence and our options. *The Behavior Analyst, 33* (2), 153–170.✤

Supplemental References: Special Section on the Human Response to Climate Change (in order of appearance)

Heward, W. L. & Chance, P. (2010). Introduction: Dealing with what is. *The Behavior Analyst, 33* (2), 145–151.

Thompson, L. (2010). Climate change: The evidence and our options. *The Behavior Analyst, 33* (2), 153–170.

Keller, J. J. (2010). The recycling solution: How I increased recycling on Dilworth Road. *The Behavior Analyst, 33* (2), 171–173.

Layng, T. V. J. (2010). Buying green. *The Behavior Analyst, 33* (2), 175–177.

Malott, R. W. (2010). I'll save the world from global warming—tomorrow: Using procrastination management to combat global warming. *The Behavior Analyst, 33* (2), 179–180.

Neuringer, A. & Oleson, K. C. (2010). Helping for change. *The Behavior Analyst, 33* (2), 181–184.

Pritchard, J. (2010). Virtual rewards for driving green. *The Behavior Analyst, 33* (2), 185–187.

Nevin, J. A. (2010). The power of cooperation. *The Behavior Analyst, 33* (2), 189–191.

Twyman, J. S. (2010). TerraKids: An interactive web site where kids learn about saving the environment. *The Behavior Analyst, 33* (2), 193–196.

Chance, P. & Heward, W. L. (2010). Climate Change: Meeting the challenge. *The Behavior Analyst, 33* (2), 197–206.℘

Appendix

*N*ote: Professor Fred Skinner provided this short article at the request of Stephen F. Ledoux and Carl D. Cheney. They were working on their 1987 book, *Grandpa Fred's Baby Tender, or Why and How We Built Our Aircribs* (Canton, NY: ABCs). The point of their book was to make accurate information on Aircribs more widely available. To support this point, Dr. Skinner contributed his article. The Aircrib design that their book followed was a new one that Dr. Skinner had composed and modeled but not actually constructed. Ledoux and Cheney placed Dr. Skinner's article as the Foreword in their book. Their book is now available as a free download (at www.behaviorology.org with more photos there than on the last page of this appendix). Again, to make accurate information on Aircribs more widely available, Skinner's Foreword also appears here. This helps clarify a development in the history of the natural science of behavior, and brings to the attention of a new generation of readers this enduring behaviorological contribution to childcare.❧

The First Baby Tender

B. F. Skinner

I designed what we called the baby tender as a laborsaving device. We wanted to have a second child, but my wife said she rather hated the chores of the first year or two. I suggested that we simplify the care of a baby. All that was needed during the early months was a clean, comfortable, warm, and safe place for the baby, and that was the point of the baby tender. I started to build it about the time we started the baby, and in spite of war–time shortages finished it just before our daughter Deborah was born.

As soon as she came home from the hospital, we put her in the baby tender. We discovered immediately that the labor we saved was far less important than the advantages for her. She slept on a tightly stretched canvas covered with a sheet (later replaced with a single plastic cloth that felt rather like linen). There were no nightclothes, sheets, or blankets, and she wore only a diaper. There was no danger that she would smother, as there occasionally is in a standard crib. She breathed clean air, which we humidified and maintained at just the right temperature. She was free of colds for many years, and I am inclined to think that it was due primarily to the warm humid air she breathed as a child. In the winter in a northern climate a house is about 30 degrees below body temperature, and the air the baby breathes is chilled further by evaporation from moist surfaces in the air passages. It is possible that the superficial layers of the bronchi and lungs grow as much as 40 degrees below body temperature, and that could make a great difference. The species originated in the tropics, where warm, moist air was standard, and there may not have been enough time for further evolutionary changes.

The space was quiet, and Deborah was free to move about and take comfortable positions at any time of day or night. She soon began to exercise much more vigorously than would have been possible in a standard crib, and she grew very strong. Our pediatrician commented on her unusual strength. Her skin stayed dry, and she never had any diaper rash. She never objected to being put into the baby tender and almost never cried.

Her rapid physical development was matched by behavioral gains. She was free to explore all parts of the space and there was a large window through which she could watch life around her. At one point she seemed to pass through a phase in which she used her feet prehensilely. Another couple who made and used a baby tender sent us a photograph of their baby holding its bottle with its feet while it drank. I made toys which Deborah used very early. By pulling a ring that hung from the ceiling, she produced a whistle. By twisting a T–bar that hung from the ceiling, she made small banners spin. Later, by pulling a ring she operated a music box, tone by tone.

She was not socially isolated. She was taken out for feeding and play, of course, and we could allow the neighborhood children to talk and gesture to her through the window without passing on their viruses. The labor we saved not only made it easy for us to treat her affectionately but encouraged us to spend more time with her. She spent a lot of time outside the baby tender, especially as she grew older. Eventually she slept in it only at night and for naps.

During her second and third years, when we could predict her bowel movements, she slept without clothing. Urine passed through the plastic cloth (which could be quickly washed and dried) into a tray to be thrown out the next morning. She learned to postpone urination, in part, I think, because of the consequences. Urination in a diaper is immediately followed by a pleasurable warmth; it is only after several minutes that a damp diaper grows cold and uncomfortable. Without a diaper urination immediately moistens the skin and chills it. Deborah began to go for long periods of time without urinating, and by the time she first slept in a bed she had learned to keep herself dry and never wet her bed. All the supposed psychological problems connected with toilet training were avoided.

I have seen many young people who spent part of their first years in similar spaces, and most of them were rather tall and strong. It would be extraordinary if those first years of rapid growth could have made that kind of difference, but it is certainly something worth exploring further.

The response to my article in *The Ladies Home Journal,* written when Deborah was nine–months–old, drew hundreds of letters asking where a "baby tender" could be purchased or how one could be made. I sent out hundreds of crude instructions. There were only a very few critical letters. I have never found anyone who, upon seeing a baby in an Aircrib, did not immediately think it was a wonderful idea. But misunderstandings began to spring up and were widely circulated. The *Journal* had given my article the title "Baby In A Box" and some of the misunderstanding came from a confusion with the equipment used in operant research. Misunderstandings are still common. Here is a sample from an article published by a reputable psychologist: "In the late '40s, Professor Skinner invented the 'air–crib,' a Skinner box for babies. It was a large, soundproof, germproof, air–conditioned box for giving children mechanical care for the first two years of life." Every statement in the passage is wrong. I designed and built the box in 1944. It is not an experimental apparatus. It is not soundproof; Deborah was shielded from loud noises, but we could hear her at all times. It is not germproof, although it was a kind of shield against sudden large doses of infection. "Air–conditioned" suggests cooling, but the air was only warmed. It is no more mechanical than a standard crib, and there was nothing mechanical about the care we gave our child. Deborah may have spent a bit more time in the Aircrib than she would have spent in a standard crib, because she was freer and more comfortable there, but in her second year she merely slept in it, at night and for naps. (Perhaps I should add that rumors that she committed suicide or became psychotic are equally wrong. Now 43 [in 1987] she is a happily married, talented artist and writer.)

It is possible to build a better world for a baby and the baby tender was a step in that direction.

B. F. Skinner
January, 1987.❧

Endnote [included in the 1987 book]: The Skinners' first daughter, Julie, is also happily married, is engaging in a successful career as a Behaviorologist, and has used an Aircrib with her own children (see www.bfskinner.org).

Endnote (for this Appendix): In a 1988 interview, Skinners' first daughter, Dr. Julie Vargas, cleared up several persistent misunderstandings about the aircrib. The other interviewees were Drs. Lawrence Fraley, Stephen Ledoux, and Ernie Vargas, Julie's husband. You can find the interview, which is entitled, "August 1988 Public Radio interview of the organizers of the first behaviorology convention," in (2013) *Journal of Behaviorology, 16* (1), 15–20. You can also order a CD recording of it through the TIBI website (www.behaviorology.org).

References (with some annotations)

Skinner, B. F. (1945, October). Baby in a box. *The Ladies Home Journal.* Also in Skinner, B. F. (1999). *Cumulative Record—Definitive Edition* (pp. 613–620). Cambridge, MA: B. F. Skinner Foundation.☙

Photographs

of the Aircrib as described in the 1987 book by Ledoux & Cheney

(more photographs in color are at www.behaviorology.org)

⇐ **The Aircrib,** and its controls ⇑, as described in the 1987 book...

⇐ **The Sleep Section** as described in the 1987 book...

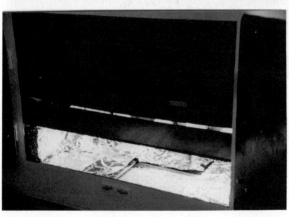

⇐ **The Climate Section** as described in the 1987 book...

Photos by Stephen F. Ledoux

Technical Glossary

Definitions of some behaviorology terms and phrases

Scientific technical terms and phrases prove valuable for readers as precise behavior–controlling stimuli. How does that happen? After appropriate conditioning the author's writing–behavior products condition a reader's expanded repertoire, which leads to the occurrence of more effective reader behavior. That is, the appearance of particular disciplinary terms or phrases affects the reader's nervous–system structure in ways that evoke a *particular* response or combination of responses or chain of responses to the terms and phrases. When a related set of scientific disciplinary terms evokes consistent responses from a group of people, that group defines a scientific disciplinary verbal community that shares a technical vocabulary. An accumulation of such defined terms and phrases results in a glossary.

This is a glossary of some of the most common terms and phrases that the scientific verbal community of behaviorologists uses at this time. These definitions may change as research developments accrue, and they may also differ somewhat from those used at different points in various chapters of this book. Such differences happen because, across the chapter coverage, topics expand through enhanced repetition and so accumulate descriptive parts into increasingly comprehensive definitions. In any case, and as an indication of the value of your expanding behaviorological repertoire, observe that both the whole *as well as the parts* of each of these alphabetically–listed definitions evoke appropriate and helpful responses.♣

ANTECEDENT — An event that occurs *before* some other event. (See POSTCEDENT.)

AVERSER — (See PUNISHER.)

AVOIDANCE — (See ESCAPE.)

BACKWARD CHAINING — The procedure of taking advantage of the DUAL FUNCTION OF A STIMULUS to build chains of responses by linking them with stimuli, that is, by strengthening the function of the evocative stimulus for a response and then using that stimulus, which has also become a conditioned reinforcer, to strengthen another response before it, and repeating this process *from end to beginning* until the chain is complete (with each stimulus functioning both as an evocative stimulus for the next response and as a conditioned reinforcing stimulus for each added "previous" response) and also by always proceeding to the end of the chain regardless of where in the chain a practice run begins.

BEHAVIOR — (See RESPONSE.)

BEHAVIORISM — (See RADICAL BEHAVIORISM.)

BEHAVIOROLOGY — Minimally, the natural science of environment–behavior functional relations and the beneficial interventions derived therefrom, both for understanding and improving both normal and problematic behaviors.

BEHAVIOROLOGICAL ENGINEERING PROCEDURE / INTERVENTION — (See PROCEDURE.)

BEHAVIORAL PROCESS — (See PROCESS.)

CHAINING — (See BACKWARD CHAINING.)

CLASS (See RESPONSE CLASS or STIMULUS CLASS) — A group the members of which share something in common.

COINCIDENTAL REINFORCER / REINFORCEMENT — Reinforcers / reinforcement that occurs through a functional chain of natural events that is otherwise unrelated to the behavior that the occurrence of these reinforcers affects; conversely, the effects of the behavior on the environment did not include the production of these reinforcers. (See SUPERSTITIOUS BEHAVIOR.)

CONDITIONED — The pairing (i.e., the occurring at about the same time) of stimulus events such that the neutral (i.e., non–functional) stimulus events start functioning like the already functional stimulus events (e.g., an operant S^r or a respondent cs). (Operant synonym: "secondary.") (Also see UNCONDITIONED.)

CONTINGENCIES — The term for the dependencies (i.e., the relationships) between the natural, real dependent variables of behavior and all of the natural, real **in**dependent variables of which behavior is a function (i.e., the single word that can cover all of the natural–science "causes" of behavior).

CONTINGENCIES OF REINFORCEMENT — The generic term that refers to *all* behavioral contingencies at once, including the many types of contingencies that do not involve reinforcement.

CUMULATIVE COMPLEXITY — (See LAW... OF...)

DIFFERENTIAL REINFORCEMENT — The term for response class members producing different consequences such that one member of a response class earns reinforcement while other class members go without reinforcement leading to their extinction. (See SHAPING.)

DISCRIMINATION — An older and less precise term for EVOCATION, due to the potential agential implications of "discrimination."

DISCRIMINATIVE STIMULUS — An older and less precise term for EVOCATIVE STIMULUS, due to the potential agential implications of "discriminative."

DUAL FUNCTION OF A STIMULUS — A stimulus serving two functions as a result of being regularly present when a REINFORCER follows a response; such a stimulus comes to function both as a conditioned reinforcing stimulus (for the previous behavior) and as an EVOCATIVE STIMULUS (for the next behavior).

ENVIRONMENT — The natural domain (on both sides of the skin) that the existence of theoretically measurable independent variables defines in behavior–controlling functional relations.

ENVIRONMENTS, EXTERNAL and INTERNAL — The environments that occur on each side of the scientifically unimportant boundary that the skin of the behaving organism specifies. (Also see ENVIRONMENT.)

ESCAPE (and AVOIDANCE) — In oversimplifed terms, escape refers to behavior that results in the termination of currently occurring unconditioned punishers (i.e., unconditioned aversive stimuli). Traditionally we differentiate between escape and avoidance on the basis of whether the unconditioned aversive stimulus is present, as in currently affecting the nervous system, or not; when the behavior terminates *conditioned* punishers, or accompanies a delay or prevention of the occurrence of unconditioned punishers, we can call it, and we traditionally have called it, "avoidance."

EVOCATION — The term that either indicates (a) the process of a particular stimulus evoking a particular response while other present stimuli have less or no such effect, at least on that particular response, or indicates (b) the process by which a particular stimulus *comes more readily to evoke* a particular response, because that stimulus has been reliably present when reinforcers follow that response; in either case, the change occurs in the physiology that mediates the behavior rather than in the stimulus itself, which remains unchanged.

EVOCATIVE STIMULUS — A stimulus that evokes a response, because it is a stimulus in the presence of which a reinforcer follows a response (which tends to make that stimulus also a conditioned reinforcer). (See DUAL FUNCTION OF A STIMULUS.)

EXTINCTION [operant] — A behavior change process in which the previously operating reinforcers cease to accompany or follow members of a response class. As a result a decrease occurs in the rate or relative frequency of that behavior across subsequent occasions. (See FORGETTING; also see PRECLUSION.)

EXTINCTION, RESPONDENT — A behavior change process in which the unconditioned stimulus, that previously occurred at least occasionally with the conditioned stimulus, ceases to occur, with the result that the occurring conditioned stimulus gradually ceases to elicit the response and returns to the status of a neutral stimulus with respect to the response.

FADING — The gradual change of or in stimuli that results in the transfer of stimulus control from one stimulus or dimension of a stimulus to another stimulus or dimension of a stimulus.

FORGETTING — A behavior change process in which the evocative stimuli have not occurred and so, as a result of the accumulation of normal degradation in physiological structures, the evocative stimuli *cannot* evoke the response, which therefore cannot occur. (See PRECLUSION; also see EXTINCTION.)

FUNCTION–ALTERING STIMULUS — A stimulus whose presence alters the function of other present stimuli.

FUNCTIONAL RELATION — The relation in which a particular dependent variable changes in some orderly, systematic way when orderly, systematic changes in a particular independent variable occur.

GENERALIZATION (STIMULUS and RESPONSE) — A STIMULUS CONTROL process with two parts; in stimulus generalization *different stimuli* evoke the same response, while in response generalization the same stimulus evokes *different responses.*

GENERALIZED (conditioned) REINFORCERS — Conditioned reinforcing stimulus events that function regardless of satiation or deprivation due to pairing with (i.e., occurring at the same time as) *several* other reinforcers. (See CONDITIONED REINFORCERS.)

INTERMITTENT REINFORCEMENT — Term for reinforcers following only some responses. (See SCHEDULES OF REINFORCEMENT.)

INTERVENTION — (See PROCEDURE.)

LAW (or theory) OF CUMULATIVE COMPLEXITY — *The natural physical/ chemical interactions of matter and energy sometimes result in more complex structures and functions that endure and naturally interact further, resulting in an accumulating complexity.* (An origin of life—*any* origin of life—is an outcome of the *Law of Cumulative Complexity.* On this planet other examples of this law include (a) the vast range of life forms available for study, (b) the vast range of available chemical compounds and materials including those that natural science has discovered and developed, (c) the vast range of behaviors that contingencies cause (e.g., the behaviors of natural scientists that contingencies induce, leading to the discoveries and developments that help us all—through verbal *behaviors*—better understand and deal with the world), and the interrelations of physiology and behaviorology. All these are cumulatively complex; all are entirely natural.)

NATURAL EVENT — An event that is definable in terms of time, distance, mass, temperature, charge, and/or a few other properties taken into account by theoretical physicists. Measurable physical properties define a natural event, which occurs only as the culmination of a sequential history of similarly definable events; natural events cannot occur spontaneously (i.e., magically, mystically, or superstitiously).

NATURALISM — The name for the general philosophy of science that informs all natural sciences, from the traditional natural sciences (e.g., physics, chemistry, biology) to behaviorology, by providing systematic verbal supplementary controlling stimuli, for science–related behaviors, in the form of assumptions that the results from experimental research and successful engineering applications have conditioned over the last several centuries (all in opposition to the convenient, anytime inventing of mystically or superstitiously grounded assumptions). (Also, see RADICAL BEHAVIORISM.)

OPERANT — A relation in which a response operates on the environment in a manner that provides energy feedback into the nervous system changing it in ways that we see as *increased or decreased* rates of that kind of response on future occasions *controlled by increased or decreased evocative effects* of the stimuli that evoked the response in the first place.

POSTCEDENT — An event that occurs *after* some other event. (See ANTECEDENT.)

POSTCEDENT ENVIRONMENT — The environment as it exists beginning immediately after a response.

PRECLUSION — A behavior change *procedure* (i.e., intervention) in which interveners prevent the evocative stimuli from occurring and so, as a result of the accumulation of normal degradation in physiological structures, the evocative stimuli *cannot* evoke the response, which therefore cannot occur. (See FORGETTING; also see EXTINCTION.)

PREMACK PRINCIPLE — A high probability behavior can reinforce a low probability behavior. (AKA "Grandma's Principle.")

PROCEDURE (Behaviorological Engineering Procedure/Intervention) — The procedural *prescription* of a behavioral process by which a functional change in environmental events (e.g., an added reinforcement procedure) produces a change in responding; the functional change occurs in an organism's internal or external environment, in the presence of facilitation by the evoked responding of one or more other organisms, which could include the subject, as these are all natural parts of the environment. (See PROCESS.)

PROCESS (Behavioral) — Any change in environmental events that produces a change in responding. The change in behavior is a function of the change in the environmental events (e.g., an added reinforcement procedure). Behavioral processes occur in an organism's internal or external environment, in the absence *or* presence of facilitation by the evoked responding of one or more other organisms, which could include the subject, as these are all natural parts of the environment. (See PROCEDURE.)

PUNISHER — A relative change in the environment (i.e., a *stimulus*) that provides an energy change at receptor cells during or immediately after a response (with reducing effects as the time between these events increases) that results in a *decrease* in the RATE or RELATIVE FREQUENCY of the behavior across subsequent occasions. Can be UNCONDITIONED (i.e., operative due to genetically produced neural structure) or CONDITIONED (i.e., operative due to pairing with—occurring at the same time as—other punishers; the pairing changes the neural structures, not the stimuli). (See OPERANT.)

PUNISHER: ADDED PUNISHER — The addition (e.g., presentation) of a stimulus in the environment during or immediately after a response that results in a decrease in the rate or relative frequency of the behavior across subsequent occasions.

PUNISHER: SUBTRACTED PUNISHER — The subtraction (e.g., withdrawal or termination) of a stimulus in the environment during or immediately following a response that results in a decrease in the rate or relative frequency of the behavior across subsequent occasions.

PUNISHMENT — A behavior change process in which a punisher occurs during or immediately after a response and results in a decrease in the rate or relative frequency of that behavior across subsequent occasions. Can be UNCONDITIONED or CONDITIONED...

PUNISHMENT: ADDED PUNISHMENT — A behavior change process involving an added punisher occurring in the environment during or immediately following a response that results in a decrease in the rate or relative frequency of the behavior across subsequent occasions.

PUNISHMENT: SUBTRACTED PUNISHMENT — A behavior change process involving a subtracted punisher occurring in the environment during or immediately following a response that results in a decrease in the rate or relative frequency of the behavior across subsequent occasions.

RADICAL BEHAVIORISM — The philosophy of science that B. F. Skinner introduced, and his scientific colleagues and students developed, that extends naturalism, the general philosophy of science in the traditional natural sciences (e.g., physics, chemistry, biology) to inform the study of behavior, including human nature and human behavior, under the more recently emerged (in the early twentieth century) natural science of behavior that we now call behaviorology.

REFLEX — A stimulus AND the response that it elicits.

REINFORCER — A relative change in the environment (i.e., a *stimulus*) that provides an energy change at receptor cells during or immediately after a response (with reducing effects as the time between these events increases) that results in an *increase* in the RATE or RELATIVE FREQUENCY of the behavior across subsequent occasions. Can be UNCONDITIONED (i.e., operative due to genetically produced neural structure) or CONDITIONED (i.e., operative due to pairing with—occurring at the same time as—other reinforcers; the pairing changes the neural structures, not the stimuli). (See OPERANT.)

REINFORCER: ADDED REINFORCER — The addition (e.g., presentation) of a stimulus in the environment during or immediately after a response that results in an increase in the rate or relative frequency of the behavior across subsequent occasions.

REINFORCER: SUBTRACTED REINFORCER — The subtraction (e.g., withdrawal or termination) of a stimulus in the environment during or immediately following a response that results in an increase in the rate or relative frequency of the behavior across subsequent occasions.

REINFORCEMENT — A behavior change process in which a reinforcer occurs during or immediately after a response and results in an increase in either the rate or relative frequency of that behavior across subsequent occasions. Can be UNCONDITIONED or CONDITIONED...

REINFORCEMENT: ADDED REINFORCEMENT — A behavior change process involving an added reinforcer occurring in the environment during or immediately following a response that results in an increase in the rate or relative frequency of the behavior across subsequent occasions.

REINFORCEMENT: SUBTRACTED REINFORCEMENT — A behavior change process involving a subtracted reinforcer occurring in the environment during or immediately following a response that results in a increase in the rate or relative frequency of the behavior across subsequent occasions.

RESISTANCE TO EXTINCTION — An effect of intermittent reinforcement schedules, in which the more a schedule resembles extinction the more responding on that schedule resists extinguishing after reinforcement stops; that is, as the number of unreinforced responses on the schedules increases, the number of responses during extinction, after reinforcement stops, also increases.

RESPONDENT — A relation involving a response that an unconditioned stimulus or a conditioned stimulus elicits.

RESPONSE — *EITHER* any covert or overt *innervated* muscular movements of an organism resulting from energy transfers within the organism that other energy changes—initially from beyond the affected body parts—evoke or elicit, *OR* patterns of neural activity resulting from energy transfers within the organism that other energy changes— initially from beyond the affected body parts—evoke or elicit; in all cases *responses are instances of behavior.*

RESPONSE CLASS — A group or variety of responses that have in common either the same eliciting stimulus, for respondent behavior, or the same effect on the environment (e.g., each class member produces the same environmental change) for operant behavior.

RESPONSE GENERALIZATION — (See GENERALIZATION.)

(RESPONSE) **RATE** — The quotient when a count of responses is divided by a count of time units during which the responses occurred:

$$\frac{\text{Count (dividend)}}{\text{Time (divisor)}} = \text{Rate (quotient)}$$

(RESPONSE) RELATIVE FREQUENCY — The quotient derived from the ratio of fulfilled opportunities to respond to total opportunities to respond. The result may be expressed as a percentage (by multiplying the quotient by 100).

$$\frac{\text{Fulfilled opportunities (dividend)}}{\text{Total opportunities (divisor)}} = \text{Relative Frequency (quotient)}$$

SCHEDULES OF REINFORCEMENT — Aside from the CRF (continuous reinforcement) schedule in which a reinforcer follows every response, "Schedules of Reinforcement" refers to the possible patterns of intermittent reinforcers (i.e., when reinforcers follow only some responses). In the four most common patterns, either reinforcers can occur based on the *ratio* of responses to reinforcers (i.e., on how many responses occur with respect to each reinforcer that occurs) or reinforcers can occur based on the *interval* of the amount of time that elapses since the last reinforcer occurred, during which time no reinforcers are available, before a reinforcer follows the next occurring response. Both ratio schedules and interval schedules can be fixed or variable. On *fixed–ratio* (FR) schedules a reinforcer follows each fixed number of responses, while on *variable–ratio* (VR) schedules a reinforcer follows a number of responses, with that number varying around some average. On *fixed–interval* (FI) schedules a reinforcer follows the first response that occurs after each fixed–duration interval elapses, while on *variable–interval* (VI) schedules a reinforcer follows the first response that occurs after each interval elapses, with the duration of each interval varying around some average. Numerous other types of reinforcement schedules are possible (e.g., mixed, multiple, chained, tandem, and concurrent) including time schedules (fixed or variable, FT or VT) which involve only intervals of time elapsing before each reinforcer occurs, regardless of responding; any reinforcers that follow responses on time schedules are coincidental, and we call any increase in responding superstitious.

SHAPING (& DIFFERENTIAL REINFORCEMENT) — Shaping is the gradual change in behavior that occurs when the criteria for the successive approximations of behavior that earn reinforcement gradually shift along some dimension of the behavior thereby producing a more refined behavior or a very different behavior or both. When shaping occurs as part of an intervention, we can say that it involves repeatedly applying the differential reinforcement procedure for each change in the reinforcement criteria. DIFFERENTIAL REINFORCEMENT involves response class members producing different consequences such that one member of a response class earns reinforcement while other class members go without reinforcement leading to their extinction.

STIMULUS — An internal or external environmental event—an energy change at receptor cells—that affects responding (e.g., elicits or evokes or consequates a response).

STIMULUS CLASS — A group of different stimuli that share some characteristic in common (e.g., a pile of different pictures in which each picture somehow shows people, and a pile of different pictures in which no picture shows people, would comprise *two* stimulus classes).

STIMULUS CONTROL — The interaction of an environmental event with the nervous system that compels the mediation of a behavior. We describe the stimulus as exerting functional control by evoking the response (see EVOCATIVE STIMULUS), or by having some other effect on another stimulus (see FUNCTION–ALTERING STIMULUS). We include respondent elicitation under stimulus control, and even as a kind of evocation. Also encompasses stimulus and response GENERALIZATION.

STIMULUS GENERALIZATION — (See GENERALIZATION.)

SUPERSTITIOUS BEHAVIOR (& COINCIDENTAL REINFORCERS) — *Superstitious behavior* is the term we use for the behavior when the occurrence of one or more coincidental reinforcers conditions (i.e., generates or maintains) a behavior. *Coincidental reinforcers* are reinforcers that occur through a functional chain of natural events that is otherwise unrelated to the behavior that the occurrence of these reinforcers affects; conversely, the effects of the behavior on the environment did not include the production of these reinforcers.

TIME OUT — A procedure (i.e., an intervention) in which the interveners are under contingencies compelling arrangements for the subject to be in a setting that disallows contact with any reinforcers (or with as few reinforcers as possible) for a short time (on the order of 30 seconds to one minute).

UNCONDITIONED — Stimulus events that already function due to nervous–system structure that genes—rather than conditioning processes—produce (e.g., an operant S^R or a respondent UCS). (Operant synonym: "primary.") (Also see CONDITIONED.) ℘

A Short & Occasionally Annotated Bibliography

This bibliography contains a range of behaviorological–science references to help interested readers expand their repertoire. Beyond that, you can find a more extensive bibliography in my 2014 textbook, *Running Out of Time—Introducing Behavoirology to Help Solve Global Problems,* which is itself a good next book. It is most easily available from the distributor, **Direct Book Services,** at 800–776–2665. They will likely answer the phone with "Dogwise," because their most popular speciality involves books about our canine friends; several of these books already specifically apply the laws of behavior that *Running Out of Time…* systematically introduces (even more so than the present book).

This bibliography includes books, and articles from accessible behaviorology journals, that may interest readers as next steps in getting to know behaviorology better. Due to space constraints, only a handful of items appear here from journals that are not explicitly behaviorology journals. The journal, *Behaviorology Today* (ISSN 1536–6669) peer reviewed most articles only minimally for its first 14 volumes, while fully peer reviewing only the occasional article that it then explicitly so labeled. However, beginning with Volume 15, Number 1 (Spring 2012) *Behaviorology Today* fully peer reviewed *all* articles. Also, beginning with Volume 16, Number 1 (Spring 2013) the name became *Journal of Behaviorology* (ISSN 2331–0774). All issues of *Behaviorology Today* and *Journal of Behaviorology* are available at www.behaviorology.org which is the website of The International Behaviorology Institute. (Only some of the references that appear at the end of various chapters also appear here.)♣

Bjork, D. W. (1993). *B. F. Skinner: A Life.* New York: Basic Books.

Cautela, J. R. & Ishaq, W. (Eds.). (1996). *Contemporary Issues in Behavior Therapy: Improving the Human Condition.* New York: Plenum.

Cheney, C. D. (1991). The source and control of behavior. In W. Ishaq (Ed.). *Human Behavior in Today's World* (pp. 73–86). New York: Praeger.

Comunidad Los Horcones. (1986). Behaviorology: An integrative denomination. *The Behavior Analyst, 9,* 227–228.

Epstein, R. (1996). *Cognition, Creativity, and Behavior.* Westport, CT: Praeger.

Ferreira, J. B. (2012). Progressive neural emotional therapy (PNET): A behaviorological analysis. *Behaviorology Today, 15* (2), 3–9.

Fraley, L. E. (1987). The cultural mission of behaviorology. *The Behavior Analyst, 10,* 123–126. Also, see Fraley, 2012a, 2013.

Fraley, L. E. (1988). Introductory comments: Behaviorology and cultural materialism. *The Behavior Analyst, 11,* 159–160.

Fraley, L. E. (1992). The religious psychology student in a behaviorology course. *Behaviorological Commentaries, Serial No. 2*, 18–21. This paper also appeared (2009) in *Behaviorology Today, 12* (2), 11–13.

Fraley, L. E. (2002). The discipline of behaviorology and the postulate of determinism. *Behaviorology Today, 5* (1), 45–49.

Fraley, L. E. (2003). The strategic misdefining of the natural sciences within universities. *Behaviorology Today, 6* (1), 15–38.

Fraley, L. E. (2008). *General Behavoirology—The Natural Science of Human Behavior.* Canton, NY: ABCs.

Fraley, L. E. (2012a). *Dignified Dying—A Behaviorological Thanatology.* Canton, NY: ABCs.

Fraley, L. E. (2012b). The evolution of a discipline and our next step. *Behaviorology Today, 15* (1), 23–28.

Fraley, L. E. (2013). *Behaviorological Rehabilitation and the Criminal Justice System.* Canton, NY: ABCs.

Fraley, L. E. & Ledoux, S. F. (2015). Origins, status, and mission of behaviorology. In S. F. Ledoux. *Origins and Components of Behaviorology—Third Edition* (pp. 33–169). Ottawa, CANADA: BehaveTech Publishing. This multi–chapter paper, originally published in 1992, later also appeared across 2006–2008 in these five parts in *Behaviorology Today:* Chapters 1 & 2: *9* (2), 13–32. Chapter 3: *10* (1), 15–25. Chapter 4: *10* (2), 9–33. Chapter 5: *11* (1), 3–30. Chapters 6 & 7: *11* (2), 3–17.

Fraley, L. E. & Vargas, E. A. (1986). Separate disciplines: The study of behavior and the study of the psyche. *The Behavior Analyst, 9,* 47–59.

Harris, M. (1974). *Cow, Pigs, Wars, and Witches: The Riddles of Culture.* New York: Random House.

Harris, M. (1977). *Cannibals and Kings: The Origins of Culture.* New York: Random House.

Harris, M. (1979). *Cultural Materialism: The Struggle for a Science of Culture.* New York: Random House.

Ishaq, W. (Ed.). (1991). *Human Behavior in Today's World.* New York: Praeger.

Johnson, P. R. (2012). A behaviorological approach to management of neuroleptic–induced tardive dyskinesia: Progressive neural emotional therapy (PNET). *Behaviorology Today, 15* (2), 11–25.

Latham, G. I. (1994). *The Power of Positive Parenting.* Logan, UT: P & T ink.

Latham, G. I. (1998). *Keys to Classroom Management.* Logan, UT: P & T ink.

Latham, G. I. (1999). *Parenting with Love.* Salt Lake City, UT: Bookcraft.

Latham, G. I. (2002). *Behind the Schoolhouse Door: Managing Chaos with Science, Skills, and Strategies.* Logan, UT: P & T ink. This book includes two earlier pieces: *Eight Skills Every Teacher Should Have* and *Management, Not Discipline: A Wakeup Call for Educators.*

Latham, G. I. (2015). China through the eyes of a behaviorologist. In S. F. Ledoux. *Origins and Components of Behaviorology—Third Edition* (pp. 297–

302). Ottawa, CANADA: BehaveTech Publishing. This paper also appeared (2002) in *Behaviorology Today, 5* (1), 17–20.

Ledoux, S. F. (1985). Designing a new *Walden Two*–inspired community. *Communities—Journal of Cooperation, No. 66,* Spring (April), 28–32, 84.

Ledoux, S. F. (1990–2015). Several books of study questions (from Canton, NY: ABCs), one with other authors, for various texts, including Skinner's *Walden Two,* Wyatt's *The Millennium Man,* Latham's *Keys to Classroom Management,* Latham's *The Power of Positive Parenting,* Maurice et al's *Behavioral Intervention for Young Children with Autism,* Ledoux's *Running Out of Time—Introducing Behaviorology to Help Solve Global Problem,* and Ledoux's *Origins and Components of Behaviorology—Third Edition.* See BOOKS at www.behaviorology.org to find some details, including study question books by other authors, such as J. O'Heare, and L. Raymond.

Ledoux, S. F. (2002). A parable of past scribes and present possibilities. *Behaviorology Today, 5* (1), 60–64. This is a parable on the 20–year, billion–dollar American education research effort called *Project Follow Through,* the outcomes of which the American education establishment tends to ignore, to the detriment of students, teachers, schools, and communities across the country and even around the world (see Watkins, 1997).

Ledoux, S. F. (2002). Carl Sagan is right again: A review of *The Millennium Man. Behaviorology Today, 5* (2), 23–25.

Ledoux, S. F. (2002). Defining natural sciences. *Behaviorology Today, 5* (1), 34–36.

Ledoux, S. F. (2012). Behaviorism at 100. *American Scientist, 100* (1), 60–65. This article extends B. F. Skinner's 1963 article "Behaviorism at fifty." The Editor introduced this article with excerpts, on pages 54–59, which he listed as an "*American Scientist* Centennial Classic 1957" that came from Skinner's 1957 *American Scientist* article "The experimental analysis of behavior." (Skinner's complete 1957 paper was also available online.)

Ledoux, S. F. (2012). Behaviorism at 100 unabridged. *Behaviorology Today, 15* (1), 3–22. With *Behaviorology Today* becoming fully peer–reviewed with this issue, this fully peer–reviewed version of the paper that originally appeared in *American Scientist* included the material set aside at the last moment to make more room for the Skinner article excerpts that accompanied the original article. *American Scientist* posted this paper online along with the original version (also, find it at www.behaviorology.org).

Ledoux, S. F. (2014). *Running Out of Time—Introducing Behaviorology to Help Solve Global Problem.* Ottawa, CANADA: BehaveTech Publishing.

Ledoux, S. F. (2015). *Origins and Components of Behaviorology—Third Edition.* Ottawa, CANADA: BehaveTech Publishing.

Ledoux, S. F. (2015). Appendix 3 Addendum: Curricular courses and resources after 25 years (1990–2015). In S. F. Ledoux. *Origins and Components of Behaviorology—Third Edition* (pp. 314–326). Ottawa, CANADA: BehaveTech Publishing. This paper reports the status of behaviorology courses and curricula at the time of the author's retirement.

Ledoux, S. F. (2015). Behaviorology in China: A status report. In S. F. Ledoux. *Origins and Components of Behaviorology—Third Edition* (pp. 187–198). Ottawa, CANADA: BehaveTech Publishing. A Chinese translation appeared (2002) in *Behaviorology Today, 5* (1), 37–44. This paper also appeared in English (2009) in *Behaviorology Today, 12* (2), 3–10.

Ledoux, S. F. & Cheney, C. D. (1987). *Grandpa Fred's Baby Tender or Why and How we Built our Aircribs.* Canton, NY: ABCs. See the Appendix for this book's Foreword, "The first baby tender," by B. F. Skinner. This book is available from www.behaviorology.org as a free download.

Ledoux, S. F., Hallatt, D., & Hallatt, T. (2014). An interview on behaviorology supporting a sustainable society. *Journal of Behaviorology, 17* (1), 3–12. You can purchase a somewhat *visually* boring, two–hour–long DVD of this interview of the author from the author or from www.behaviorology.org.

Ledoux, S. F. & O'Heare, J. (2015). Elements of the ongoing history of the behaviorology discipline. In S. F. Ledoux. *Origins and Components of Behaviorology—Third Edition* (pp. 259–296). Ottawa, CANADA: BehaveTech Publishing.

Lloyd, K. E. (1985). Behavioral anthropology: A review of Marvin Harris' *Cultural Materialism. Journal of the Experimental Analysis of Behavior, 43,* 279–287.

Logue, A. W. (1988). A behaviorist's biologist: Review of Philip J. Pauly's *Controlling Life: Jacques Loeb and the Engineering Ideal in Biology. The Behavior Analyst, 11,* 205–207.

Malott, R. W. (1988). Rule–governed behavior and behavioral anthropology. *The Behavior Analyst, 11,* 181–203.

Moore, J. (1984). On privacy, causes, and contingencies. *The Behavior Analyst, 7,* 3–16.

O'Heare, J. (2014). *The Science and Technology of Dog Training.* Ottawa, CANADA: BehaveTech Publishing.

O'Heare, J. (2015). *The Science and Technology of Animal Training.* Ottawa, CANADA: BehaveTech Publishing.

Peterson, N. & Ledoux, S. F. (2014). *An Introduction to Verbal Behavior—Second Edition.* Canton, NY: ABCs. Going through this self–study book enables the reader to go through Skinner's *Verbal Behavior* book more easily.

Sidman, M. (1960). *Tactics of Scientific Research.* New York: Basic Books. Authors Cooperative, in Boston MA, republished this book in 1988.

Sidman, M. (2001). *Coercion and its Fallout—Revised Edition.* Boston, MA: Authors Cooperative.

Sidman, M. (2003). Reinforcement in diplomacy: More effective than coercion. *Behaviorology Today, 6* (2), 30–35.

Skinner, B. F. (1938). *The Behavior of Organisms.* New York: Appleton–Century–Crofts. Seventh printing, 1966, with special preface: Englewood Cliffs, NJ: Prentice–Hall. The B. F. Skinner Foundation (www.bfskinner.org) in Cambridge, MA, republished this book in 1991.

Skinner, B. F. (1948). *Walden Two.* New York: Macmillan. In 1976 Macmillan issued a new paperback edition with Skinner's introductory essay, *"Walden*

Two Revisited." This novel provides a fictional description of a culture the design of which is based on behaviorological science. While this story is as relevant today as when it was written, the author often used the word "psychology" in this novel to denote the natural science of behavior into which he was trying, at the time, to turn the traditional field of psychology. Since these usages can confuse readers today, less confusion occurs if they substitute "behaviorology" for these usages.

Skinner, B. F. (1953). *Science and Human Behavior.* New York: Macmillan. The Free Press, New York, published a paperback edition in 1965.

Skinner, B. F. (1957). The experimental analysis of behavior. *American Scientist, 45* (4), 343–371.

Skinner, B. F. (1957). *Verbal Behavior.* New York: Appleton–Century–Crofts. The B. F. Skinner Foundation (www.bfskinner.org) in Cambridge, MA, republished this book in 1992.

Skinner, B. F. (1963). Behaviorism at fifty. *Science, 140,* 951–958.

Skinner, B. F. (1966). The phylogeny and ontogeny of behavior. *Science, 153,* 1205–1213.

Skinner, B. F. (1968). *The Technology of Teaching.* New York: Appleton–Century–Crofts. The B. F. Skinner Foundation (www.bfskinner.org) in Cambridge, MA, republished this book in 2003.

Skinner, B. F. (1969). *Contingencies of Reinforcement: A Theoretical Analysis.* New York: Appleton–Century–Crofts. The B. F. Skinner Foundation (www.bfskinner.org) in Cambridge, MA, republished this book in 2013.

Skinner, B. F. (1971). *Beyond Freedom and Dignity.* New York: Knopf.

Skinner, B. F. (1972). *Cumulative Record: A Selection of Papers (Third Edition).* New York: Appleton–Century–Crofts. The B. F. Skinner Foundation (www.bfskinner.org) in Cambridge, MA, republished this book as "... *Definitive Edition,"* in 1999. Several of the papers in this collection contain material concerning Skinner's invention of various pieces of equipment. The paper, "A case history in scientific method"—on pages 108–131 in the *Definitive Edition*—is particularly relevant to the invention of the **Cumulative Recorder.**

Skinner, B. F. (1974). *About Behaviorism.* New York: Knopf. With the legitimate time–frame excuse of appearing before the 1987 adoption of the term behaviorology, this book portrays radical behaviorism as the philosophy of "behavior analysis." Back then this could be accurate, because behavior analysts were then moving to become an independent natural science under the behavior analysis label. Sadly the efforts under *that* label were ultimately unsuccessful, because psychology claimed the behavior analysis label, without the kind of strenuous objections from behavior analysts that might have stopped it. Thus, psychology succeeded in making the term "behavior analysis" a part of the non–natural psychology discipline, something that cost behavior analysts some of the credibility they had—as natural scientists of behavior moving to separate from psychology—with traditional natural scientists (which, unfortunately, makes their contributions to solving global problems more difficult to provide). Meanwhile, we use *behaviorology* to denote the independent natural science of behavior. Nevertheless, this book is a wonderfully thorough and readable treatment of radical behaviorism, the philosophy of science of behaviorology.

Skinner, B. F. (1976). *Particulars of My Life*. New York: Knopf. This is the first of three volumes in Skinner's autobiography.

Skinner, B. F. (1978). *Reflections on Behaviorism and Society.* Englewood Cliffs, NJ: Prentice–Hall.

Skinner, B. F. (1979). *The Shaping of a Behaviorist*. New York: Knopf. This is the second of three volumes in Skinner's autobiography, and includes some interesting material about his equipment inventions.

Skinner, B. F. (1983). *A Matter of Consequences*. New York: Knopf. This is the third of three volumes in Skinner's autobiography.

Skinner, B. F. (1987). *Upon Further Reflection*. Englewood Cliffs, NJ: Prentice–Hall.

Skinner, B. F. (1987). The first baby tender. In S. F. Ledoux & C. D. Cheney. *Grandpa Fred's Baby Tender or why and how we built our aircribs* (pp. iii–v). Canton, NY: ABCs. This paper also appeared (2004) in *Behaviorology Today,* 7 (1), 3–4. See the Appendix. Also see Ledoux & Cheney, 1987.

Skinner, B. F. (1989). *Recent Issues in the Analysis of Behavior.* Columbus, OH: Merrill.

Skinner, B. F. & Vaughan, M. E. (1983). *Enjoy Old Age*. New York: W.W. Norton.

Thompson, L. (2010). Climate change: The evidence and our options. *The Behavior Analyst, 33* (2), 153–170.

Vargas, E. A. (1987). "Separate disciplines" is another name for survival. *The Behavior Analyst, 10,* 119–121.

Vargas, E. A. (1991). Behaviorology: Its paradigm. In W. Ishaq (Ed.). *Human Behavior in Today's World* (pp. 139–147). New York: Praeger.

Vargas, E. A. (2013). The importance of form in Skinner's analysis of verbal behavior and a further step. *The Analysis of Verbal Behavior, 29,* 167–183. This paper helps us appreciate the value of traditional linguists' work.

Watkins, C. L. (1997). *Project Follow Through: A Case Study of Contingencies Influencing Instructional Practices of the Educational Establishment.* Cambridge, MA: Cambridge Center for Behavioral Studies.

West, R. P. & Hamerlynck, L. A. (Eds.). (1992). *Designs for Excellence in Education: The Legacy of B. F. Skinner (Limited Edition).* Longmont, CO: Sopris West.

Wyatt, W. J. (1997). *The Millennium Man.* Hurricane, WV: Third Millennium Press. In this novel the author often used the term "behavior analysis" as the name for the natural science of behavior to show that it is different from, and not any kind of, psychology. Today, however, psychology claims "behavior analysis" as part of itself. To avoid confusion readers today should substitute "behaviorology" when "behavior analysis" appears in this book.

Wyatt, W. J., Hawkins, R. P., & Davis, P. (1986). Behaviorism: Are reports of its death exaggerated? *The Behavior Analyst, 9,* 101–105.℘

Index

$\mathcal{T}$his Index features a *selected* set of topical entries, not a comprehensive set. In addition the page numbers only represent content of some substantive nature rather than every appearance of the entry in the book.❧

interval reinforcement schedules
203–208

$\mathcal{J}$

Johnson, M. 174
Journal of Behaviorology (JOB) 8, 397
*Journal of the Experimental Analysis
of Behavior* (JEAB) 145

K

Keenan, B. 220
knowing (also see *awareness,
consciousness*) 330, 332–335,
338–339, 358–367
Kurland, A. 174

$\mathcal{L}$

language (also see *verbal behavior*)
47, 51–53, 283
language–related fictional
explanations, other 51–53
Larsen, D. 382
Latham, G. 222–227, 231, 277
Law of Cumulative Complexity
17–18, 30, 50, 85, 339, 342, 368
Ledoux, S. F. 18, 33, 45, 134, 158,
164, 192, 201, 212, 214, 228,
249, 281, 293, 311, 342, 371,
385, 395, 397, 429–432
Lee, D. 12
life 341–344
life, meaning of 346–347
Limits to Growth, The (also see *global
problems/solutions*) 378
linguistics 51–53, 284–288
linguistic behavior (see *verbal behavior*)
listener and speaker in *verbal
behavior* 289–290

Loeb, J. 385
logic coercion trap 224
love 100–104
Luce, S. 293

$\mathcal{M}$

MacCorquodale, K. 285
Malott, R. xix, 134, 167, 186–187,
193, 220
Mark Twain 368
Maurice, C. 293
maxi and mini deities 55, 69,
104, 317
McIntyre, L. 375
Meadows, D. H. 378
Meadows, D. L. 378
meaning of life 346–347
meaning of "nature" and "natural"
141–142
mediator and verbalizer in verbal
behavior 290
medical ethics (also see *ethics*)
355–356
memories 17
method of successive approximation
169–177
methodology 33, 200–201, 214
Meyerson, L. 194, 198
Michael, J. 194, 198
Michaelangelo 245
Millennium Man, The 136, 304
mini and maxi deities 55, 69,
104, 317
Mlodinow, L. 17, 359
modifier, contingency 247–248, 250
modifier, environment 242, 248
morals 305, 314–317, 379–381
Morrow, J. 259*ff*
motor behavior 87–88
multi–term contingencies 131–132
muscular behavior 87, 273, 275, 283

T

College 2012 portrait Photo courtesy of SUNY–Canton

About the Author

*A*fter completing the usual four years of high school at St. Pius X Seminary (in Galt, CA) the author, Dr. Stephen Ledoux (pronounced "la–dew") graduated in June, 1968, and then attended St. Patrick's College seminary (in Mountain View, CA) for two years. He began his professional activities in the early 1970s after earning his B.A. and M.A. degrees, in 1972 and 1973 respectively, at *California State University, Sacramento.* At various times he has taught courses in behaviorology, education, English, and psychology, all (except English) at both the undergraduate and graduate levels, with behaviorology taught at the high school level as well (at St. Pius X in 1972). After several years of university teaching, he returned to full–time study, and earned his ph.d. from *Western Michigan University,* in 1982, in The Experimental Analysis of Behavior. (He says that, in part, the ph.d. seemed necessary, because his California m.a. degree was signed by "Ronald Reagan" and "James Bond"...)

Prof. Ledoux has held positions both at home and abroad. For four years, 1975–1979, he taught in Australia, starting as a "Tutor" at the *University of Queensland* in Brisbane, and then as a "Lecturer" at the *Gippsland Institute of Advanced Education* near Melbourne. He also taught in China, both in 1979 (at *Xi'an Jiaotong University* where he was their first foreign English teacher) and in 1990–1991 (at *Xi'an Foreign Languages University*) as part of a faculty exchange. Since 1982 he taught at the Canton campus of the *State University of New York* from which, after 33 years, he retired with Emeritus status in 2015.

Over the last several decades, Dr. Ledoux prepared numerous publications and presentations, and was involved in many other ways in professional work. With colleague Dr. Carl Cheney, of Utah State University, Logan, he worked extensively in the early 1980s with the concept and function of the *Aircrib* (a.k.a. the "baby tender") an invention of the late Prof. B. F. Skinner (see the Appendix). For several decades he has been very involved with the movement formally establishing the independent discipline of behaviorology, including a three–year term (1988–1991) as the first elected president of *The International Behaviorology Association.* In 1997 he published two books, *Origins and Components of Behaviorology* (which went into its third edition in 2015) and, with his spouse, Dr. Nelly Case (a professor of the Crane School of Music at suny–Potsdam) as the primary author, a book about the year they taught in China, with their then five–year–old son, entitled *The Panda and Monkey King Christmas—A Family's Year in China.*

In 1998 Prof. Ledoux was elected Chair of the Board of Directors of *The International Behaviorology Institute* (tibi) which is a non–profit educational corporation that he helped establish to support behaviorologists by providing training in behaviorology. In 2001 he declined to continue as Chair so that he would have the time to fulfill appointments as the Editor both of the tibi journal *Behaviorology Today,* which he edited through volume 14, and of the tibi web site (www.behaviorology.org) which he still edits. After also completing a half dozen study–question books for various textbooks, in 2005 he accepted the task of editing Lawrence Fraley's 1,600–page, three–course text, *General Behaviorology: The Natural Science of Human Behavior,* which was completed and published in late 2008. This was the first extensive and systematic text of the independent behaviorology discipline, and an inspiration for both the present book as well his earlier (2014) textbook (entitled *Running Out of Time—Introducing Behaviorology to Help Solve Global Problems,* and published by BehaveTech Publishing in Ottawa, Canada). Dr. Ledoux also edited two other books by Prof. Fraley, *Dignified Dying—A Behaviorological Thanatology* (2012) and *Behaviorological Rehabilitation and the Criminal Justice System* (2013). (You can find full references for all these books in the Bibliography.)

In 2012 Prof. Ledoux updated B. F. Skinner's 1963 article, Behaviorism at Fifty, by publishing his article, Behaviorism at 100, in *American Scientist.* Two months later *Behaviorology Today* (now called *Journal of Behaviorology*) published a longer, peer–reviewed version of this article using the title, Behaviorism at 100 Unabridged. His next steps were the *Running Out of Time…* book, and the present book.

Beyond teaching and writing, Dr. Ledoux's professional interests include verbal behavior (especially as applied to language teaching), pedagogical effectiveness, and the experimental analysis of simultaneously evoked and simultaneously selected human operants. His hobbies involve star gazing, meteorites, and arts from China, Japan, and Southwest Native Americans. With his children grown, he retired to Los Alamos, NM, with his spouse.☙

Photo courtesy of SUNY–Canton

The author, Stephen F. Ledoux, Ph.D., in January 2012, upon receiving the volume 100, number 1, issue of *American Scientist* containing his "Behaviorism at 100" article.

Behaviorology in its immediate scientific neighborhood

"The Green Book" [cover color]